LINEAR ALGEBRA – THEOREMS AND APPLICATIONS

Edited by **Hassan Abid Yasser**

INTECHOPEN.COM

Linear Algebra – Theorems and Applications
http://dx.doi.org/10.5772/3107
Edited by Hassan Abid Yasser

Contributors

Matsuo Sato, Francesco Aldo Costabile, Elisabetta Longo, A. Amparan, S. Marcaida, I. Zaballa, Taro Kimura, Ricardo L. Soto, Daniel N. Miller, Raymond A. de Callafon, P. Cervantes, L.F. González, F.J. Ortiz, A.D. García, Pattrawut Chansangiam, Jadranka Mićić, Josip Pečarić, Abdulhadi Aminu, Mohammad Poursina, Imad M. Khan, Kurt S. Anderson

Published by InTech

Janeza Trdine 9, 51000 Rijeka, Croatia

Edition 2014

Publishing Process Manager Marijan Polic
Technical Editor InTech DTP team
Cover InTech Design team

Additional hard copies can be obtained from orders@intechopen.com

Linear Algebra – Theorems and Applications, Edited by Hassan Abid Yasser
p. cm.
ISBN 978-953-51-0669-2

INTECH

open science | open minds

Contents

Preface

The core of linear algebra is essential to every mathematician, and we not only treat this core, but add material that is essential to mathematicians in specific fields. This book is for advanced researchers. We presume you are already familiar with elementary linear algebra and that you know how to multiply matrices, solve linear systems, etc. We do not treat elementary material here, though we occasionally return to elementary material from a more advanced standpoint to show you what it really means. We have written a book that we hope will be broadly useful. In a few places we have succumbed to temptation and included material that is not quite so well known, but which, in our opinion, should be. We hope that you will be enlightened not only by the specific material in the book but also by its style of argument. We also hope this book will serve as a valuable reference throughout your mathematical career.

Chapter 1 reviews the metric Hermitian 3-algebra, which has been playing important roles recently in sting theory. It is classified by using a correspondence to a class of the super Lie algebra. It also reviews the Lie and Hermitian 3-algebra models of M-theory. Chapter 2 deals with algebraic analysis of Appell polynomials. It presents the determinantal approaches of Appell polynomials and the related topics, where many classical and non-classical examples are presented. Chapter 3 reviews a universal relation between combinatorics and the matrix model, and discusses its relation to the gauge theory. Chapter 4 covers the nonnegative matrices that have been a source of interesting and challenging mathematical problems. They arise in many applications such as: communications systems, biological systems, economics, ecology, computer sciences, machine learning, and many other engineering systems. Chapter 5 presents the central theory behind realization-based system identification and connects the theory to many tools in linear algebra, including the QR-decomposition, the singular value decomposition, and linear least-squares problems. Chapter 6 presents a novel iterative-recursive algorithm for computing GI for block matrices in the context of wireless MIMO communication systems within RFC. Chapter 7 deals with the development of the theory of operator means. It setups basic notations and states some background about operator monotone functions which play important roles in the theory of operator means. Chapter 8 studies a general formulation of Jensen's operator inequality for a continuous field of self-adjoint operators and a field of positive linear mappings. The aim of chapter 9 is to present a system of linear equation and inequalities in max-algebra. Max-algebra is an analogue of linear algebra developed on a pair of operations extended to matrices and vectors. Chapter 10 covers an efficient algorithm for the coarse to fine scale transition in multi-flexible-body systems with application to biomolecular systems that are modeled as articulated bodies and undergo discontinuous changes in the model definition. Finally, chapter 11 studies the structure of matrices defined over arbitrary fields whose elements are rational functions with no poles at infinity and prescribed finite poles. Complete systems of invariants are provided for each one of these equivalence relations and the relationship between both systems of invariants is clarified. This result can be seen as an extension of the classical theorem on pole assignment by Rosenbrock.

Dr. Hassan Abid Yasser
College of Science
University of Thi-Qar, Thi-Qar
Iraq

Chapter 1

3-Algebras in String Theory

Matsuo Sato

Additional information is available at the end of the chapter

http://dx.doi.org/10.5772/46480

1. Introduction

In this chapter, we review 3-algebras that appear as fundamental properties of string theory. 3-algebra is a generalization of Lie algebra; it is defined by a tri-linear bracket instead of by a bi-linear bracket, and satisfies fundamental identity, which is a generalization of Jacobi identity [1–3]. We consider 3-algebras equipped with invariant metrics in order to apply them to physics.

It has been expected that there exists M-theory, which unifies string theories. In M-theory, some structures of 3-algebras were found recently. First, it was found that by using $u(N) \oplus u(N)$ Hermitian 3-algebra, we can describe a low energy effective action of N coincident supermembranes [4–8], which are fundamental objects in M-theory.

With this as motivation, 3-algebras with invariant metrics were classified [9–22]. Lie 3-algebras are defined in real vector spaces and tri-linear brackets of them are totally anti-symmetric in all the three entries. Lie 3-algebras with invariant metrics are classified into $\mathcal{A}_4$ algebra, and Lorentzian Lie 3-algebras, which have metrics with indefinite signatures. On the other hand, Hermitian 3-algebras are defined in Hermitian vector spaces and their tri-linear brackets are complex linear and anti-symmetric in the first two entries, whereas complex anti-linear in the third entry. Hermitian 3-algebras with invariant metrics are classified into $u(N) \oplus u(M)$ and $sp(2N) \oplus u(1)$ Hermitian 3-algebras.

Moreover, recent studies have indicated that there also exist structures of 3-algebras in the Green-Schwartz supermembrane action, which defines full perturbative dynamics of a supermembrane. It had not been clear whether the total supermembrane action including fermions has structures of 3-algebras, whereas the bosonic part of the action can be described by using a tri-linear bracket, called Nambu bracket [23, 24], which is a generalization of Poisson bracket. If we fix to a light-cone gauge, the total action can be described by using Poisson bracket, that is, only structures of Lie algebra are left in this gauge [25]. However, it was shown under an approximation that the total action can be described by Nambu bracket if we fix to a semi-light-cone gauge [26]. In this gauge, the eleven dimensional space-time of M-theory is manifest in the supermembrane action, whereas only ten dimensional part is manifest in the light-cone gauge.

The BFSS matrix theory is conjectured to describe an infinite momentum frame (IMF) limit of M-theory [27] and many evidences were found. The action of the BFSS matrix theory can be obtained by replacing Poisson bracket with a finite dimensional Lie algebra's bracket in the supermembrane action in the light-cone gauge. Because of this structure, only variables that represent the ten dimensional part of the eleven-dimensional space-time are manifest in the BFSS matrix theory. Recently, 3-algebra models of M-theory were proposed [26, 28, 29], by replacing Nambu bracket with finite dimensional 3-algebras' brackets in an action that is shown, by using an approximation, to be equivalent to the semi-light-cone supermembrane action. All the variables that represent the eleven dimensional space-time are manifest in these models. It was shown that if the DLCQ limit of the 3-algebra models of M-theory is taken, they reduce to the BFSS matrix theory [26, 28], as they should [30–35].

2. Definition and classification of metric Hermitian 3-algebra

In this section, we will define and classify the Hermitian 3-algebras equipped with invariant metrics.

2.1. General structure of metric Hermitian 3-algebra

The metric Hermitian 3-algebra is a map $V \times V \times V \rightarrow V$ defined by $(x, y, z) \mapsto [x, y; z]$, where the 3-bracket is complex linear in the first two entries, whereas complex anti-linear in the last entry, equipped with a metric $< x, y >$, satisfying the following properties:
the fundamental identity

$$[[x,y;z],v;w] = [[x,v;w],y;z] + [x,[y,v;w];z] - [x,y;[z,w;v]] \tag{1}$$

the metric invariance

$$< [x,v;w],y > - < x,[y,w;v] >= 0 \tag{2}$$

and the anti-symmetry

$$[x,y;z] = -[y,x;z] \tag{3}$$

for

$$x,y,z,v,w \in V \tag{4}$$

The Hermitian 3-algebra generates a symmetry, whose generators $D(x,y)$ are defined by

$$D(x,y)z := [z,x;y] \tag{5}$$

From (1), one can show that $D(x,y)$ form a Lie algebra,

$$[D(x,y),D(v,w)] = D(D(x,y)v,w) - D(v,D(y,x)w) \tag{6}$$

There is an one-to-one correspondence between the metric Hermitian 3-algebra and a class of metric complex super Lie algebras [19]. Such a class satisfies the following conditions among complex super Lie algebras $S = S_0 \oplus S_1$, where S_0 and S_1 are even and odd parts, respectively. S_1 is decomposed as $S_1 = V \oplus \bar{V}$, where V is an unitary representation of S_0: for $a \in S_0$, $u, v \in V$,

$$[a,u] \in V \tag{7}$$

and

$$< [a,u],v > + < u,[a^*,v] > = 0 \tag{8}$$

$\bar{v} \in \bar{V}$ is defined by

$$\bar{v} = < \ ,v > \tag{9}$$

The super Lie bracket satisfies

$$[V,V] = 0, \quad [\bar{V},\bar{V}] = 0 \tag{10}$$

From the metric Hermitian 3-algebra, we obtain the class of the metric complex super Lie algebra in the following way. The elements in S_0, V, and $\bar{V}$ are defined by (5), (4), and (9), respectively. The algebra is defined by (6) and

$$\begin{aligned}
&[D(x,y),z] := D(x,y)z = [z,x;y] \\
&[D(x,y),\bar{z}] := -D(\bar{y},x)z = -[z,\bar{y};x] \\
&[x,\bar{y}] := D(x,y) \\
&[x,y] := 0 \\
&[\bar{x},\bar{y}] := 0
\end{aligned} \tag{11}$$

One can show that this algebra satisfies the super Jacobi identity and (7)-(10) as in [19].

Inversely, from the class of the metric complex super Lie algebra, we obtain the metric Hermitian 3-algebra by

$$[x,y;z] := \alpha[[y,\bar{z}],x] \tag{12}$$

where α is an arbitrary constant. One can also show that this algebra satisfies (1)-(3) for (4) as in [19].

2.2. Classification of metric Hermitian 3-algebra

The classical Lie super algebras satisfying (7)-(10) are $A(m-1,n-1)$ and $C(n+1)$. The even parts of $A(m-1,n-1)$ and $C(n+1)$ are $u(m) \oplus u(n)$ and $sp(2n) \oplus u(1)$, respectively. Because the metric Hermitian 3-algebra one-to-one corresponds to this class of the super Lie algebra, the metric Hermitian 3-algebras are classified into $u(m) \oplus u(n)$ and $sp(2n) \oplus u(1)$ Hermitian 3-algebras.

First, we will construct the $u(m) \oplus u(n)$ Hermitian 3-algebra from $A(m-1,n-1)$, according to the relation in the previous subsection. $A(m-1,n-1)$ is simple and is obtained by dividing $sl(m,n)$ by its ideal. That is, $A(m-1,n-1) = sl(m,n)$ when $m \neq n$ and $A(n-1,n-1) = sl(n,n)/\lambda 1_{2n}$.

Real $sl(m,n)$ is defined by

$$\begin{pmatrix} h_1 & c \\ ic^{\dagger} & h_2 \end{pmatrix} \tag{13}$$

where h_1 and h_2 are $m \times m$ and $n \times n$ anti-Hermite matrices and c is an $n \times m$ arbitrary complex matrix. Complex $sl(m,n)$ is a complexification of real $sl(m,n)$, given by

$$\begin{pmatrix} \alpha & \beta \\ \gamma & \delta \end{pmatrix} \tag{14}$$

where α, β, γ, and δ are $m \times m$, $n \times m$, $m \times n$, and $n \times n$ complex matrices that satisfy

$$\mathrm{tr}\alpha = \mathrm{tr}\delta \tag{15}$$

Complex $A(m-1,n-1)$ is decomposed as $A(m-1,n-1) = S_0 \oplus V \oplus \bar{V}$, where

$$\begin{pmatrix} \alpha & 0 \\ 0 & \delta \end{pmatrix} \in S_0$$
$$\begin{pmatrix} 0 & \beta \\ 0 & 0 \end{pmatrix} \in V$$
$$\begin{pmatrix} 0 & 0 \\ \gamma & 0 \end{pmatrix} \in \bar{V} \tag{16}$$

(9) is rewritten as $V \to \bar{V}$ defined by

$$B = \begin{pmatrix} 0 & \beta \\ 0 & 0 \end{pmatrix} \mapsto B^\dagger = \begin{pmatrix} 0 & 0 \\ \beta^\dagger & 0 \end{pmatrix} \tag{17}$$

where $B \in V$ and $B^\dagger \in \bar{V}$. (12) is rewritten as

$$[X, Y; Z] = \alpha[[Y, Z^\dagger], X] = \alpha \begin{pmatrix} 0 & yz^\dagger x - xz^\dagger y \\ 0 & 0 \end{pmatrix} \tag{18}$$

for

$$X = \begin{pmatrix} 0 & x \\ 0 & 0 \end{pmatrix} \in V$$
$$Y = \begin{pmatrix} 0 & y \\ 0 & 0 \end{pmatrix} \in V$$
$$Z = \begin{pmatrix} 0 & z \\ 0 & 0 \end{pmatrix} \in V \tag{19}$$

As a result, we obtain the $u(m) \oplus u(n)$ Hermitian 3-algebra,

$$[x, y; z] = \alpha(yz^\dagger x - xz^\dagger y) \tag{20}$$

where x, y, and z are arbitrary $n \times m$ complex matrices. This algebra was originally constructed in [8].

Inversely, from (20), we can construct complex $A(m-1,n-1)$. (5) is rewritten as

$$D(x, y) = (xy^\dagger, y^\dagger x) \in S_0 \tag{21}$$

(6) and (11) are rewritten as

$$\begin{aligned}
&[(xy^\dagger, y^\dagger x), (x'y'^\dagger, y'^\dagger x')] = ([xy^\dagger, x'y'^\dagger], [y^\dagger x, y'^\dagger x']) \\
&[(xy^\dagger, y^\dagger x), z] = xy^\dagger z - zy^\dagger x \\
&[(xy^\dagger, y^\dagger x), w^\dagger] = y^\dagger x w^\dagger - w^\dagger x y^\dagger \\
&[x, y^\dagger] = (xy^\dagger, y^\dagger x) \\
&[x, y] = 0 \\
&[x^\dagger, y^\dagger] = 0
\end{aligned} \tag{22}$$

This algebra is summarized as

$$\left[\begin{pmatrix} xy^\dagger & z \\ w^\dagger & y^\dagger x \end{pmatrix}, \begin{pmatrix} x'y'^\dagger & z' \\ w'^\dagger & y'^\dagger x' \end{pmatrix}\right] \tag{23}$$

which forms complex $A(m-1, n-1)$.

Next, we will construct the $sp(2n) \oplus u(1)$ Hermitian 3-algebra from $C(n+1)$. Complex $C(n+1)$ is decomposed as $C(n+1) = S_0 \oplus V \oplus \bar{V}$. The elements are given by

$$\begin{pmatrix} \alpha & 0 & 0 & 0 \\ 0 & -\alpha & 0 & 0 \\ 0 & 0 & a & b \\ 0 & 0 & c & -a^T \end{pmatrix} \in S_0$$
$$\begin{pmatrix} 0 & 0 & x_1 & x_2 \\ 0 & 0 & 0 & 0 \\ 0 & x_2^T & 0 & 0 \\ 0 & -x_1^T & 0 & 0 \end{pmatrix} \in V$$
$$\begin{pmatrix} 0 & 0 & 0 & 0 \\ 0 & 0 & y_1 & y_2 \\ y_2^T & 0 & 0 & 0 \\ -y_1^T & 0 & 0 & 0 \end{pmatrix} \in \bar{V} \tag{24}$$

where α is a complex number, a is an arbitrary $n \times n$ complex matrix, b and c are $n \times n$ complex symmetric matrices, and x_1, x_2, y_1 and y_2 are $n \times 1$ complex matrices. (9) is rewritten as $V \to \bar{V}$ defined by $B \mapsto \bar{B} = UB^*U^{-1}$, where $B \in V$, $\bar{B} \in \bar{V}$ and

$$U = \begin{pmatrix} 0 & 1 & 0 & 0 \\ 1 & 0 & 0 & 0 \\ 0 & 0 & 0 & \mathbf{1} \\ 0 & 0 & -\mathbf{1} & 0 \end{pmatrix} \tag{25}$$

Explicitly,

$$B = \begin{pmatrix} 0 & 0 & x_1 & x_2 \\ 0 & 0 & 0 & 0 \\ 0 & x_2^T & 0 & 0 \\ 0 & -x_1^T & 0 & 0 \end{pmatrix} \mapsto \bar{B} = \begin{pmatrix} 0 & 0 & 0 & 0 \\ 0 & 0 & x_2^* & -x_1^* \\ -x_1^\dagger & 0 & 0 & 0 \\ -x_2^\dagger & 0 & 0 & 0 \end{pmatrix} \tag{26}$$

(12) is rewritten as

$$\begin{aligned} [X, Y; Z] &:= \alpha[[Y, \bar{Z}], X] \\ &= \alpha \left[\left[\begin{pmatrix} 0 & 0 & y_1 & y_2 \\ 0 & 0 & 0 & 0 \\ 0 & y_2^T & 0 & 0 \\ 0 & -y_1^T & 0 & 0 \end{pmatrix}, \begin{pmatrix} 0 & 0 & 0 & 0 \\ 0 & 0 & z_2^* & -z_1^* \\ -z_1^\dagger & 0 & 0 & 0 \\ -z_2^\dagger & 0 & 0 & 0 \end{pmatrix}\right], \begin{pmatrix} 0 & 0 & x_1 & x_2 \\ 0 & 0 & 0 & 0 \\ 0 & x_2^T & 0 & 0 \\ 0 & -x_1^T & 0 & 0 \end{pmatrix}\right] \\ &= \alpha \begin{pmatrix} 0 & 0 & w_1 & w_2 \\ 0 & 0 & 0 & 0 \\ 0 & w_2^T & 0 & 0 \\ 0 & -w_1^T & 0 & 0 \end{pmatrix} \end{aligned} \tag{27}$$

for

$$X = \begin{pmatrix} 0 & 0 & x_1 & x_2 \\ 0 & 0 & 0 & 0 \\ 0 & x_2^T & 0 & 0 \\ 0 & -x_1^T & 0 & 0 \end{pmatrix} \in V$$

$$Y = \begin{pmatrix} 0 & 0 & y_1 & y_2 \\ 0 & 0 & 0 & 0 \\ 0 & y_2^T & 0 & 0 \\ 0 & -y_1^T & 0 & 0 \end{pmatrix} \in V$$

$$Z = \begin{pmatrix} 0 & 0 & z_1 & z_2 \\ 0 & 0 & 0 & 0 \\ 0 & z_2^T & 0 & 0 \\ 0 & -z_1^T & 0 & 0 \end{pmatrix} \in V \tag{28}$$

where w_1 and w_2 are given by

$$(w_1, w_2) = -(y_1 z_1^\dagger + y_2 z_2^\dagger)(x_1, x_2) + (x_1 z_1^\dagger + x_2 z_2^\dagger)(y_1, y_2) + (x_2 y_1^T - x_1 y_2^T)(z_2^*, -z_1^*) \tag{29}$$

As a result, we obtain the $sp(2n) \oplus u(1)$ Hermitian 3-algebra,

$$[x, y; z] = \alpha((y \odot \tilde{z})x + (\tilde{z} \odot x)y - (x \odot y)\tilde{z}) \tag{30}$$

for $x = (x_1, x_2)$, $y = (y_1, y_2)$, $z = (z_1, z_2)$, where x_1, x_2, y_1, y_2, z_1, and z_2 are n-vectors and

$$\begin{aligned} \tilde{z} &= (z_2^*, -z_1^*) \\ a \odot b &= a_1 \cdot b_2 - a_2 \cdot b_1 \end{aligned} \tag{31}$$

3. 3-algebra model of M-theory

In this section, we review the fact that the supermembrane action in a semi-light-cone gauge can be described by Nambu bracket, where structures of 3-algebra are manifest. The 3-algebra Models of M-theory are defined based on the semi-light-cone supermembrane action. We also review that the models reduce to the BFSS matrix theory in the DLCQ limit.

3.1. Supermembrane and 3-algebra model of M-theory

The fundamental degrees of freedom in M-theory are supermembranes. The action of the covariant supermembrane action in M-theory [36] is given by

$$\begin{aligned} S_{M2} = \int d^3\sigma \Big(\sqrt{-G} + \frac{i}{4}\epsilon^{\alpha\beta\gamma}\bar{\Psi}\Gamma_{MN}\partial_\alpha\Psi(\Pi_\beta{}^M\Pi_\gamma{}^N + \frac{i}{2}\Pi_\beta{}^M\bar{\Psi}\Gamma^N\partial_\gamma\Psi \\ -\frac{1}{12}\bar{\Psi}\Gamma^M\partial_\beta\Psi\bar{\Psi}\Gamma^N\partial_\gamma\Psi)\Big) \end{aligned} \tag{32}$$

where $M, N = 0, \cdots, 10$, $\alpha, \beta, \gamma = 0, 1, 2$, $G_{\alpha\beta} = \Pi_\alpha{}^M \Pi_{\beta M}$ and $\Pi_\alpha{}^M = \partial_\alpha X^M - \frac{i}{2}\bar{\Psi}\Gamma^M\partial_\alpha\Psi$. Ψ is a $SO(1,10)$ Majorana fermion.

This action is invariant under dynamical supertransformations,

$$\begin{aligned}\delta\Psi &= \epsilon \\ \delta X^M &= -i\bar{\Psi}\Gamma^M\epsilon\end{aligned} \tag{33}$$

These transformations form the $\mathcal{N}=1$ supersymmetry algebra in eleven dimensions,

$$[\delta_1,\delta_2]X^M = -2i\epsilon_1\Gamma^M\epsilon_2$$

$$[\delta_1,\delta_2]\Psi = 0 \tag{34}$$

The action is also invariant under the κ-symmetry transformations,

$$\begin{aligned}\delta\Psi &= (1+\Gamma)\kappa(\sigma) \\ \delta X^M &= i\bar{\Psi}\Gamma^M(1+\Gamma)\kappa(\sigma)\end{aligned} \tag{35}$$

where

$$\Gamma = \frac{1}{3!\sqrt{-G}}\epsilon^{\alpha\beta\gamma}\Pi^L_\alpha\Pi^M_\beta\Pi^N_\gamma\Gamma_{LMN} \tag{36}$$

If we fix the κ-symmetry (35) of the action by taking a semi-light-cone gauge [26][1]

$$\Gamma^{012}\Psi = -\Psi \tag{37}$$

we obtain a semi-light-cone supermembrane action,

$$\begin{aligned}S_{M2} = \int d^3\sigma\Big(\sqrt{-G} + \frac{i}{4}\epsilon^{\alpha\beta\gamma}(\bar{\Psi}\Gamma_{\mu\nu}\partial_\alpha\Psi(\Pi_\beta{}^\mu\Pi_\gamma{}^\nu + \frac{i}{2}\Pi_\beta{}^\mu\bar{\Psi}\Gamma^\nu\partial_\gamma\Psi - \frac{1}{12}\bar{\Psi}\Gamma^\mu\partial_\beta\Psi\bar{\Psi}\Gamma^\nu\partial_\gamma\Psi) \\ +\bar{\Psi}\Gamma_{IJ}\partial_\alpha\Psi\partial_\beta X^I\partial_\gamma X^J)\Big)\end{aligned} \tag{38}$$

where $G_{\alpha\beta} = h_{\alpha\beta} + \Pi_\alpha{}^\mu\Pi_{\beta\mu}$, $\Pi_\alpha{}^\mu = \partial_\alpha X^\mu - \frac{i}{2}\bar{\Psi}\Gamma^\mu\partial_\alpha\Psi$, and $h_{\alpha\beta} = \partial_\alpha X^I\partial_\beta X_I$.

In [26], it is shown under an approximation up to the quadratic order in $\partial_\alpha X^\mu$ and $\partial_\alpha\Psi$ but exactly in X^I, that this action is equivalent to the continuum action of the 3-algebra model of M-theory,

$$\begin{aligned}S_{cl} = \int d^3\sigma\sqrt{-g}\Big(&-\frac{1}{12}\{X^I,X^J,X^K\}^2 - \frac{1}{2}(A_{\mu ab}\{\varphi^a,\varphi^b,X^I\})^2 \\ &-\frac{1}{3}E^{\mu\nu\lambda}A_{\mu ab}A_{\nu cd}A_{\lambda ef}\{\varphi^a,\varphi^c,\varphi^d\}\{\varphi^b,\varphi^e,\varphi^f\} + \frac{1}{2}\Lambda \\ &-\frac{i}{2}\bar{\Psi}\Gamma^\mu A_{\mu ab}\{\varphi^a,\varphi^b,\Psi\} + \frac{i}{4}\bar{\Psi}\Gamma_{IJ}\{X^I,X^J,\Psi\}\Big)\end{aligned} \tag{39}$$

where $I,J,K = 3,\cdots,10$ and $\{\varphi^a,\varphi^b,\varphi^c\} = \epsilon^{\alpha\beta\gamma}\partial_\alpha\varphi^a\partial_\beta\varphi^b\partial_\gamma\varphi^c$ is the Nambu-Poisson bracket. An invariant symmetric bilinear form is defined by $\int d^3\sigma\sqrt{-g}\varphi^a\varphi^b$ for complete basis φ^a in three dimensions. Thus, this action is manifestly VPD covariant even when the world-volume metric is flat. X^I is a scalar and Ψ is a $SO(1,2)\times SO(8)$ Majorana-Weyl fermion

[1] Advantages of a semi-light-cone gauges against a light-cone gauge are shown in [37–39]

satisfying (37). $E^{\mu\nu\lambda}$ is a Levi-Civita symbol in three dimensions and Λ is a cosmological constant.

The continuum action of 3-algebra model of M-theory (39) is invariant under 16 dynamical supersymmetry transformations,

$$\begin{aligned}
\delta X^I &= i\bar{\epsilon}\Gamma^I\Psi \\
\delta A_\mu(\sigma,\sigma') &= \frac{i}{2}\bar{\epsilon}\Gamma_\mu\Gamma_I(X^I(\sigma)\Psi(\sigma') - X^I(\sigma')\Psi(\sigma)), \\
\delta\Psi &= -A_{\mu ab}\{\varphi^a,\varphi^b,X^I\}\Gamma^\mu\Gamma_I\epsilon - \frac{1}{6}\{X^I,X^J,X^K\}\Gamma_{IJK}\epsilon
\end{aligned} \tag{40}$$

where $\Gamma_{012}\epsilon = -\epsilon$. These supersymmetries close into gauge transformations on-shell,

$$\begin{aligned}
[\delta_1,\delta_2]X^I &= \Lambda_{cd}\{\varphi^c,\varphi^d,X^I\} \\
[\delta_1,\delta_2]A_{\mu ab}\{\varphi^a,\varphi^b,\quad\} &= \Lambda_{ab}\{\varphi^a,\varphi^b,A_{\mu cd}\{\varphi^c,\varphi^d,\quad\}\} \\
&\quad - A_{\mu ab}\{\varphi^a,\varphi^b,\Lambda_{cd}\{\varphi^c,\varphi^d,\quad\}\} + 2i\bar{\epsilon}_2\Gamma^\nu\epsilon_1 O^A_{\mu\nu} \\
[\delta_1,\delta_2]\Psi &= \Lambda_{cd}\{\varphi^c,\varphi^d,\Psi\} + (i\bar{\epsilon}_2\Gamma^\mu\epsilon_1\Gamma_\mu - \frac{i}{4}\bar{\epsilon}_2\Gamma^{KL}\epsilon_1\Gamma_{KL})O^\Psi
\end{aligned} \tag{41}$$

where gauge parameters are given by $\Lambda_{ab} = 2i\bar{\epsilon}_2\Gamma^\mu\epsilon_1 A_{\mu ab} - i\bar{\epsilon}_2\Gamma_{JK}\epsilon_1 X^J_a X^K_b$. $O^A_{\mu\nu} = 0$ and $O^\Psi = 0$ are equations of motions of $A_{\mu\nu}$ and Ψ, respectively, where

$$\begin{aligned}
O^A_{\mu\nu} &= A_{\mu ab}\{\varphi^a,\varphi^b,A_{\nu cd}\{\varphi^c,\varphi^d,\quad\}\} - A_{\nu ab}\{\varphi^a,\varphi^b,A_{\mu cd}\{\varphi^c,\varphi^d,\quad\}\} \\
&\quad + E_{\mu\nu\lambda}(-\{X^I,A^\lambda_{ab}\{\varphi^a,\varphi^b,X_I\},\quad\} + \frac{i}{2}\{\bar{\Psi},\Gamma^\lambda\Psi,\quad\}) \\
O^\Psi &= -\Gamma^\mu A_{\mu ab}\{\varphi^a,\varphi^b,\Psi\} + \frac{1}{2}\Gamma_{IJ}\{X^I,X^J,\Psi\}
\end{aligned} \tag{42}$$

(41) implies that a commutation relation between the dynamical supersymmetry transformations is

$$\delta_2\delta_1 - \delta_1\delta_2 = 0 \tag{43}$$

up to the equations of motions and the gauge transformations.

This action is invariant under a translation,

$$\delta X^I(\sigma) = \eta^I, \qquad \delta A^\mu(\sigma,\sigma') = \eta^\mu(\sigma) - \eta^\mu(\sigma') \tag{44}$$

where η^I are constants.

The action is also invariant under 16 kinematical supersymmetry transformations

$$\tilde{\delta}\Psi = \tilde{\epsilon} \tag{45}$$

and the other fields are not transformed. $\tilde{\epsilon}$ is a constant and satisfy $\Gamma_{012}\tilde{\epsilon} = \tilde{\epsilon}$. $\tilde{\epsilon}$ and ϵ should come from sixteen components of thirty-two $\mathcal{N} = 1$ supersymmetry parameters in eleven dimensions, corresponding to eigen values ± 1 of Γ_{012}, respectively. This $\mathcal{N} = 1$ supersymmetry consists of remaining 16 target-space supersymmetries and transmuted 16 κ-symmetries in the semi-light-cone gauge [25, 26, 40].

A commutation relation between the kinematical supersymmetry transformations is given by

$$\tilde{\delta}_2\tilde{\delta}_1 - \tilde{\delta}_1\tilde{\delta}_2 = 0 \tag{46}$$

A commutator of dynamical supersymmetry transformations and kinematical ones acts as

$$\begin{aligned}(\tilde{\delta}_2\delta_1 - \delta_1\tilde{\delta}_2)X^I(\sigma) &= i\bar{\epsilon}_1\Gamma^I\tilde{\epsilon}_2 \equiv \eta_0^I \\ (\tilde{\delta}_2\delta_1 - \delta_1\tilde{\delta}_2)A^\mu(\sigma,\sigma') &= \frac{i}{2}\bar{\epsilon}_1\Gamma^\mu\Gamma_I(X^I(\sigma) - X^I(\sigma'))\tilde{\epsilon}_2 \equiv \eta_0^\mu(\sigma) - \eta_0^\mu(\sigma')\end{aligned} \tag{47}$$

where the commutator that acts on the other fields vanishes. Thus, the commutation relation is given by

$$\tilde{\delta}_2\delta_1 - \delta_1\tilde{\delta}_2 = \delta_\eta \tag{48}$$

where δ_η is a translation.

If we change a basis of the supersymmetry transformations as

$$\begin{aligned}\delta' &= \delta + \tilde{\delta} \\ \tilde{\delta}' &= i(\delta - \tilde{\delta})\end{aligned} \tag{49}$$

we obtain

$$\begin{aligned}\delta_2'\delta_1' - \delta_1'\delta_2' &= \delta_\eta \\ \tilde{\delta}_2'\tilde{\delta}_1' - \tilde{\delta}_1'\tilde{\delta}_2' &= \delta_\eta \\ \tilde{\delta}_2'\delta_1' - \delta_1'\tilde{\delta}_2' &= 0\end{aligned} \tag{50}$$

These thirty-two supersymmetry transformations are summarised as $\Delta = (\delta', \tilde{\delta}')$ and (50) implies the $\mathcal{N} = 1$ supersymmetry algebra in eleven dimensions,

$$\Delta_2\Delta_1 - \Delta_1\Delta_2 = \delta_\eta \tag{51}$$

3.2. Lie 3-algebra models of M-theory

In this and next subsection, we perform the second quantization on the continuum action of the 3-algebra model of M-theory: By replacing the Nambu-Poisson bracket in the action (39) with brackets of finite-dimensional 3-algebras, Lie and Hermitian 3-algebras, we obtain the Lie and Hermitian 3-algebra models of M-theory [26, 28], respectively. In this section, we review the Lie 3-algebra model.

If we replace the Nambu-Poisson bracket in the action (39) with a completely antisymmetric real 3-algebra's bracket [21, 22],

$$\begin{aligned}\int d^3\sigma\sqrt{-g} &\to \langle\ \rangle \\ \{\varphi^a, \varphi^b, \varphi^c\} &\to [T^a, T^b, T^c]\end{aligned} \tag{52}$$

we obtain the Lie 3-algebra model of M-theory [26, 28],

$$\begin{aligned}S_0 = \Big\langle &-\frac{1}{12}[X^I, X^J, X^K]^2 - \frac{1}{2}(A_{\mu ab}[T^a, T^b, X^I])^2 \\ &-\frac{1}{3}E^{\mu\nu\lambda}A_{\mu ab}A_{\nu cd}A_{\lambda ef}[T^a, T^c, T^d][T^b, T^e, T^f] \\ &-\frac{i}{2}\bar{\Psi}\Gamma^\mu A_{\mu ab}[T^a, T^b, \Psi] + \frac{i}{4}\bar{\Psi}\Gamma_{IJ}[X^I, X^J, \Psi]\Big\rangle\end{aligned} \tag{53}$$

We have deleted the cosmological constant Λ, which corresponds to an operator ordering ambiguity, as usual as in the case of other matrix models [27, 41].

This model can be obtained formally by a dimensional reduction of the $\mathcal{N} = 8$ BLG model [4–6],

$$\begin{aligned} S_{\mathcal{N}=8BLG} = \int d^3x \Big\langle &-\frac{1}{12}[X^I, X^J, X^K]^2 - \frac{1}{2}(D_\mu X^I)^2 - E^{\mu\nu\lambda}\big(\frac{1}{2}A_{\mu ab}\partial_\nu A_{\lambda cd}T^a[T^b, T^c, T^d] \\ &+\frac{1}{3}A_{\mu ab}A_{\nu cd}A_{\lambda ef}[T^a, T^c, T^d][T^b, T^e, T^f]\big) \\ &+\frac{i}{2}\bar{\Psi}\Gamma^\mu D_\mu \Psi + \frac{i}{4}\bar{\Psi}\Gamma_{IJ}[X^I, X^J, \Psi]\Big\rangle \end{aligned} \tag{54}$$

The formal relations between the Lie (Hermitian) 3-algebra models of M-theory and the $\mathcal{N} = 8$ ($\mathcal{N} = 6$) BLG models are analogous to the relation among the $\mathcal{N} = 4$ super Yang-Mills in four dimensions, the BFSS matrix theory [27], and the IIB matrix model [41]. They are completely different theories although they are related to each others by dimensional reductions. In the same way, the 3-algebra models of M-theory and the BLG models are completely different theories.

The fields in the action (53) are spanned by the Lie 3-algebra T^a as $X^I = X^I_a T^a$, $\Psi = \Psi_a T^a$ and $A^\mu = A^\mu_{ab} T^a \otimes T^b$, where $I = 3, \cdots, 10$ and $\mu = 0, 1, 2$. $<>$ represents a metric for the 3-algebra. Ψ is a Majorana spinor of SO(1,10) that satisfies $\Gamma_{012}\Psi = \Psi$. $E^{\mu\nu\lambda}$ is a Levi-Civita symbol in three-dimensions.

Finite dimensional Lie 3-algebras with an invariant metric is classified into four-dimensional Euclidean $\mathcal{A}_4$ algebra and the Lie 3-algebras with indefinite metrics in [9–11, 21, 22]. We do not choose $\mathcal{A}_4$ algebra because its degrees of freedom are just four. We need an algebra with arbitrary dimensions N, which is taken to infinity to define M-theory. Here we choose the most simple indefinite metric Lie 3-algebra, so called the Lorentzian Lie 3-algebra associated with $u(N)$ Lie algebra,

$$\begin{aligned} &[T^{-1}, T^a, T^b] = 0 \\ &[T^0, T^i, T^j] = [T^i, T^j] = f^{ij}{}_k T^k \\ &[T^i, T^j, T^k] = f^{ijk} T^{-1} \end{aligned} \tag{55}$$

where $a = -1, 0, i$ ($i = 1, \cdots, N^2$). T^i are generators of $u(N)$. A metric is defined by a symmetric bilinear form,

$$< T^{-1}, T^0 > = -1 \tag{56}$$

$$< T^i, T^j > = h^{ij} \tag{57}$$

and the other components are 0. The action is decomposed as

$$\begin{aligned} S = \mathrm{Tr}(&-\frac{1}{4}(x_0^K)^2[x^I, x^J]^2 + \frac{1}{2}(x_0^I[x_I, x^J])^2 - \frac{1}{2}(x_0^I b_\mu + [a_\mu, x^I])^2 - \frac{1}{2}E^{\mu\nu\lambda} b_\mu [a_\nu, a_\lambda] \\ &+ i\bar{\psi}_0 \Gamma^\mu b_\mu \psi - \frac{i}{2}\bar{\psi}\Gamma^\mu [a_\mu, \psi] + \frac{i}{2} x_0^I \bar{\psi}\Gamma_{IJ}[x^J, \psi] - \frac{i}{2}\bar{\psi}_0 \Gamma_{IJ}[x^I, x^J]\psi) \end{aligned} \tag{58}$$

where we have renamed $X_0^I \to x_0^I$, $X_i^I T^i \to x^I$, $\Psi_0 \to \psi_0$, $\Psi_i T^i \to \psi$, $2A_{\mu 0i} T^i \to a_\mu$, and $A_{\mu ij}[T^i, T^j] \to b_\mu$. a_μ correspond to the target coordinate matrices X^μ, whereas b_μ are auxiliary fields.

In this action, T^{-1} mode; X_{-1}^I, Ψ_{-1} or A_{-1a}^μ does not appear, that is they are unphysical modes. Therefore, the indefinite part of the metric (56) does not exist in the action and the Lie 3-algebra model of M-theory is ghost-free like a model in [42]. This action can be obtained by a dimensional reduction of the three-dimensional $\mathcal{N} = 8$ BLG model [4–6] with the same 3-algebra. The BLG model possesses a ghost mode because of its kinetic terms with indefinite signature. On the other hand, the Lie 3-algebra model of M-theory does not possess a kinetic term because it is defined as a zero-dimensional field theory like the IIB matrix model [41].

This action is invariant under the translation

$$\delta x^I = \eta^I, \qquad \delta a^\mu = \eta^\mu \tag{59}$$

where η^I and η^μ belong to $u(1)$. This implies that eigen values of x^I and a^μ represent an eleven-dimensional space-time.

The action is also invariant under 16 kinematical supersymmetry transformations

$$\tilde{\delta}\psi = \tilde{\epsilon} \tag{60}$$

and the other fields are not transformed. $\tilde{\epsilon}$ belong to $u(1)$ and satisfy $\Gamma_{012}\tilde{\epsilon} = \tilde{\epsilon}$. $\tilde{\epsilon}$ and ϵ should come from sixteen components of thirty-two $\mathcal{N} = 1$ supersymmetry parameters in eleven dimensions, corresponding to eigen values ± 1 of Γ_{012}, respectively, as in the previous subsection.

A commutation relation between the kinematical supersymmetry transformations is given by

$$\tilde{\delta}_2\tilde{\delta}_1 - \tilde{\delta}_1\tilde{\delta}_2 = 0 \tag{61}$$

The action is invariant under 16 dynamical supersymmetry transformations,

$$\begin{aligned}
&\delta X^I = i\bar{\epsilon}\Gamma^I\Psi \\
&\delta A_{\mu ab}[T^a, T^b, \quad] = i\bar{\epsilon}\Gamma_\mu\Gamma_I[X^I, \Psi, \quad] \\
&\delta\Psi = -A_{\mu ab}[T^a, T^b, X^I]\Gamma^\mu\Gamma_I\epsilon - \frac{1}{6}[X^I, X^J, X^K]\Gamma_{IJK}\epsilon
\end{aligned} \tag{62}$$

where $\Gamma_{012}\epsilon = -\epsilon$. These supersymmetries close into gauge transformations on-shell,

$$\begin{aligned}
&[\delta_1, \delta_2]X^I = \Lambda_{cd}[T^c, T^d, X^I] \\
&[\delta_1, \delta_2]A_{\mu ab}[T^a, T^b, \quad] = \Lambda_{ab}[T^a, T^b, A_{\mu cd}[T^c, T^d, \quad]] \\
&\qquad - A_{\mu ab}[T^a, T^b, \Lambda_{cd}[T^c, T^d, \quad]] + 2i\bar{\epsilon}_2\Gamma^\nu\epsilon_1 O_{\mu\nu}^A \\
&[\delta_1, \delta_2]\Psi = \Lambda_{cd}[T^c, T^d, \Psi] + (i\bar{\epsilon}_2\Gamma^\mu\epsilon_1\Gamma_\mu - \frac{i}{4}\bar{\epsilon}_2\Gamma^{KL}\epsilon_1\Gamma_{KL})O^\Psi
\end{aligned} \tag{63}$$

where gauge parameters are given by $\Lambda_{ab} = 2i\bar{\epsilon}_2\Gamma^{\mu}\epsilon_1 A_{\mu ab} - i\bar{\epsilon}_2\Gamma_{JK}\epsilon_1 X^J_a X^K_b$. $O^A_{\mu\nu} = 0$ and $O^{\Psi} = 0$ are equations of motions of $A_{\mu\nu}$ and Ψ, respectively, where

$$\begin{aligned} O^A_{\mu\nu} &= A_{\mu ab}[T^a, T^b, A_{\nu cd}[T^c, T^d, \quad]] - A_{\nu ab}[T^a, T^b, A_{\mu cd}[T^c, T^d, \quad]] \\ &\quad + E_{\mu\nu\lambda}(-[X^I, A^{\lambda}_{ab}[T^a, T^b, X_I], \quad] + \frac{i}{2}[\bar{\Psi}, \Gamma^{\lambda}\Psi, \quad]) \\ O^{\Psi} &= -\Gamma^{\mu} A_{\mu ab}[T^a, T^b, \Psi] + \frac{1}{2}\Gamma_{IJ}[X^I, X^J, \Psi] \end{aligned} \tag{64}$$

(63) implies that a commutation relation between the dynamical supersymmetry transformations is

$$\delta_2\delta_1 - \delta_1\delta_2 = 0 \tag{65}$$

up to the equations of motions and the gauge transformations.

The 16 dynamical supersymmetry transformations (62) are decomposed as

$$\begin{aligned} \delta x^I &= i\bar{\epsilon}\Gamma^I\psi \\ \delta x^I_0 &= i\bar{\epsilon}\Gamma^I\psi_0 \\ \delta x^I_{-1} &= i\bar{\epsilon}\Gamma^I\psi_{-1} \\ \\ \delta\psi &= -(b_{\mu}x^I_0 + [a_{\mu}, x^I])\Gamma^{\mu}\Gamma_I\epsilon - \frac{1}{2}x^I_0[x^J, x^K]\Gamma_{IJK}\epsilon \\ \delta\psi_0 &= 0 \\ \delta\psi_{-1} &= -\mathrm{Tr}(b_{\mu}x^I)\Gamma^{\mu}\Gamma_I\epsilon - \frac{1}{6}\mathrm{Tr}([x^I, x^J]x^K)\Gamma_{IJK}\epsilon \\ \\ \delta a_{\mu} &= i\bar{\epsilon}\Gamma_{\mu}\Gamma_I(x^I_0\psi - \psi_0 x^I) \\ \delta b_{\mu} &= i\bar{\epsilon}\Gamma_{\mu}\Gamma_I[x^I, \psi] \\ \delta A_{\mu-1i} &= i\bar{\epsilon}\Gamma_{\mu}\Gamma_I\frac{1}{2}(x^I_{-1}\psi_i - \psi_{-1}x^I_i) \\ \delta A_{\mu-10} &= i\bar{\epsilon}\Gamma_{\mu}\Gamma_I\frac{1}{2}(x^I_{-1}\psi_0 - \psi_{-1}x^I_0) \end{aligned} \tag{66}$$

and thus a commutator of dynamical supersymmetry transformations and kinematical ones acts as

$$\begin{aligned} (\tilde{\delta}_2\delta_1 - \delta_1\tilde{\delta}_2)x^I &= i\bar{\epsilon}_1\Gamma^I\tilde{\epsilon}_2 \equiv \eta^I \\ (\tilde{\delta}_2\delta_1 - \delta_1\tilde{\delta}_2)a^{\mu} &= i\bar{\epsilon}_1\Gamma^{\mu}\Gamma_I x^I_0\tilde{\epsilon}_2 \equiv \eta^{\mu} \\ (\tilde{\delta}_2\delta_1 - \delta_1\tilde{\delta}_2)A^{\mu}_{-1i}T^i &= \frac{1}{2}i\bar{\epsilon}_1\Gamma^{\mu}\Gamma_I x^I_{-1}\tilde{\epsilon}_2 \end{aligned} \tag{67}$$

where the commutator that acts on the other fields vanishes. Thus, the commutation relation for physical modes is given by

$$\tilde{\delta}_2\delta_1 - \delta_1\tilde{\delta}_2 = \delta_{\eta} \tag{68}$$

where δ_{η} is a translation.

(61), (65), and (68) imply the $\mathcal{N} = 1$ supersymmetry algebra in eleven dimensions as in the previous subsection.

3.3. Hermitian 3-algebra model of M-theory

In this subsection, we study the Hermitian 3-algebra models of M-theory [26]. Especially, we study mostly the model with the $u(N) \oplus u(N)$ Hermitian 3-algebra (20).

The continuum action (39) can be rewritten by using the triality of $SO(8)$ and the $SU(4) \times U(1)$ decomposition [8, 43, 44] as

$$\begin{aligned} S_{cl} = \int d^3\sigma \sqrt{-g}\Big(&-V - A_{\mu ba}\{Z^A, T^a, T^b\} A^{\mu}_{dc}\{Z_A, T^c, T^d\} \\ &+\frac{1}{3}E^{\mu\nu\lambda} A_{\mu ba} A_{\nu dc} A_{\lambda fe}\{T^a, T^c, T^d\}\{T^b, T^f, T^e\} \\ &+i\bar{\psi}^A \Gamma^\mu A_{\mu ba}\{\psi_A, T^a, T^b\} + \frac{i}{2}E_{ABCD}\bar{\psi}^A\{Z^C, Z^D, \psi^B\} - \frac{i}{2}E^{ABCD} Z_D\{\bar{\psi}_A, \psi_B, Z_C\} \\ &-i\bar{\psi}^A\{\psi_A, Z^B, Z_B\} + 2i\bar{\psi}^A\{\psi_B, Z^B, Z_A\}\Big) \end{aligned} \tag{69}$$

where fields with a raised A index transform in the 4 of SU(4), whereas those with lowered one transform in the $\bar{4}$. $A_{\mu ba}$ ($\mu = 0, 1, 2$) is an anti-Hermitian gauge field, Z^A and Z_A are a complex scalar field and its complex conjugate, respectively. ψ_A is a fermion field that satisfies

$$\Gamma^{012}\psi_A = -\psi_A \tag{70}$$

and ψ^A is its complex conjugate. $E^{\mu\nu\lambda}$ and E^{ABCD} are Levi-Civita symbols in three dimensions and four dimensions, respectively. The potential terms are given by

$$\begin{aligned} V &= \frac{2}{3}\Upsilon^{CD}_B \Upsilon^B_{CD} \\ \Upsilon^{CD}_B &= \{Z^C, Z^D, Z_B\} - \frac{1}{2}\delta^C_B\{Z^E, Z^D, Z_E\} + \frac{1}{2}\delta^D_B\{Z^E, Z^C, Z_E\} \end{aligned} \tag{71}$$

If we replace the Nambu-Poisson bracket with a Hermitian 3-algebra's bracket [19, 20],

$$\begin{aligned} &\int d^3\sigma\sqrt{-g} \rightarrow \langle\ \rangle \\ &\{\varphi^a, \varphi^b, \varphi^c\} \rightarrow [T^a, T^b; \bar{T}^{\bar{c}}] \end{aligned} \tag{72}$$

we obtain the Hermitian 3-algebra model of M-theory [26],

$$\begin{aligned} S = \Big\langle &-V - A_{\mu\bar{b}a}[Z^A, T^a; \bar{T}^{\bar{b}}] A^{\mu}_{\bar{d}c}\overline{[Z_A, T^c; \bar{T}^{\bar{d}}]} + \frac{1}{3}E^{\mu\nu\lambda} A_{\mu\bar{b}a} A_{\nu\bar{d}c} A_{\lambda\bar{f}e}[T^a, T^c; \bar{T}^{\bar{d}}]\overline{[T^b, T^f; \bar{T}^{\bar{e}}]} \\ &+i\bar{\psi}^A\Gamma^\mu A_{\mu\bar{b}a}[\psi_A, T^a; \bar{T}^{\bar{b}}] + \frac{i}{2}E_{ABCD}\bar{\psi}^A[Z^C, Z^D; \bar{\psi}^B] - \frac{i}{2}E^{ABCD}\bar{Z}_D[\bar{\psi}_A, \psi_B; \bar{Z}_C] \\ &-i\bar{\psi}^A[\psi_A, Z^B; \bar{Z}_B] + 2i\bar{\psi}^A[\psi_B, Z^B; \bar{Z}_A]\Big\rangle \end{aligned} \tag{73}$$

where the cosmological constant has been deleted for the same reason as before. The potential terms are given by

$$\begin{aligned} V &= \frac{2}{3}\Upsilon^{CD}_B \overline{\Upsilon}^B_{CD} \\ \Upsilon^{CD}_B &= [Z^C, Z^D; \bar{Z}_B] - \frac{1}{2}\delta^C_B[Z^E, Z^D; \bar{Z}_E] + \frac{1}{2}\delta^D_B[Z^E, Z^C; \bar{Z}_E] \end{aligned} \tag{74}$$

This matrix model can be obtained formally by a dimensional reduction of the $\mathcal{N}=6$ BLG action [8], which is equivalent to ABJ(M) action [7, 45][2],

$$S_{\mathcal{N}=6BLG} = \int d^3x \Big\langle -V - D_\mu Z^A \overline{D^\mu Z_A} + E^{\mu\nu\lambda}\big(\frac{1}{2}A_{\mu\bar{c}b}\partial_\nu A_{\lambda\bar{d}a}\bar{T}^{\bar{d}}[T^a, T^b; \bar{T}^{\bar{c}}] + \frac{1}{3}A_{\mu\bar{b}a}A_{\nu\bar{d}c}A_{\lambda\bar{f}e}[T^a, T^c; \bar{T}^{\bar{d}}]\overline{[T^b, T^f; \bar{T}^{\bar{e}}]}\big) - i\bar{\psi}^A\Gamma^\mu D_\mu\psi_A + \frac{i}{2}E_{ABCD}\bar{\psi}^A[Z^C, Z^D; \psi^B] - \frac{i}{2}E^{ABCD}\bar{Z}_D[\bar{\psi}_A, \psi_B; \bar{Z}_C] - i\bar{\psi}^A[\psi_A, Z^B; \bar{Z}_B] + 2i\bar{\psi}^A[\psi_B, Z^B; \bar{Z}_A]\Big\rangle \tag{75}$$

The Hermitian 3-algebra models of M-theory are classified into the models with $u(m) \oplus u(n)$ Hermitian 3-algebra (20) and $sp(2n) \oplus u(1)$ Hermitian 3-algebra (30). In the following, we study the $u(N) \oplus u(N)$ Hermitian 3-algebra model. By substituting the $u(N) \oplus u(N)$ Hermitian 3-algebra (20) to the action (73), we obtain

$$S = \mathrm{Tr}\Big(-\frac{(2\pi)^2}{k^2}V - (Z^A A^R_\mu - A^L_\mu Z^A)(Z^A A^{R\mu} - A^{L\mu}Z^A)^\dagger - \frac{k}{2\pi}\frac{i}{3}E^{\mu\nu\lambda}(A^R_\mu A^R_\nu A^R_\lambda - A^L_\mu A^L_\nu A^L_\lambda) - \bar{\psi}^A\Gamma^\mu(\psi_A A^R_\mu - A^L_\mu\psi_A) + \frac{2\pi}{k}(iE_{ABCD}\bar{\psi}^A Z^C\psi^{\dagger B}Z^D - iE^{ABCD}Z^\dagger_D\bar{\psi}^\dagger{}_A Z^\dagger_C\psi_B - i\bar{\psi}^A\psi_A Z^\dagger_B Z^B + i\bar{\psi}^A Z^B Z^\dagger_B\psi_A + 2i\bar{\psi}^A\psi_B Z^\dagger_A Z^B - 2i\bar{\psi}^A Z^B Z^\dagger_A\psi_B)\Big) \tag{76}$$

where $A^R_\mu \equiv -\frac{k}{2\pi}iA_{\mu\bar{b}a}T^{\dagger\bar{b}}T^a$ and $A^L_\mu \equiv -\frac{k}{2\pi}iA_{\mu\bar{b}a}T^aT^{\dagger\bar{b}}$ are $N \times N$ Hermitian matrices. In the algebra, we have set $\alpha = \frac{2\pi}{k}$, where k is an integer representing the Chern-Simons level. We choose $k=1$ in order to obtain 16 dynamical supersymmetries. V is given by

$$V = +\frac{1}{3}Z^\dagger_A Z^A Z^\dagger_B Z^B Z^\dagger_C Z^C + \frac{1}{3}Z^A Z^\dagger_A Z^B Z^\dagger_B Z^C Z^\dagger_C + \frac{4}{3}Z^\dagger_A Z^B Z^\dagger_C Z^A Z^\dagger_B Z^C - Z^\dagger_A Z^A Z^\dagger_B Z^C Z^\dagger_C Z^B - Z^A Z^\dagger_A Z^B Z^\dagger_C Z^C Z^\dagger_B \tag{77}$$

By redefining fields as

$$Z^A \to \left(\frac{k}{2\pi}\right)^{\frac{1}{3}} Z^A$$
$$A^\mu \to \left(\frac{2\pi}{k}\right)^{\frac{1}{3}} A^\mu$$
$$\psi^A \to \left(\frac{k}{2\pi}\right)^{\frac{1}{6}} \psi^A \tag{78}$$

we obtain an action that is independent of Chern-Simons level:

$$S = \mathrm{Tr}\Big(-V - (Z^A A^R_\mu - A^L_\mu Z^A)(Z^A A^{R\mu} - A^{L\mu}Z^A)^\dagger - \frac{i}{3}E^{\mu\nu\lambda}(A^R_\mu A^R_\nu A^R_\lambda - A^L_\mu A^L_\nu A^L_\lambda) - \bar{\psi}^A\Gamma^\mu(\psi_A A^R_\mu - A^L_\mu\psi_A) + iE_{ABCD}\bar{\psi}^A Z^C\psi^{\dagger B}Z^D - iE^{ABCD}Z^\dagger_D\bar{\psi}^\dagger{}_A Z^\dagger_C\psi_B - i\bar{\psi}^A\psi_A Z^\dagger_B Z^B + i\bar{\psi}^A Z^B Z^\dagger_B\psi_A + 2i\bar{\psi}^A\psi_B Z^\dagger_A Z^B - 2i\bar{\psi}^A Z^B Z^\dagger_A\psi_B\Big) \tag{79}$$

[2] The authors of [46–49] studied matrix models that can be obtained by a dimensional reduction of the ABJM and ABJ gauge theories on S^3. They showed that the models reproduce the original gauge theories on S^3 in planar limits.

as opposed to three-dimensional Chern-Simons actions.

If we rewrite the gauge fields in the action as $A^L_\mu = A_\mu + b_\mu$ and $A^R_\mu = A_\mu - b_\mu$, we obtain

$$\begin{aligned} S = \mathrm{Tr}\Big(&-V + ([A_\mu, Z^A] + \{b_\mu, Z^A\})([A^\mu, Z_A] - \{b^\mu, Z_A\}) + iE^{\mu\nu\lambda}(\frac{2}{3}b_\mu b_\nu b_\lambda + 2A_\mu A_\nu b_\lambda) \\ &+\bar{\psi}^A\Gamma^\mu([A_\mu, \psi_A] + \{b_\mu, \psi_A\}) + iE_{ABCD}\bar{\psi}^A Z^C \psi^{\dagger B} Z^D - iE^{ABCD} Z^\dagger_D \bar{\psi}^\dagger{}_A Z^\dagger_C \psi_B \\ &-i\bar{\psi}^A \psi_A Z^\dagger_B Z^B + i\bar{\psi}^A Z^B Z^\dagger_B \psi_A + 2i\bar{\psi}^A \psi_B Z^\dagger_A Z^B - 2i\bar{\psi}^A Z^B Z^\dagger_A \psi_B\Big) \end{aligned} \tag{80}$$

where $[\ ,\ ]$ and $\{\ ,\ \}$ are the ordinary commutator and anticommutator, respectively. The $u(1)$ parts of A^μ decouple because A^μ appear only in commutators in the action. b^μ can be regarded as auxiliary fields, and thus A^μ correspond to matrices X^μ that represents three space-time coordinates in M-theory. Among $N \times N$ arbitrary complex matrices Z^A, we need to identify matrices X^I ($I = 3, \cdots 10$) representing the other space coordinates in M-theory, because the model possesses not $SO(8)$ but $SU(4) \times U(1)$ symmetry. Our identification is

$$\begin{aligned} Z^A &= iX^{A+2} - X^{A+6}, \\ X^I &= \hat{X}^I - ix^I \mathbf{1} \end{aligned} \tag{81}$$

where $\hat{X}^I$ and x^I are $su(N)$ Hermitian matrices and real scalars, respectively. This is analogous to the identification when we compactify ABJM action, which describes N M2 branes, and obtain the action of N D2 branes [7, 50, 51]. We will see that this identification works also in our case. We should note that while the $su(N)$ part is Hermitian, the $u(1)$ part is anti-Hermitian. That is, an eigen-value distribution of X^μ, Z^A, and not X^I determine the spacetime in the Hermitian model. In order to define light-cone coordinates, we need to perform Wick rotation: $a^0 \to -ia^0$. After the Wick rotation, we obtain

$$A^0 = \hat{A}^0 - ia^0 \mathbf{1} \tag{82}$$

where $\hat{A}^0$ is a $su(N)$ Hermitian matrix.

3.4. DLCQ Limit of 3-algebra model of M-theory

It was shown that M-theory in a DLCQ limit reduces to the BFSS matrix theory with matrices of finite size [30–35]. This fact is a strong criterion for a model of M-theory. In [26, 28], it was shown that the Lie and Hermitian 3-algebra models of M-theory reduce to the BFSS matrix theory with matrices of finite size in the DLCQ limit. In this subsection, we show an outline of the mechanism.

DLCQ limit of M-theory consists of a light-cone compactification, $x^- \approx x^- + 2\pi R$, where $x^\pm = \frac{1}{\sqrt{2}}(x^{10} \pm x^0)$, and Lorentz boost in x^{10} direction with an infinite momentum. After appropriate scalings of fields [26, 28], we define light-cone coordinate matrices as

$$\begin{aligned} X^0 &= \frac{1}{\sqrt{2}}(X^+ - X^-) \\ X^{10} &= \frac{1}{\sqrt{2}}(X^+ + X^-) \end{aligned} \tag{83}$$

We integrate out b^μ by using their equations of motion.

A matrix compactification [52] on a circle with a radius R imposes the following conditions on X^- and the other matrices Y:

$$\begin{aligned} X^- - (2\pi R)1 &= U^\dagger X^- U \\ Y &= U^\dagger Y U \end{aligned} \tag{84}$$

where U is a unitary matrix. In order to obtain a solution to (84), we need to take $N \to \infty$ and consider matrices of infinite size [52]. A solution to (84) is given by $X^- = \bar{X}^- + \tilde{X}^-$, $Y = \tilde{Y}$ and

$$U = \begin{pmatrix} \ddots & \ddots & & & \\ & 0 & 1 & 0 & \\ & & 0 & 1 & \\ & & & 0 & 1 \\ & 0 & & 0 & \ddots \\ & & & & \ddots \end{pmatrix} \otimes 1_{n\times n} \in U(N) \tag{85}$$

Backgrounds $\bar{X}^-$ are

$$\bar{X}^- = -T^3 \bar{x}_0^- T^0 - (2\pi R)\mathrm{diag}(\cdots, s-1, s, s+1, \cdots) \otimes \mathbf{1}_{n\times n} \tag{86}$$

in the Lie 3-algebra case, whereas

$$\bar{X}^- = -i(T^3 \bar{x}^-)\mathbf{1} - i(2\pi R)\mathrm{diag}(\cdots, s-1, s, s+1, \cdots) \otimes \mathbf{1}_{n\times n} \tag{87}$$

in the Hermitian 3-algebra case. A fluctuation $\tilde{x}$ that represents $u(N)$ parts of $\tilde{X}^-$ and $\tilde{Y}$ is

$$\begin{pmatrix} \ddots & \ddots & \ddots & & & & \\ \ddots & \tilde{x}(0) & \tilde{x}(1) & \tilde{x}(2) & & & \ddots \\ \ddots & \tilde{x}(-1) & \tilde{x}(0) & \tilde{x}(1) & \tilde{x}(2) & & \\ & \tilde{x}(-2) & \tilde{x}(-1) & \tilde{x}(0) & \tilde{x}(1) & \tilde{x}(2) & \\ & & \tilde{x}(-2) & \tilde{x}(-1) & \tilde{x}(0) & \tilde{x}(1) & \tilde{x}(2) \\ & & & \tilde{x}(-2) & \tilde{x}(-1) & \tilde{x}(0) & \tilde{x}(1) \ \ddots \\ & \ddots & & & \tilde{x}(-2) & \tilde{x}(-1) & \tilde{x}(0) \ \ddots \\ & & & & \ddots & \ddots & \ddots \end{pmatrix} \tag{88}$$

Each $\tilde{x}(s)$ is a $n \times n$ matrix, where s is an integer. That is, the (s, t)-th block is given by $\tilde{x}_{s,t} = \tilde{x}(s-t)$.

We make a Fourier transformation,

$$\tilde{x}(s) = \frac{1}{2\pi\tilde{R}} \int_0^{2\pi\tilde{R}} d\tau x(\tau) e^{is\frac{\tau}{\tilde{R}}} \tag{89}$$

where $x(\tau)$ is a $n \times n$ matrix in one-dimension and $R\tilde{R} = 2\pi$. From (86)-(89), the following identities hold:

$$\sum_t \tilde{x}_{s,t}\tilde{x'}_{t,u} = \frac{1}{2\pi\tilde{R}}\int_0^{2\pi\tilde{R}} d\tau\, x(\tau)x'(\tau)e^{i(s-u)\frac{\tau}{\tilde{R}}}$$

$$\mathrm{tr}(\sum_{s,t} \tilde{x}_{s,t}\tilde{x'}_{t,s}) = V\frac{1}{2\pi\tilde{R}}\int_0^{2\pi\tilde{R}} d\tau\, \mathrm{tr}(x(\tau)x'(\tau))$$

$$[\bar{x}^-, \tilde{x}]_{s,t} = \frac{1}{2\pi\tilde{R}}\int_0^{2\pi\tilde{R}} d\tau\, \partial_\tau x(\tau)e^{i(s-t)\frac{\tau}{\tilde{R}}} \tag{90}$$

where tr is a trace over $n \times n$ matrices and $V = \sum_s 1$.

Next, we boost the system in x^{10} direction:

$$\tilde{X}'^+ = \frac{1}{T}\tilde{X}^+$$

$$\tilde{X}'^- = T\tilde{X}^- \tag{91}$$

The DLCQ limit is achieved when $T \to \infty$, where the "novel Higgs mechanism" [51] is realized. In $T \to \infty$, the actions of the 3-algebra models of M-theory reduce to that of the BFSS matrix theory [27] with matrices of finite size,

$$S = \frac{1}{g^2}\int_{-\infty}^{\infty} d\tau \mathrm{tr}(\frac{1}{2}(D_0 x^P)^2 - \frac{1}{4}[x^P, x^Q]^2 + \frac{1}{2}\bar{\psi}\Gamma^0 D_0\psi - \frac{i}{2}\bar{\psi}\Gamma^P[x_P, \psi]) \tag{92}$$

where $P, Q = 1, 2, \cdots, 9$.

3.5. Supersymmetric deformation of Lie 3-algebra model of M-theory

A supersymmetric deformation of the Lie 3-algebra Model of M-theory was studied in [53] (see also [54–56]). If we add mass terms and a flux term,

$$S_m = \left\langle -\frac{1}{2}\mu^2(X^I)^2 - \frac{i}{2}\mu\bar{\Psi}\Gamma_{3456}\Psi + H_{IJKL}[X^I, X^J, X^K]X^L \right\rangle \tag{93}$$

such that

$$H_{IJKL} = \begin{cases} -\frac{\mu}{6}\epsilon_{IJKL} & (I, J, K, L = 3, 4, 5, 6 \text{ or } 7, 8, 9, 10) \\ 0 & (\text{otherwise}) \end{cases} \tag{94}$$

to the action (53), the total action $S_0 + S_m$ is invariant under dynamical 16 supersymmetries,

$$\delta X^I = i\bar{\epsilon}\Gamma^I\Psi$$

$$\delta A_{\mu ab}[T^a, T^b, \quad] = i\bar{\epsilon}\Gamma_\mu\Gamma_I[X^I, \Psi, \quad]$$

$$\delta\Psi = -\frac{1}{6}[X^I, X^J, X^K]\Gamma_{IJK}\epsilon - A_{\mu ab}[T^a, T^b, X^I]\Gamma^\mu\Gamma_I\epsilon + \mu\Gamma_{3456}X^I\Gamma_I\epsilon \tag{95}$$

From this action, we obtain various interesting solutions, including fuzzy sphere solutions [53].

4. Conclusion

The metric Hermitian 3-algebra corresponds to a class of the super Lie algebra. By using this relation, the metric Hermitian 3-algebras are classified into $u(m) \oplus u(n)$ and $sp(2n) \oplus u(1)$ Hermitian 3-algebras.

The Lie and Hermitian 3-algebra models of M-theory are obtained by second quantizations of the supermembrane action in a semi-light-cone gauge. The Lie 3-algebra model possesses manifest $\mathcal{N} = 1$ supersymmetry in eleven dimensions. In the DLCQ limit, both the models reduce to the BFSS matrix theory with matrices of finite size as they should.

Acknowledgements

We would like to thank T. Asakawa, K. Hashimoto, N. Kamiya, H. Kunitomo, T. Matsuo, S. Moriyama, K. Murakami, J. Nishimura, S. Sasa, F. Sugino, T. Tada, S. Terashima, S. Watamura, K. Yoshida, and especially H. Kawai and A. Tsuchiya for valuable discussions.

Author details

Matsuo Sato
Hirosaki University, Japan

5. References

[1] V. T. Filippov, n-Lie algebras, Sib. Mat. Zh. 26, No. 6, (1985) 126140.
[2] N. Kamiya, A structure theory of Freudenthal-Kantor triple systems, J. Algebra 110 (1987) 108.
[3] S. Okubo, N. Kamiya, Quasi-classical Lie superalgebras and Lie supertriple systems, Comm. Algebra 30 (2002) no. 8, 3825.
[4] J. Bagger, N. Lambert, Modeling Multiple M2's, Phys. Rev. D75 (2007) 045020.
[5] A. Gustavsson, Algebraic structures on parallel M2-branes, Nucl. Phys. B811 (2009) 66.
[6] J. Bagger, N. Lambert, Gauge Symmetry and Supersymmetry of Multiple M2-Branes, Phys. Rev. D77 (2008) 065008.
[7] O. Aharony, O. Bergman, D. L. Jafferis, J. Maldacena, N=6 superconformal Chern-Simons-matter theories, M2-branes and their gravity duals, JHEP 0810 (2008) 091.
[8] J. Bagger, N. Lambert, Three-Algebras and N=6 Chern-Simons Gauge Theories, Phys. Rev. D79 (2009) 025002.
[9] J. Figueroa-O'Farrill, G. Papadopoulos, Pluecker-type relations for orthogonal planes, J. Geom. Phys. 49 (2004) 294.
[10] G. Papadopoulos, M2-branes, 3-Lie Algebras and Plucker relations, JHEP 0805 (2008) 054.
[11] J. P. Gauntlett, J. B. Gutowski, Constraining Maximally Supersymmetric Membrane Actions, JHEP 0806 (2008) 053.
[12] D. Gaiotto, E. Witten, Janus Configurations, Chern-Simons Couplings, And The Theta-Angle in N=4 Super Yang-Mills Theory, arXiv:0804.2907[hep-th].
[13] K. Hosomichi, K-M. Lee, S. Lee, S. Lee, J. Park, N=5,6 Superconformal Chern-Simons Theories and M2-branes on Orbifolds, JHEP 0809 (2008) 002.

[14] M. Schnabl, Y. Tachikawa, Classification of N=6 superconformal theories of ABJM type, arXiv:0807.1102[hep-th].
[15] J. Gomis, G. Milanesi, J. G. Russo, Bagger-Lambert Theory for General Lie Algebras, JHEP 0806 (2008) 075.
[16] S. Benvenuti, D. Rodriguez-Gomez, E. Tonni, H. Verlinde, N=8 superconformal gauge theories and M2 branes, JHEP 0901 (2009) 078.
[17] P.-M. Ho, Y. Imamura, Y. Matsuo, M2 to D2 revisited, JHEP 0807 (2008) 003.
[18] M. A. Bandres, A. E. Lipstein, J. H. Schwarz, Ghost-Free Superconformal Action for Multiple M2-Branes, JHEP 0807 (2008) 117.
[19] P. de Medeiros, J. Figueroa-O'Farrill, E. Me'ndez-Escobar, P. Ritter, On the Lie-algebraic origin of metric 3-algebras, Commun. Math. Phys. 290 (2009) 871.
[20] S. A. Cherkis, V. Dotsenko, C. Saeman, On Superspace Actions for Multiple M2-Branes, Metric 3-Algebras and their Classification, Phys. Rev. D79 (2009) 086002.
[21] P.-M. Ho, Y. Matsuo, S. Shiba, Lorentzian Lie (3-)algebra and toroidal compactification of M/string theory, arXiv:0901.2003 [hep-th].
[22] P. de Medeiros, J. Figueroa-O'Farrill, E. Mendez-Escobar, P. Ritter, Metric 3-Lie algebras for unitary Bagger-Lambert theories, JHEP 0904 (2009) 037.
[23] Y. Nambu, Generalized Hamiltonian dynamics, Phys. Rev. D7 (1973) 2405.
[24] H. Awata, M. Li, D. Minic, T. Yoneya, On the Quantization of Nambu Brackets, JHEP 0102 (2001) 013.
[25] B. de Wit, J. Hoppe, H. Nicolai, On the Quantum Mechanics of Supermembranes, Nucl. Phys. B305 (1988) 545.
[26] M. Sato, Model of M-theory with Eleven Matrices, JHEP 1007 (2010) 026.
[27] T. Banks, W. Fischler, S.H. Shenker, L. Susskind, M Theory As A Matrix Model: A Conjecture, Phys. Rev. D55 (1997) 5112.
[28] M. Sato, Supersymmetry and the Discrete Light-Cone Quantization Limit of the Lie 3-algebra Model of M-theory, Phys. Rev. D85 (2012), 046003.
[29] M. Sato, Zariski Quantization as Second Quantization, arXiv:1202.1466 [hep-th].
[30] L. Susskind, Another Conjecture about M(atrix) Theory, hep-th/9704080.
[31] A. Sen, D0 Branes on T^n and Matrix Theory, Adv. Theor. Math. Phys. 2 (1998) 51.
[32] N. Seiberg, Why is the Matrix Model Correct?, Phys. Rev. Lett. 79 (1997) 3577.
[33] J. Polchinski, M-Theory and the Light Cone, Prog. Theor. Phys. Suppl. 134 (1999) 158.
[34] J. Polchinski, String Theory Vol. 2: Superstring Theory and Beyond, Cambridge University Press, Cambridge, UK (1998).
[35] K. Becker, M. Becker, J. H. Schwarz, String Theory and M-theory, Cambridge University Press, Cambridge, UK (2007).
[36] E. Bergshoeff, E. Sezgin, P.K. Townsend, Supermembranes and Eleven-Dimensional Supergravity, Phys. Lett. B189 (1987) 75.
[37] S. Carlip, Loop Calculations For The Green-Schwarz Superstring, Phys. Lett. B186 (1987) 141.
[38] R.E. Kallosh, Quantization of Green-Schwarz Superstring, Phys. Lett. B195 (1987) 369.
[39] Y. Kazama, N. Yokoi, Superstring in the plane-wave background with RR flux as a conformal field theory, JHEP 0803 (2008) 057.
[40] T. Banks, N. Seiberg, S. Shenker, Branes from Matrices, Nucl. Phys. B490 (1997) 91.
[41] N. Ishibashi, H. Kawai, Y. Kitazawa, A. Tsuchiya, A Large-N Reduced Model as Superstring, Nucl. Phys. B498 (1997) 467.
[42] M. Sato, Covariant Formulation of M-Theory, Int. J. Mod. Phys. A24 (2009) 5019.

[43] H. Nishino, S. Rajpoot, Triality and Bagger-Lambert Theory, Phys. Lett. B671 (2009) 415.
[44] A. Gustavsson, S-J. Rey, Enhanced N=8 Supersymmetry of ABJM Theory on R(8) and R(8)/Z(2), arXiv:0906.3568 [hep-th].
[45] O. Aharony, O. Bergman, D. L. Jafferis, Fractional M2-branes, JHEP 0811 (2008) 043.
[46] M. Hanada, L. Mannelli, Y. Matsuo, Large-N reduced models of supersymmetric quiver, Chern-Simons gauge theories and ABJM, arXiv:0907.4937 [hep-th].
[47] G. Ishiki, S. Shimasaki, A. Tsuchiya, Large N reduction for Chern-Simons theory on S^3, Phys. Rev. D80 (2009) 086004.
[48] H. Kawai, S. Shimasaki, A. Tsuchiya, Large N reduction on group manifolds, arXiv:0912.1456 [hep-th].
[49] G. Ishiki, S. Shimasaki, A. Tsuchiya, A Novel Large-N Reduction on S^3: Demonstration in Chern-Simons Theory, arXiv:1001.4917 [hep-th].
[50] Y. Pang, T. Wang, From N M2's to N D2's, Phys. Rev. D78 (2008) 125007.
[51] S. Mukhi, C. Papageorgakis, M2 to D2, JHEP 0805 (2008) 085.
[52] W. Taylor, D-brane field theory on compact spaces, Phys. Lett. B394 (1997) 283.
[53] J. DeBellis, C. Saemann, R. J. Szabo, Quantized Nambu-Poisson Manifolds in a 3-Lie Algebra Reduced Model, JHEP 1104 (2011) 075.
[54] M. M. Sheikh-Jabbari, Tiny Graviton Matrix Theory: DLCQ of IIB Plane-Wave String Theory, A Conjecture , JHEP 0409 (2004) 017.
[55] J. Gomis, A. J. Salim, F. Passerini, Matrix Theory of Type IIB Plane Wave from Membranes, JHEP 0808 (2008) 002.
[56] K. Hosomichi, K. Lee, S. Lee, Mass-Deformed Bagger-Lambert Theory and its BPS Objects, Phys.Rev. D78 (2008) 066015.

Chapter 2

Algebraic Theory of Appell Polynomials with Application to General Linear Interpolation Problem

Francesco Aldo Costabile and Elisabetta Longo

Additional information is available at the end of the chapter

http://dx.doi.org/10.5772/46482

1. Introduction

In 1880 P. E. Appell ([1]) introduced and widely studied sequences of n-degree polynomials

$$A_n(x), \; n = 0, 1, \ldots \tag{1}$$

satisfying the differential relation

$$DA_n(x) = nA_{n-1}(x), \; n = 1, 2, \ldots \tag{2}$$

Sequences of polynomials, verifying the (2), nowadays called Appell polynomials, have been well studied because of their remarkable applications not only in different branches of mathematics ([2, 3]) but also in theoretical physics and chemistry ([4, 5]). In 1936 an initial bibliography was provided by Davis ([6, p. 25]). In 1939 Sheffer ([7]) introduced a new class of polynomials which extends the class of Appell polynomials; he called these polynomials of type zero, but nowadays they are called Sheffer polynomials. Sheffer also noticed the similarities between Appell polynomials and the umbral calculus, introduced in the second half of the 19th century with the work of such mathematicians as Sylvester, Cayley and Blissard (for examples, see [8]). The Sheffer theory is mainly based on formal power series. In 1941 Steffensen ([9]) published a theory on Sheffer polynomials based on formal power series too. However, these theories were not suitable as they did not provide sufficient computational tools. Afterwards Mullin, Roman and Rota ([10–12]), using operators method, gave a beautiful theory of umbral calculus, including Sheffer polynomials. Recently, Di Bucchianico and Loeb ([13]) summarized and documented more than five hundred old and new findings related to Appell polynomial sequences. In last years attention has centered on finding a novel representation of Appell polynomials. For instance, Lehemer ([14]) illustrated six different approaches to representing the sequence of Bernoulli polynomials, which is a special case of Appell polynomial sequences. Costabile ([15, 16]) also gave a new form of Bernoulli polynomials, called determinantal form, and later these ideas have been extended to Appell polynomial sequences. In fact, in 2010, Costabile and Longo ([17]) proposed an algebraic and elementary approach to Appell polynomial sequences. At the same time, Yang and Youn ([18]) also gave an algebraic approach, but with different methods. The approach to Appell polynomial sequences via linear algebra is an easily comprehensible mathematical tool, specially for non-specialists; that is very good because many polynomials arise in physics, chemistry and engineering. The present work concerns with these topics and it is organized as follows: in Section 2 we mention the Appell method ([1]); in Section 3 we provide the determinantal approach ([17]) and prove the equivalence with other definitions; in Section 4 classical and non-classical examples are given; in Section 5, by using elementary tools of linear algebra, general properties of Appell polynomials are provided; in Section 6 we mention Appell polynomials of second kind ([19, 20]) and, in Section 7 two classical examples are given; in Section 8 we provide an application to general linear interpolation problem([21]), giving, in Section 9, some examples; in Section 10 the Yang and Youn approach ([18]) is sketched; finally, in Section 11 conclusions close the work.

2. The Appell approach

Let $\{A_n(x)\}_n$ be a sequence of n-degree polynomials satisfying the differential relation (2). Then we have

Remark 1. *There is a one-to-one correspondence of the set of such sequences $\{A_n(x)\}_n$ and the set of numerical sequences $\{\alpha_n\}_n$, $\alpha_0 \neq 0$ given by the explicit representation*

$$A_n(x) = \alpha_n + \binom{n}{1}\alpha_{n-1}x + \binom{n}{2}\alpha_{n-2}x^2 + \cdots + \alpha_0 x^n,\ n = 0, 1, ... \tag{3}$$

Equation (3), in particular, shows explicitly that for each $n \geq 1$ the polynomial $A_n(x)$ is completely determined by $A_{n-1}(x)$ and by the choice of the constant of integration α_n.

Remark 2. *Given the formal power series*

$$a(h) = \alpha_0 + \frac{h}{1!}\alpha_1 + \frac{h^2}{2!}\alpha_2 + \cdots + \frac{h^n}{n!}\alpha_n + \cdots, \quad \alpha_0 \neq 0, \tag{4}$$

with α_i $i = 0, 1, ...$ real coefficients, the sequence of polynomials, $A_n(x)$, determined by the power series expansion of the product $a(h)e^{hx}$, i.e.

$$a(h)e^{hx} = A_0(x) + \frac{h}{1!}A_1(x) + \frac{h^2}{2!}A_2(x) + \cdots + \frac{h^n}{n!}A_n(x) + \cdots, \tag{5}$$

satisfies (2).

The function $a(h)$ is said, by Appell, 'generating function' of the sequence $\{A_n(x)\}_n$.

Appell also noticed various examples of sequences of polynomials verifying (2).

He also considered ([1]) an application of these polynomial sequences to linear differential equations, which is out of this context.

3. The determinantal approach

Let be $\beta_i \in \mathbb{R},\ i = 0, 1, ...,$ with $\beta_0 \neq 0$.

We give the following

Definition 1. *The polynomial sequence defined by*

$$\begin{cases} A_0(x) = \frac{1}{\beta_0}, \\ A_n(x) = \frac{(-1)^n}{(\beta_0)^{n+1}} \begin{vmatrix} 1 & x & x^2 & \cdots\cdots & x^{n-1} & x^n \\ \beta_0 & \beta_1 & \beta_2 & \cdots\cdots & \beta_{n-1} & \beta_n \\ 0 & \beta_0 & \binom{2}{1}\beta_1 & \cdots\cdots & \binom{n-1}{1}\beta_{n-2} & \binom{n}{1}\beta_{n-1} \\ 0 & 0 & \beta_0 & \cdots\cdots & \binom{n-1}{2}\beta_{n-3} & \binom{n}{2}\beta_{n-2} \\ \vdots & & & \ddots & \vdots & \vdots \\ \vdots & & & & \ddots\ \vdots & \vdots \\ 0 & \cdots & \cdots & \cdots\ 0 & \beta_0 & \binom{n}{n-1}\beta_1 \end{vmatrix}, \quad n = 1, 2, ... \end{cases} \tag{6}$$

is called Appell polynomial sequence for β_i.

Then we have

Theorem 1. *If $A_n(x)$ is the Appell polynomial sequence for β_i the differential relation (2) holds.*

Proof. Using the properties of linearity we can differentiate the determinant (6), expand the resulting determinant with respect to the first column and recognize the factor $A_{n-1}(x)$ after multiplication of the i-th row by $i-1,\ i = 2, ..., n$ and j-th column by $\frac{1}{j},\ j = 1, ..., n$. □

Theorem 2. *If $A_n(x)$ is the Appell polynomial sequence for β_i we have the equality (3) with*

$$\alpha_0 = \frac{1}{\beta_0}, \tag{7}$$

$$\alpha_i = \frac{(-1)^i}{(\beta_0)^{i+1}} \begin{vmatrix} \beta_1 & \beta_2 & \cdots\cdots & \beta_{i-1} & \beta_i \\ \beta_0 & \binom{2}{1}\beta_1 & \cdots\cdots & \binom{i-1}{1}\beta_{i-2} & \binom{i}{1}\beta_{i-1} \\ 0 & \beta_0 & \cdots\cdots & \binom{i-1}{2}\beta_{i-3} & \binom{i}{2}\beta_{i-2} \\ \vdots & & \ddots & \vdots & \vdots \\ \vdots & & \ddots & \vdots & \vdots \\ 0 & \cdots & \cdots\ 0 & \beta_0 & \binom{i}{i-1}\beta_1 \end{vmatrix} =$$

$$= -\frac{1}{\beta_0}\sum_{k=0}^{i-1}\binom{i}{k}\beta_{i-k}\alpha_k, \qquad i = 1, 2, ..., n. \tag{8}$$

Proof. From (6), by expanding the determinant $A_n(x)$ with respect to the first row, we obtain the (3) with α_i given by (7) and the determinantal form in (8); this is a determinant of an upper Hessenberg matrix of order i ([16]), then setting $\overline{\alpha}_i = (-1)^i (\beta_0)^{i+1} \alpha_i$ for $i = 1, 2, ..., n$, we have

$$\overline{\alpha}_i = \sum_{k=0}^{i-1} (-1)^{i-k-1} h_{k+1,i} q_k(i) \overline{\alpha}_k, \tag{9}$$

where:

$$h_{l,m} = \begin{cases} \beta_m & \text{for } l = 1, \\ \binom{m}{l-1}\beta_{m-l+1} & \text{for } 1 < l \leq m+1, \\ 0 & \text{for } l > m+1, \end{cases} \quad l, m = 1, 2, ..., i, \tag{10}$$

$$q_k(i) = \prod_{j=k+2}^{i} h_{j,j-1} = (\beta_0)^{i-k-1}, \quad k = 0, 1, ..., i-2, \tag{11}$$

$$q_{i-1}(i) = 1. \tag{12}$$

By virtue of the previous setting, (9) implies

$$\begin{aligned} \overline{\alpha}_i &= \sum_{k=0}^{i-2} (-1)^{i-k-1} \binom{i}{k} \beta_{i-k} (\beta_0)^{i-k-1} \overline{\alpha}_k + \binom{i}{i-1} \beta_1 \overline{\alpha}_{i-1} = \\ &= (-1)^i (\beta_0)^{i+1} \left(-\frac{1}{\beta_0} \sum_{k=0}^{i-1} \binom{i}{k} \beta_{i-k} \alpha_k \right), \end{aligned}$$

and the proof is concluded. □

Remark 3. *We note that (7) and (8) are equivalent to*

$$\sum_{k=0}^{i} \binom{i}{k} \beta_{i-k} \alpha_k = \begin{cases} 1 & i = 0 \\ 0 & i > 0 \end{cases} \tag{13}$$

and that for each sequence of Appell polynomials there exist two sequences of numbers α_i and β_i related by (13).

Corollary 1. *If $A_n(x)$ is the Appell polynomial sequence for β_i we have*

$$A_n(x) = \sum_{j=0}^{n} \binom{n}{j} A_{n-j}(0) x^j, \; n = 0, 1, ... \tag{14}$$

Proof. Follows from Theorem 2 being

$$A_i(0) = \alpha_i, \quad i = 0, 1, ..., n. \tag{15}$$

□

Remark 4. *For computation we can observe that α_n is a n-order determinant of a particular upper Hessenberg form and it's known that the algorithm of Gaussian elimination without pivoting for computing the determinant of an upper Hessenberg matrix is stable ([22, p. 27]).*

Theorem 3. *If $a(h)$ is the function defined in (4) and $A_n(x)$ is the polynomial sequence defined by (5), setting*

$$\begin{cases} \beta_0 = \frac{1}{\alpha_0}, \\ \beta_n = -\frac{1}{\alpha_0}\left(\sum_{k=1}^{n}\binom{n}{k}\alpha_k\beta_{n-k}\right), \quad n = 1,2,..., \end{cases} \tag{16}$$

we have that $A_n(x)$ satisfies the (6), i.e. $A_n(x)$ is the Appell polynomial sequence for β_i.

Proof. Let be

$$b(h) = \beta_0 + \frac{h}{1!}\beta_1 + \frac{h^2}{2!}\beta_2 + \cdots + \frac{h^n}{n!}\beta_n + \cdots \tag{17}$$

with β_n as in (16). Then we have $a(h)\,b(h) = 1$, where the product is intended in the Cauchy sense, i.e.:

$$a(h)\,b(h) = \sum_{n=0}^{\infty}\sum_{k=0}^{n}\binom{n}{k}\alpha_k\beta_{n-k}\frac{h^n}{n!}.$$

Let us multiply both hand sides of equation

$$a(h)e^{hx} = \sum_{n=0}^{\infty} A_n(x)\frac{h^n}{n!} \tag{18}$$

for $\frac{1}{a(h)}$ and, in the same equation, replace functions e^{hx} and $\frac{1}{a(h)}$ by their Taylor series expansion at the origin; then (18) becomes

$$\sum_{n=0}^{\infty}\frac{x^n h^n}{n!} = \sum_{n=0}^{\infty} A_n(x)\frac{h^n}{n!}\sum_{n=0}^{\infty}\frac{h^n}{n!}\beta_n. \tag{19}$$

By multiplying the series on the left hand side of (19) according to the Cauchy-product rules, previous equality leads to the following system of infinite equations in the unknown $A_n(x)$, $n = 0,1,...$

$$\begin{cases} A_0(x)\,\beta_0 = 1, \\ A_0(x)\,\beta_1 + A_1(x)\,\beta_0 = x, \\ A_0(x)\,\beta_2 + \binom{2}{1}A_1(x)\,\beta_1 + A_2(x)\,\beta_0 = x^2, \\ \vdots \\ A_0(x)\,\beta_n + \binom{n}{1}A_1(x)\,\beta_{n-1} + ... + A_n(x)\,\beta_0 = x^n, \\ \vdots \end{cases} \tag{20}$$

From the first one of (20) we obtain the first one of (6). Moreover, the special form of the previous system (lower triangular) allows us to work out the unknown $A_n(x)$ operating with the first $n+1$ equations, only by applying the Cramer rule:

$$A_n(x) = \frac{1}{(\beta_0)^{n+1}} \begin{vmatrix} \beta_0 & 0 & 0 & \cdots & 0 & 1 \\ \beta_1 & \beta_0 & 0 & \cdots & 0 & x \\ \beta_2 & \binom{2}{1}\beta_1 & \beta_0 & \cdots & 0 & x^2 \\ \vdots & & & \ddots & & \vdots \\ \beta_{n-1} & \binom{n-1}{1}\beta_{n-2} & \cdots & \cdots & \beta_0 & x^{n-1} \\ \beta_n & \binom{n}{1}\beta_{n-1} & \cdots & \cdots & \binom{n}{n-1}\beta_1 & x^n \end{vmatrix}.$$

By transposition of the previous, we have

$$A_n(x) = \frac{1}{(\beta_0)^{n+1}} \begin{vmatrix} \beta_0 & \beta_1 & \beta_2 & \cdots & \beta_{n-1} & \beta_n \\ 0 & \beta_0 & \binom{2}{1}\beta_1 & \cdots & \binom{n-1}{1}\beta_{n-2} & \binom{n}{1}\beta_{n-1} \\ 0 & 0 & \beta_0 & & & \vdots \\ \vdots & & & \ddots & & \vdots \\ 0 & 0 & 0 & \cdots & \beta_0 & \binom{n}{n-1}\beta_1 \\ 1 & x & x^2 & \cdots & x^{n-1} & x^n \end{vmatrix}, \quad n = 1, 2, ..., \tag{21}$$

that is exactly the second one of (6) after n circular row exchanges: more precisely, the i-th row moves to the $(i+1)$-th position for $i = 1, \ldots, n-1$, the n-th row goes to the first position. □

Definition 2. *The function $a(h)\, e^{hx}$, as in (4) and (5), is said 'generating function' of the Appell polynomial sequence $A_n(x)$ for β_i.*

Theorems 1, 2, 3 concur to assert the validity of following

Theorem 4 (Circular). *If $A_n(x)$ is the Appell polynomial sequence for β_i we have*

$$(6) \Rightarrow (2) \Rightarrow (3) \Rightarrow (5) \Rightarrow (6).$$

Proof.

(6)⇒(2): Follows from Theorem 1.

(2)⇒(3): Follows from Theorem 2, or more simply by direct integration of the differential equation (2).

(3)⇒(5): Follows ordering the Cauchy product of the developments $a(h)$ and e^{hx} with respect to the powers of h and recognizing polynomials $A_n(x)$, expressed in form (3), as coefficients of $\frac{h^n}{n!}$.

(5)⇒(6): Follows from Theorem 3.

□

Remark 5. *In virtue of the Theorem 4, any of the relations (2), (3), (5), (6) can be assumed as definition of Appell polynomial sequences.*

4. Examples of Appell polynomial sequences

The following are classical examples of Appell polynomial sequences.

a) Bernoulli polynomials ([17, 23]):

$$\beta_i = \frac{1}{i+1}, \quad i = 0, 1, ..., \tag{22}$$

$$a(h) = \frac{h}{e^h - 1}; \tag{23}$$

b) Euler polynomials ([17, 23]):

$$\beta_0 = 1, \quad \beta_i = \frac{1}{2}, \quad i = 1, 2, ..., \tag{24}$$

$$a(h) = \frac{2}{e^h + 1}; \tag{25}$$

c) Normalized Hermite polynomials ([17, 24]):

$$\beta_i = \frac{1}{\sqrt{\pi}} \int_{-\infty}^{+\infty} e^{-x^2} x^i dx = \begin{cases} 0 & \text{for } i \text{ odd} \\ \frac{(i-1)(i-3)\cdot\dots\cdot 3\cdot 1}{2^{\frac{i}{2}}} & \text{for } i \text{ even} \end{cases}, \quad i = 0, 1, ..., \tag{26}$$

$$a(h) = e^{-\frac{h^2}{4}}; \tag{27}$$

d) Laguerre polynomials ([17, 24]):

$$\beta_i = \int_0^{+\infty} e^{-x} x^i dx = \Gamma(i+1) = i!, \quad i = 0, 1, ..., \tag{28}$$

$$a(h) = 1 - h; \tag{29}$$

The following are non-classical examples of Appell polynomial sequences.

e) Generalized Bernoulli polynomials

- with Jacobi weight ([17]):

$$\beta_i = \int_0^1 (1-x)^\alpha x^\beta x^i dx = \frac{\Gamma(\alpha+1)\Gamma(\beta+i+1)}{\Gamma(\alpha+\beta+i+2)}, \quad \alpha, \beta > -1, \quad i = 0, 1, ..., \tag{30}$$

$$a(h) = \frac{1}{\int_0^1 (1-x)^\alpha x^\beta e^{hx} dx}; \tag{31}$$

- of order k ([11]):

$$\beta_i = \left(\frac{1}{i+1}\right)^k, \quad k \text{ integer}, \quad i = 0, 1, ..., \tag{32}$$

$$a(h) = \left(\frac{h}{e^h - 1}\right)^k; \tag{33}$$

f) Central Bernoulli polynomials ([25]):

$$\beta_{2i} = \frac{1}{i+1},$$
$$\beta_{2i+1} = 0, \quad i = 0, 1, ..., \tag{34}$$
$$a(h) = \frac{h}{\sinh(h)}; \tag{35}$$

g) Generalized Euler polynomials ([17]):

$$\beta_0 = 1,$$
$$\beta_i = \frac{w_1}{w_1 + w_2}, \quad w_1, w_2 > 0, \quad i = 1, 2, ..., \tag{36}$$
$$a(h) = \frac{w_1 + w_2}{w_1 e^h + w_2}; \tag{37}$$

h) Generalized Hermite polynomials ([17]):

$$\beta_i = \frac{1}{\sqrt{\pi}} \int_{-\infty}^{+\infty} e^{-|x|^\alpha} x^i dx$$
$$= \begin{cases} 0 & \text{for } i \text{ odd} \\ \frac{2}{\alpha\sqrt{\pi}} \Gamma\left(\frac{i+1}{\alpha}\right) & \text{for } i \text{ even} \end{cases}, \quad \begin{matrix} i = 0, 1, ..., \\ \alpha > 0, \end{matrix} \tag{38}$$
$$a(h) = \frac{\sqrt{\pi}}{\int_{-\infty}^{\infty} e^{-|x|^\alpha} e^{hx} dx}; \tag{39}$$

i) Generalized Laguerre polynomials ([17]):

$$\beta_i = \int_0^{+\infty} e^{-\alpha x} x^i dx$$
$$= \frac{\Gamma(i+1)}{\alpha^{i+1}} = \frac{i!}{\alpha^{i+1}}, \quad \alpha > 0, \quad i = 0, 1, ..., \tag{40}$$
$$a(h) = \alpha - h. \tag{41}$$

5. General properties of Appell polynomials

By elementary tools of linear algebra we can prove the general properties of Appell polynomials.

Let $A_n(x)$, $n = 0, 1, ...$, be a polynomial sequence and $\beta_i \in \mathbb{R}$, $i = 0, 1, ...$, with $\beta_0 \neq 0$.

Theorem 5 (Recurrence). *$A_n(x)$ is the Appell polynomial sequence for β_i if and only if*

$$A_n(x) = \frac{1}{\beta_0} \left(x^n - \sum_{k=0}^{n-1} \binom{n}{k} \beta_{n-k} A_k(x) \right), \quad n = 1, 2, ... \tag{42}$$

Proof. Follows observing that the following holds:

$$A_n(x) = \frac{(-1)^n}{(\beta_0)^{n+1}} \begin{vmatrix} 1 & x & x^2 & \cdots\cdots & x^{n-1} & x^n \\ \beta_0 & \beta_1 & \beta_2 & \cdots\cdots & \beta_{n-1} & \beta_n \\ 0 & \beta_0 & \binom{2}{1}\beta_1 & \cdots\cdots & \binom{n-1}{1}\beta_{n-2} & \binom{n}{1}\beta_{n-1} \\ 0 & 0 & \beta_0 & \cdots\cdots & \binom{n-1}{2}\beta_{n-3} & \binom{n}{2}\beta_{n-2} \\ \vdots & & & \ddots & \vdots & \vdots \\ \vdots & & & \ddots & \vdots & \vdots \\ 0 & \cdots & \cdots & \cdots\ 0 & \beta_0 & \binom{n}{n-1}\beta_1 \end{vmatrix} =$$

$$= \frac{1}{\beta_0}\left(x^n - \sum_{k=0}^{n-1}\binom{n}{k}\beta_{n-k}A_k(x)\right), \quad n = 1, 2, \ldots \tag{43}$$

In fact, if $A_n(x)$ is the Appell polynomial sequence for β_i, from (6), we can observe that $A_n(x)$ is a determinant of an upper Hessenberg matrix of order $n+1$ ([16]) and, proceeding as in Theorem 2, we can obtain the (43). □

Corollary 2. *If $A_n(x)$ is the Appell polynomial sequence for β_i then*

$$x^n = \sum_{k=0}^{n}\binom{n}{k}\beta_{n-k}A_k(x), \quad n = 0, 1, \ldots \tag{44}$$

Proof. Follows from (42). □

Corollary 3. *Let $\mathcal{P}_n$ be the space of polynomials of degree $\leq n$ and $\{A_n(x)\}_n$ be an Appell polynomial sequence, then $\{A_n(x)\}_n$ is a basis for $\mathcal{P}_n$.*

Proof. If we have

$$P_n(x) = \sum_{k=0}^{n} a_{n,k}x^k, \quad a_{n,k} \in \mathbb{R}, \tag{45}$$

then, by Corollary 2, we get

$$P_n(x) = \sum_{k=0}^{n} a_{n,k}\sum_{j=0}^{k}\binom{k}{j}\beta_{k-j}A_j(x) = \sum_{k=0}^{n} c_{n,k}A_k(x),$$

where

$$c_{n,k} = \sum_{j=0}^{n-k}\binom{k+j}{k}a_{k+j}\beta_j. \tag{46}$$

□

Remark 6. *An alternative recurrence relation can be determined from (5) after differentiation with respect to h ([18, 26]).*

Let be $\beta_i,\ \gamma_i \in \mathbb{R},\ i = 0, 1, ...,$ with $\beta_0,\ \gamma_0 \neq 0.$
Let us consider the Appell polynomial sequences $A_n(x)$ and $B_n(x)$, $n = 0, 1, ...,$ for β_i and γ_i, respectively, and indicate with $(AB)_n(x)$ the polynomial that is obtained replacing in $A_n(x)$ the powers $x^0, x^1, ..., x^n$, respectively, with the polynomials $B_0(x), B_1(x), ..., B_n(x)$. Then we have

Theorem 6. *The sequences*

i) $\lambda A_n(x) + \mu B_n(x), \quad \lambda, \mu \in \mathbb{R},$

ii) $(AB)_n(x)$

are sequences of Appell polynomials again.

Proof. *i)* Follows from the property of linearity of determinant.

ii) Expanding the determinant $(AB)_n(x)$ with respect to the first row we obtain

$$(AB)_n(x) = \frac{(-1)^n}{(\beta_0)^{n+1}} \sum_{j=0}^{n} (-1)^j (\beta_0)^j \binom{n}{j} \overline{\alpha}_{n-j} B_j(x) = \\ = \sum_{j=0}^{n} \frac{(-1)^{n-j}}{(\beta_0)^{n-j+1}} \binom{n}{j} \overline{\alpha}_{n-j} B_j(x), \tag{47}$$

where

$$\overline{\alpha}_0 = 1,$$
$$\overline{\alpha}_i = \begin{vmatrix} \beta_1 & \beta_2 & \cdots\cdots & \beta_{i-1} & \beta_i \\ \beta_0 & \binom{2}{1}\beta_1 & \cdots\cdots & \binom{i-1}{1}\beta_{i-2} & \binom{i}{1}\beta_{i-1} \\ 0 & \beta_0 & \cdots\cdots & \binom{i-1}{2}\beta_{i-3} & \binom{i}{2}\beta_{i-2} \\ \vdots & & \ddots & \vdots & \vdots \\ \vdots & & \ddots & \vdots & \vdots \\ 0 & \cdots & \cdots\ 0 & \beta_0 & \binom{i}{i-1}\beta_1 \end{vmatrix}, \quad i = 1, 2, ..., n.$$

We observe that

$$A_i(0) = \frac{(-1)^i}{(\beta_0)^{i+1}} \overline{\alpha}_i, \quad i = 1, 2, ..., n$$

and hence (47) becomes

$$(AB)_n(x) = \sum_{j=0}^{n} \binom{n}{j} A_{n-j}(0)\, B_j(x). \tag{48}$$

Differentiating both hand sides of (48) and since $B_j(x)$ is a sequence of Appell polynomials, we deduce

$$((AB)_n(x))' = n\,(AB)_{n-1}(x). \tag{49}$$

□

Let us, now, introduce the Appell vector.

Definition 3. *If $A_n(x)$ is the Appell polynomial sequence for β_i the vector of functions $\overline{A}_n(x) = [A_0(x), ..., A_n(x)]^T$ is called Appell vector for β_i.*

Then we have

Theorem 7 (Matrix form). *Let $\overline{A}_n(x)$ be a vector of polynomial functions. Then $\overline{A}_n(x)$ is the Appell vector for β_i if and only if, putting*

$$(M)_{i,j} = \begin{cases} \binom{i}{j}\beta_{i-j} & i \geq j \\ 0 & otherwise \end{cases}, \qquad i,j = 0, ..., n, \tag{50}$$

and $X(x) = [1, x, ..., x^n]^T$ the following relation holds

$$X(x) = M\overline{A}_n(x) \tag{51}$$

or, equivalently,

$$\overline{A}_n(x) = \left(M^{-1}\right) X(x), \tag{52}$$

being M^{-1} the inverse matrix of M.

Proof. If $\overline{A}_n(x)$ is the Appell vector for β_i the result easily follows from Corollary 2.

Vice versa, observing that the matrix M defined by (50) is invertible, setting

$$\left(M^{-1}\right)_{i,j} = \begin{cases} \binom{i}{j}\alpha_{i-j} & i \geq j \\ 0 & otherwise \end{cases}, \qquad i,j = 0, ..., n, \tag{53}$$

we have the (52) and therefore the (3) and, being the coefficients α_k and β_k related by (13), we have that $A_n(x)$ is the Appell polynomial sequence for β_i. □

Theorem 8 (Connection constants). *Let $\overline{A}_n(x)$ and $\overline{B}_n(x)$ be the Appell vectors for β_i and γ_i, respectively. Then*

$$\overline{A}_n(x) = C\overline{B}_n(x), \tag{54}$$

where

$$(C)_{i,j} = \begin{cases} \binom{i}{j}c_{i-j} & i \geq j \\ 0 & otherwise \end{cases}, \qquad i,j = 0, ..., n. \tag{55}$$

with

$$c_n = \sum_{k=0}^{n} \binom{n}{k} \alpha_{n-k}\gamma_k. \tag{56}$$

Proof. From Theorem 7 we have

$$X(x) = M\overline{A}_n(x)$$

with M as in (50) or, equivalently,

$$\overline{A}_n(x) = \left(M^{-1}\right) X(x),$$

with M^{-1} as in (53).

Always from Theorem 7 we get

$$X(x) = N\overline{B}_n(x)$$

with

$$(N)_{i,j} = \begin{cases} \binom{i}{j}\gamma_{i-j} & i \geq j \\ 0 & otherwise \end{cases}, \qquad i,j = 0,...,n.$$

Then

$$\overline{A}_n(x) = M^{-1}N\overline{B}_n(x),$$

from which, setting $C = M^{-1}N$, we have the thesis. □

Theorem 9 (Inverse relations). *Let $A_n(x)$ be the Appell polynomial sequence for β_i then the following are inverse relations:*

$$\begin{cases} y_n = \sum_{k=0}^{n} \binom{n}{k} \beta_{n-k} x_k \\ x_n = \sum_{k=0}^{n} \binom{n}{k} A_{n-k}(0) y_k. \end{cases} \tag{57}$$

Proof. Let us remember that

$$A_k(0) = \alpha_k,$$

where the coefficients α_k and β_k are related by (13).

Moreover, setting $\overline{y}_n = [y_0, ..., y_n]^T$ and $\overline{x}_n = [x_0, ..., x_n]^T$, from (57) we have

$$\begin{cases} \overline{y}_n = M_1\overline{x}_n \\ \overline{x}_n = M_2\overline{y}_n \end{cases}$$

with

$$(M_1)_{i,j} = \begin{cases} \binom{i}{j}\beta_{i-j} & i \geq j \\ 0 & otherwise \end{cases}, \qquad i,j = 0,...,n, \tag{58}$$

$$(M_2)_{i,j} = \begin{cases} \binom{i}{j}\alpha_{i-j} & i \geq j \\ 0 & otherwise \end{cases}, \qquad i,j = 0,...,n, \tag{59}$$

and, from (13) we get

$$M_1M_2 = I_{n+1},$$

i.e. (57) are inverse relations. □

Theorem 10 (Inverse relation between two Appell polynomial sequences). *Let $\overline{A}_n(x)$ and $\overline{B}_n(x)$ be the Appell vectors for β_i and γ_i, respectively. Then the following are inverse relations:*

$$\begin{cases} \overline{A}_n(x) = C\overline{B}_n(x) \\ \overline{B}_n(x) = \widetilde{C}\overline{A}_n(x) \end{cases} \tag{60}$$

with

$$(C)_{i,j} = \begin{cases} \binom{i}{j} c_{i-j} & i \geq j \\ 0 & otherwise \end{cases}, \qquad \left(\widetilde{C}\right)_{i,j} = \begin{cases} \binom{i}{j} \widetilde{c}_{i-j} & i \geq j \\ 0 & otherwise \end{cases}, \quad i,j = 0,...,n, \tag{61}$$

$$c_n = \sum_{k=0}^{n} \binom{n}{k} A_{n-k}(0)\gamma_k, \quad \widetilde{c}_n = \sum_{k=0}^{n} \binom{n}{k} B_{n-k}(0)\beta_k. \tag{62}$$

Proof. Follows from Theorem 8, after observing that

$$\sum_{k=0}^{n} \binom{n}{k} c_{n-k}\widetilde{c}_k = \begin{cases} 1 & n = 0 \\ 0 & n > 0 \end{cases} \tag{63}$$

and therefore

$$C\widetilde{C} = I_{n+1}.$$

□

Theorem 11 (Binomial identity). *If $A_n(x)$ is the Appell polynomial sequence for β_i we have*

$$A_n(x+y) = \sum_{i=0}^{n} \binom{n}{i} A_i(x)\, y^{n-i}, \quad n = 0, 1, ... \tag{64}$$

Proof. Starting by the Definition 1 and using the identity

$$(x+y)^i = \sum_{k=0}^{i} \binom{i}{k} y^k x^{i-k}, \tag{65}$$

we infer

$$A_n(x+y) = \frac{(-1)^n}{(\beta_0)^{n+1}} \begin{vmatrix} 1 & (x+y)^1 & \cdots & (x+y)^{n-1} & (x+y)^n \\ \beta_0 & \beta_1 & \cdots & \beta_{n-1} & \beta_n \\ 0 & \ddots & & & \vdots \\ \vdots & & \ddots & & \vdots \\ 0 & & \cdots & \beta_0 & \beta_1\binom{n}{n-1} \end{vmatrix} =$$

$$= \sum_{i=0}^{n} y^i \frac{(-1)^{n-i}}{(\beta_0)^{n-i+1}} \begin{vmatrix} \binom{i}{i} & \binom{i+1}{i}x^1 & \binom{i+2}{i}x^2 & \cdots & \binom{n-1}{i}x^{n-i-1} & \binom{n}{i}x^{n-i} \\ \beta_0 & \beta_1\binom{i+1}{i} & \beta_2\binom{i+2}{i} & \cdots & \beta_{n-i-1}\binom{n-1}{i} & \beta_{n-i}\binom{n}{i} \\ 0 & \beta_0 & \beta_1\binom{i+2}{i+1} & \cdots & \beta_{n-i-2}\binom{n-1}{i+1} & \beta_{n-i-1}\binom{n}{i+1} \\ \vdots & & \beta_0 & & & \vdots \\ \vdots & & & \ddots & & \vdots \\ 0 & \cdots & \cdots & 0 & \beta_0 & \beta_1\binom{n}{n-1} \end{vmatrix}.$$

We divide, now, each $j-$th column, $j = 2, ..., n-i+1$, for $\binom{i+j-1}{i}$ and multiply each $h-$th row, $h = 3, ..., n-i+1$, for $\binom{i+h-2}{i}$. Thus we finally obtain

$$A_n(x+y) =$$

$$= \sum_{i=0}^{n} \frac{\binom{i+1}{i} \cdots \binom{n}{i}}{\binom{i+1}{i} \cdots \binom{n-1}{i}} y^i \frac{(-1)^{n-i}}{(\beta_0)^{n-i+1}}$$

$$\begin{vmatrix} 1 & x^1 & x^2 & \cdots & x^{n-i-1} & x^{n-i} \\ \beta_0 & \beta_1 & \beta_2 & \cdots & \beta_{n-i-1} & \beta_{n-i} \\ 0 & \beta_0 & \beta_1\binom{2}{1} & \cdots & \beta_{n-i-2}\binom{n-i-1}{1} & \beta_{n-i-1}\binom{n-i}{1} \\ \vdots & & \beta_0 & & & \vdots \\ \vdots & & & \ddots & & \vdots \\ 0 & \dots & \dots & 0 & \beta_0 & \beta_1\binom{n-i}{n-i-1} \end{vmatrix} =$$

$$= \sum_{i=0}^{n} \binom{n}{i} A_{n-i}(x) y^i = \sum_{i=0}^{n} \binom{n}{i} A_i(x) y^{n-i}.$$

□

Theorem 12 (Generalized Appell identity). *Let $A_n(x)$ and $B_n(x)$ be the Appell polynomial sequences for β_i and γ_i, respectively. Then, if $C_n(x)$ is the Appell polynomial sequence for δ_i with*

$$\begin{cases} \delta_0 = \frac{1}{C_0(0)}, \\ \delta_i = -\frac{1}{C_0(0)} \sum_{k=1}^{i} \binom{i}{k} \delta_{i-k} C_k(0), \; i = 1, ..., \end{cases} \tag{66}$$

and

$$C_i(0) = \sum_{j=0}^{i} \binom{i}{j} B_{i-j}(0) A_j(0), \tag{67}$$

where $A_i(0)$ and $B_i(0)$ are related to β_i and γ_i, respectively, by relations similar to (66), we have

$$C_n(y+z) = \sum_{k=0}^{n} \binom{n}{k} A_k(y) B_{n-k}(z). \tag{68}$$

Proof. Starting from (3) we have

$$C_n(y+z) = \sum_{k=0}^{n} \binom{n}{k} C_{n-k}(0)(y+z)^k. \tag{69}$$

Then, applying (67) and the well-known classical binomial identity, after some calculation, we obtain the thesis. □

Theorem 13 (Combinatorial identities). *Let $A_n(x)$ and $B_n(x)$ be the Appell polynomial sequences for β_i and γ_i, respectively. Then the following relations holds:*

$$\sum_{k=0}^{n}\binom{n}{k}A_k(x)B_{n-k}(-x)=\sum_{k=0}^{n}\binom{n}{k}A_k(0)B_{n-k}(0), \tag{70}$$

$$\sum_{k=0}^{n}\binom{n}{k}A_k(x)B_{n-k}(z)=\sum_{k=0}^{n}\binom{n}{k}A_k(x+z)B_{n-k}(0). \tag{71}$$

Proof. If $C_n(x)$ is the Appell polynomial sequence for δ_i defined as in (66), from the generalized Appell identity, we have

$$\sum_{k=0}^{n}\binom{n}{k}A_k(x)B_{n-k}(-x)=C_n(0)=\sum_{k=0}^{n}\binom{n}{k}A_k(0)B_{n-k}(0)$$

and

$$\sum_{k=0}^{n}\binom{n}{k}A_k(x)B_{n-k}(z)=C_n(x+z)=\sum_{k=0}^{n}\binom{n}{k}A_k(x+z)B_{n-k}(0).$$

□

Theorem 14 (Forward difference). *If $A_n(x)$ is the Appell polynomial sequence for β_i we have*

$$\Delta A_n(x)\equiv A_n(x+1)-A_n(x)=\sum_{i=0}^{n-1}\binom{n}{i}A_i(x),\quad n=0,1,... \tag{72}$$

Proof. The desired result follows from (64) with $y=1$. □

Theorem 15 (Multiplication Theorem). *Let $\overline{A}_n(x)$ be the Appell vector for β_i.*

The following identities hold:

$$\overline{A}_n(mx)=B(x)\overline{A}_n(x)\qquad n=0,1,...,\qquad m=1,2,..., \tag{73}$$

$$\overline{A}_n(mx)=M^{-1}DX(x)\qquad n=0,1,...,\qquad m=1,2,..., \tag{74}$$

where

$$(B(x))_{i,j}=\begin{cases}\binom{i}{j}(m-1)^{i-j}x^{i-j} & i\geq j\\ 0 & otherwise\end{cases},\qquad i,j=0,...,n, \tag{75}$$

$D=diag[1,m,...,m^n]$ and M^{-1} defined as in (53).

Proof. The (73) follows from (64) setting $y=x(m-1)$. In fact we get

$$A_n(mx)=\sum_{i=0}^{n}\binom{n}{i}A_i(x)(m-1)^{n-i}x^{n-i}. \tag{76}$$

The (74) follows from Theorem 7. In fact we get

$$\overline{A}_n(mx)=M^{-1}X(mx)=M^{-1}DX(x), \tag{77}$$

and

$$A_n(mx) = \sum_{i=0}^{n} \binom{n}{i} \alpha_{n-i} m^i x^i. \tag{78}$$

□

Theorem 16 (Differential equation). *If $A_n(x)$ is the Appell polynomial sequence for β_i then $A_n(x)$ satisfies the linear differential equation:*

$$\frac{\beta_n}{n!} y^{(n)}(x) + \frac{\beta_{n-1}}{(n-1)!} y^{(n-1)}(x) + ... + \frac{\beta_2}{2!} y^{(2)}(x) + \beta_1 y^{(1)}(x) + \beta_0 y(x) = x^n \tag{79}$$

Proof. From Theorem 5 we have

$$A_{n+1}(x) = \frac{1}{\beta_0} \left(x^{n+1} - \sum_{k=0}^{n} \binom{n+1}{k+1} \beta_{k+1} A_{n-k}(x) \right). \tag{80}$$

From Theorem 1 we find that

$$A'_{n+1}(x) = (n+1) A_n(x), \quad \text{and} \quad A_{n-k}(x) = \frac{A_n^{(k)}(x)}{n(n-1)...(n-k+1)}, \tag{81}$$

and replacing $A_{n-k}(x)$ in the (80) we obtain

$$A_{n+1}(x) = \frac{1}{\beta_0} \left(x^{n+1} - (n+1) \sum_{k=0}^{n} \beta_{k+1} \frac{A_n^{(k)}(x)}{(k+1)!} \right). \tag{82}$$

Differentiating both hand sides of the last one and replacing $A'_{n+1}(x)$ with $(n+1)A_n(x)$, after some calculation we obtain the thesis. □

Remark 7. *An alternative differential equation for Appell polynomial sequences can be determined by the recurrence relation referred to in Remark 6 ([18, 26]).*

6. Appell polynomial sequences of second kind

Let $f : I \subset \mathbb{R} \rightarrow \mathbb{R}$ and Δ be the finite difference operator ([23]), i.e.:

$$\Delta[f](x) = f(x+1) - f(x), \tag{83}$$

we define the finite difference operator of order i, with $i \in \mathbb{N}$, as

$$\Delta^i[f](x) = \Delta(\Delta^{i-1}[f](x)) = \sum_{j=0}^{i} (-1)^{i-j} \binom{i}{j} f(x+j), \tag{84}$$

meaning $\Delta^0 = I$ and $\Delta^1 = \Delta$, where I is the identity operator.
Let the sequence of falling factorial defined by

$$\begin{cases} (x)_0 = 1, \\ (x)_n = x(x-1)(x-2)\cdots(x-n+1), \; n = 1,2,..., \end{cases} \tag{85}$$

we give the following

Definition 4. *Let $\beta_i \in \mathbb{R}$, $i = 0, 1, \ldots$, with $\beta_0 \neq 0$. The polynomial sequence*

$$\begin{cases} \mathcal{A}_0(x) = \frac{1}{\beta_0}, \\ \mathcal{A}_n(x) = \frac{(-1)^n}{(\beta_0)^{n+1}} \begin{vmatrix} 1 & (x)_1 & (x)_2 & \cdots\cdots & (x)_{n-1} & (x)_n \\ \beta_0 & \beta_1 & \beta_2 & \cdots\cdots & \beta_{n-1} & \beta_n \\ 0 & \beta_0 & \binom{2}{1}\beta_1 & \cdots\cdots & \binom{n-1}{1}\beta_{n-2} & \binom{n}{1}\beta_{n-1} \\ 0 & 0 & \beta_0 & \cdots\cdots & \binom{n-1}{2}\beta_{n-3} & \binom{n}{2}\beta_{n-2} \\ \vdots & & \ddots & & \vdots & \vdots \\ \vdots & & & \ddots & \vdots & \vdots \\ 0 & \cdots & \cdots & \cdots\ 0 & \beta_0 & \binom{n}{n-1}\beta_1 \end{vmatrix}, \quad n = 1, 2, \ldots \end{cases} \tag{86}$$

is called Appell polynomial sequence of second kind.

Then, we have

Theorem 17. *For Appell polynomial sequences of second kind we get*

$$\Delta \mathcal{A}_n(x) = n \mathcal{A}_{n-1}(x) \quad n = 1, 2, \ldots \tag{87}$$

Proof. By the well-known relation ([23])

$$\Delta (x)_n = n (x)_{n-1}, \quad n = 1, 2, \ldots, \tag{88}$$

applying the operator Δ to the definition (86) and using the properties of linearity of Δ we have

$$\Delta \mathcal{A}_n(x) = \frac{(-1)^n}{(\beta_0)^{n+1}} \begin{vmatrix} \Delta 1 & \Delta (x)_1 & \Delta (x)_2 & \cdots\cdots & \Delta (x)_{n-1} & \Delta (x)_n \\ \beta_0 & \beta_1 & \beta_2 & \cdots\cdots & \beta_{n-1} & \beta_n \\ 0 & \beta_0 & \binom{2}{1}\beta_1 & \cdots\cdots & \binom{n-1}{1}\beta_{n-2} & \binom{n}{1}\beta_{n-1} \\ 0 & 0 & \beta_0 & \cdots\cdots & \binom{n-1}{2}\beta_{n-3} & \binom{n}{2}\beta_{n-2} \\ \vdots & & \ddots & & \vdots & \vdots \\ \vdots & & & \ddots & \vdots & \vdots \\ 0 & \cdots & \cdots & \cdots\ 0 & \beta_0 & \binom{n}{n-1}\beta_1 \end{vmatrix}, \quad n = 1, 2, \ldots \tag{89}$$

We can expand the determinant in (89) with respect to the first column and, after multiplying the i-th row by $i-1$, $i = 2, \ldots, n$ and the j-th column by $\frac{1}{j}$, $j = 1, \ldots, n$, we can recognize the factor $\mathcal{A}_{n-1}(x)$. □

We can observe that the structure of the determinant in (86) is similar to that one of the determinant in (6). In virtue of this it is possible to obtain a dual theory of Appell polynomials of first kind, in the sense that similar properties can be proven ([19]).

For example, the generating function is

$$H(x,h) = a(h)(1+h)^x, \tag{90}$$

where $a(h)$ is an invertible formal series of power.

7. Examples of Appell polynomial sequences of second kind

The following are classical examples of Appell polynomial sequences of second kind.

a) Bernoulli polynomials of second kind ([19, 23]):

$$\beta_i = \frac{(-1)^i}{i+1} i!,\ i = 0, 1, ..., \tag{91}$$

$$H(x,h) = \frac{h(1+h)^x}{ln(1+h)}; \tag{92}$$

b) Boole polynomials ([19, 23]):

$$\beta_i = \begin{cases} 1, & i = 0 \\ \frac{1}{2}, & i = 1 \\ 0, & i = 2, ... \end{cases} \tag{93}$$

$$H(x,h) = \frac{2(1+h)^x}{2+h}. \tag{94}$$

8. An application to general linear interpolation problem

Let X be the linear space of real functions defined in the interval $[0,1]$ continuous and with continuous derivatives of all necessary orders. Let L be a linear functional on X such that $L(1) \neq 0$. If in (6) and respectively in (86) we set

$$\beta_i = L(x^i), \qquad \beta_i = L((x)_i), \quad i = 0, 1, ..., \tag{95}$$

$A_n(x)$ and $\mathcal{A}_n(x)$ will be said Appell polynomial sequences of first or of second kind related to the functional L and denoted by $A_{L,n}(x)$ and $\mathcal{A}_{L,n}(x)$, respectively.

Remark 8. *The generating function of the sequence $A_{L,n}(x)$ is*

$$G(x,h) = \frac{e^{xh}}{L_x(e^{xh})}, \tag{96}$$

and for $\mathcal{A}_{L,n}(x)$ is

$$H(x,h) = \frac{(1+h)^x}{L_x((1+h)^x)}, \tag{97}$$

where L_x means that the functional L is applied to the argument as a function of x.

Proof. For $A_{L,n}(x)$ if $G(x,h) = a(h)e^{xh}$ with $\frac{1}{a(h)} = \sum_{i=0}^{\infty} \beta_i \frac{h^i}{i!}$ we have

$$G(x,t) = \frac{e^{xh}}{\frac{1}{a(h)}} = \frac{e^{xh}}{\sum\limits_{i=0}^{\infty} \beta_i \frac{h^i}{i!}} = \frac{e^{xh}}{\sum\limits_{i=0}^{\infty} L(x^i) \frac{h^i}{i!}} = \frac{e^{xh}}{L\left(\sum\limits_{i=0}^{\infty} x^i \frac{h^i}{i!}\right)} = \frac{e^{xh}}{L_x(e^{xh})}.$$

For $\mathcal{A}_{L,n}(x)$, the proof similarly follows. □

Then, we have

Theorem 18. *Let $\omega_i \in \mathbb{R}$, $i = 0, ..., n$, the polynomials*

$$P_n(x) = \sum_{i=0}^{n} \frac{\omega_i}{i!} A_{L,i}(x), \tag{98}$$

$$P_n^*(x) = \sum_{i=0}^{n} \frac{\omega_i}{i!} \mathcal{A}_{L,i}(x) \tag{99}$$

are the unique polynomials of degree less than or equal to n, such that

$$L(P_n^{(i)}) = i!\omega_i, \;\; i = 0, ..., n, \tag{100}$$

$$L(\Delta^i P_n^*) = i!\omega_i, \;\; i = 0, ..., n. \tag{101}$$

Proof. The proof follows observing that, by the hypothesis on functional L there exists a unique polynomial of degree $\leq n$ verifying (100) and , respectively, (101); moreover from the properties of $A_{L,i}(x)$ and $\mathcal{A}_{L,i}(x)$, we have

$$L(A_{L,i}^{(j)}(x)) = i(i-1)...(i-j+1)L(A_{L,i-j}(x)) = j!\binom{i}{j}\delta_{ij}, \tag{102}$$

$$L(\Delta^i \mathcal{A}_{L,i}(x)) = i(i-1)...(i-j+1)L(\mathcal{A}_{L,i-j}(x)) = j!\binom{i}{j}\delta_{ij}, \tag{103}$$

where δ_{ij} is the Kronecker symbol.

From (102) and (103) it is easy to prove that the polynomials (98) and (99) verify (100) and (101), respectively. □

Remark 9. *For every linear functional L on X, $\{A_{L,i}(x)\}$, $\{\mathcal{A}_{L,i}(x)\}$, $i = 0, ..., n$, are basis for $\mathcal{P}_n$ and, $\forall P_n(x) \in \mathcal{P}_n$, we have*

$$P_n(x) = \sum_{i=0}^{n} \frac{L(P_n^{(i)})}{i!} A_{L,i}(x), \tag{104}$$

$$P_n(x) = \sum_{i=0}^{n} \frac{L(\Delta^i P_n)}{i!} \mathcal{A}_{L,i}(x). \tag{105}$$

Let us consider a function $f \in X$. Then we have the following

Theorem 19. *The polynomials*

$$P_{L,n}[f](x) = \sum_{i=0}^{n} \frac{L(f^{(i)})}{i!} A_{L,i}(x), \tag{106}$$

$$P^*_{L,n}[f](x) = \sum_{i=0}^{n} \frac{L(\Delta^i f)}{i!} \mathcal{A}_{L,i}(x) \tag{107}$$

are the unique polynomial of degree $\leq n$ such that

$$L(P_{L,n}[f]^{(i)}) = L(f^{(i)}),\ i = 0, ..., n,$$

$$L(\Delta^i P^*_{L,n}[f]) = L(\Delta^i f),\ i = 0, ..., n.$$

Proof. Setting $\omega_i = \frac{L(f^{(i)})}{i!}$, and respectively, $\omega_i = \frac{L(\Delta^i f)}{i!}$, $i = 0, ..., n$, the result follows from Theorem 18. □

Definition 5. *The polynomials (106) and (107) are called Appell interpolation polynomial for f of first and of second kind, respectively.*

Now it is interesting to consider the estimation of the remainders

$$R_{L,n}[f](x) = f(x) - P_{L,n}[f](x),\ \forall x \in [0,1], \tag{108}$$

$$R^*_{L,n}[f](x) = f(x) - P^*_{L,n}[f](x),\ \forall x \in [0,1]. \tag{109}$$

Remark 10. *For any $f \in \mathcal{P}_n$*

$$R_{L,n}[f](x) = 0, \quad R_{L,n}[x^{n+1}] \neq 0,\ \forall x \in [0,1], \tag{110}$$

$$R^*_{L,n}[f](x) = 0, \quad R^*_{L,n}[(x)_{n+1}] \neq 0,\ \forall x \in [0,1], \tag{111}$$

i. e. the polynomial operators (106) and (107) are exact on $\mathcal{P}_n$.

For a fixed x we may consider the remainder $R_{L,n}[f](x)$ and $R^*_{L,n}[f](x)$ as linear functionals which act on f and annihilate all elements of $\mathcal{P}_n$. From Peano's Theorem ([27, p. 69]) if a linear functional has this property, then it must also have a simple representation in terms of $f^{(n+1)}$. Therefore we have

Theorem 20. *Let $f \in C^{n+1}[a,b]$, the following relations hold*

$$R_{L,n}(f,x) = \frac{1}{n!}\int_0^1 K_n(x,t) f^{(n+1)}(t)\,dt, \quad \forall x \in [0,1], \tag{112}$$

$$R^*_{L,n}(f,x) = \frac{1}{n!}\int_0^1 K^*_n(x,t) f^{(n+1)}(t)\,dt, \quad \forall x \in [0,1], \tag{113}$$

where

$$K_n(x,t) = R_{L,n}\left[(x-t)_+^n\right] = (x-t)_+^n - \sum_{i=0}^{n}\binom{n}{i} L\left((x-t)_+^{n-i}\right) A_{L,i}(x), \tag{114}$$

$$K_n^*(x,t) = R_{L,n}^*\left[(x-t)_+^n\right] = (x-t)_+^n - \sum_{i=0}^{n} \frac{L\left(\Delta^i (x-t)_+^n\right)}{i!} \mathcal{A}_{L,i}(x). \tag{115}$$

Proof. After some calculation, the results follow by Remark 10 and Peano's Theorem. □

Remark 11 (Bounds). *If $f^{(n+1)} \in \mathcal{L}^p[0,1]$ and $K_n(x,t)$, $K_n^*(x,t) \in \mathcal{L}^q[0,1]$ with $\frac{1}{p}+\frac{1}{q}=1$ then we apply the Hölder's inequality so that*

$$|R_{L,n}[f](x)| \le \frac{1}{n!}\left(\int_0^1 |K_n(x,t)|^q\,dt\right)^{\frac{1}{q}} \left(\int_0^1 \left|f^{(n+1)}(t)\right|^p dt\right)^{\frac{1}{p}},$$

$$|R_{L,n}^*[f](x)| \le \frac{1}{n!}\left(\int_0^1 |K_n^*(x,t)|^q\,dt\right)^{\frac{1}{q}} \left(\int_0^1 \left|f^{(n+1)}(t)\right|^p dt\right)^{\frac{1}{p}}.$$

The two most important cases are $p=q=2$ and $q=1$, $p=\infty$:

i) for $p=q=2$ we have the estimates

$$|R_{L,n}[f](x)| \le \sigma_n \,|||f|||, \quad |R_{L,n}^*[f](x)| \le \sigma_n^* \,|||f|||, \tag{116}$$

where

$$(\sigma_n)^2 = \left(\frac{1}{n!}\right)^2 \int_0^1 (K_n(x,t))^2\,dt, \quad (\sigma_n^*)^2 = \left(\frac{1}{n!}\right)^2 \int_0^1 (K_n^*(x,t))^2\,dt, \tag{117}$$

and

$$|||f|||^2 = \int_0^1 \left(f^{(n+1)}(t)\right)^2 dt; \tag{118}$$

ii) for $q=1$, $p=\infty$ we have that

$$|R_{L,n}[f](x)| \le \frac{1}{n!}M_{n+1}\int_0^1 |K_n(x,t)|\,dt, \quad |R_{L,n}^*[f](x)| \le \frac{1}{n!}M_{n+1}\int_0^1 |K_n^*(x,t)|\,dt, \tag{119}$$

where

$$M_{n+1} = \sup_{a\le x\le b} \left|f^{(n+1)}(x)\right|. \tag{120}$$

A further polynomial operator can be determined as follows:
for any fixed $z \in [0,1]$ we consider the polynomial

$$\overline{P}_{L,n}[f](x) \equiv f(z) + P_{L,n}[f](x) - P_{L,n}[f](z) = f(z) + \sum_{i=1}^{n} \frac{L(f^{(i)})}{i!}\left(A_{L,i}(x) - A_{L,i}(z)\right), \tag{121}$$

and, respectively,

$$\overline{P}_{L,n}^*[f](x) \equiv f(z) + P_{L,n}^*[f](x) - P_{L,n}^*[f](z) = f(z) + \sum_{i=1}^{n} \frac{L(\Delta^i f)}{i!}\left(\mathcal{A}_{L,i}(x) - \mathcal{A}_{L,i}(z)\right). \tag{122}$$

Then we have the following

Theorem 21. *The polynomials* $\overline{P}_{L,n}[f](x), \overline{P}^*_{L,n}[f](x)$ *are approximating polynomials of degree n for* $f(x)$, *i.e.:*

$$\forall x \in [0,1], \quad f(x) = \overline{P}_{L,n}[f](x) + \overline{R}_{L,n}[f](x), \tag{123}$$

$$f(x) = \overline{P}^*_{L,n}[f](x) + \overline{R}^*_{L,n}[f](x), \tag{124}$$

where

$$\overline{R}_{L,n}[f](x) = R_{L,n}[f](x) - R_{L,n}[f](z), \tag{125}$$

$$\overline{R}^*_{L,n}[f](x) = R^*_{L,n}[f](x) - R^*_{L,n}[f](z), \tag{126}$$

with

$$\overline{R}_{L,n}[x^i] = 0, \quad i = 0,..,n, \quad \overline{R}_{L,n}[x^{n+1}] \neq 0, \tag{127}$$

$$\overline{R}^*_{L,n}[(x)_i] = 0, \quad i = 0,..,n, \quad \overline{R}^*_{L,n}[(x)_{n+1}] \neq 0. \tag{128}$$

Proof. $\forall x \in [0,1]$ and for any fixed $z \in [0,1]$, from (108), we have

$$f(x) - f(z) = P_{L,n}[f](x) - P_{L,n}[f](z) + R_{L,n}[f](x) - R_{L,n}[f](z),$$

from which we get (123) and (125). The exactness of the polynomial $\overline{P}_{L,n}[f](x)$ follows from the exactness of the polynomial $P_{L,n}[f](x)$.

Proceeding in the same manner we can prove the result for the polynomial $\overline{P}^*_{L,n}[f](x)$. □

Remark 12. *The polynomials* $\overline{P}_{L,n}[f](x), \overline{P}^*_{L,n}[f](x)$ *satisfy the interpolation conditions*

$$\overline{P}_{L,n}[f](z) = f(z), \quad L(\overline{P}^{(i)}_{L,n}[f]) = L(f^{(i)}),\ i = 1,...,n, \tag{129}$$

$$\overline{P}^*_{L,n}[f](z) = f(z), \quad L(\Delta^i \overline{P}^*_{L,n}[f]) = L(\Delta^i f),\ i = 1,...,n. \tag{130}$$

9. Examples of Appell interpolation polynomials

a) Taylor interpolation and classical interpolation on equidistant points:
Assuming

$$L(f) = f(x_0), \qquad x_0 \in [0,1], \tag{131}$$

the polynomials $P_{L,n}[f](x)$ and $P^*_{L,n}[f](x)$ are, respectively, the Taylor interpolation polynomial and the classical interpolation polynomial on equidistant points;

b) Bernoulli interpolation of first and of second kind:

- Bernoulli interpolation of first kind ([15, 21]):
  Assuming

$$L(f) = \int_0^1 f(x)dx, \tag{132}$$

  the interpolation polynomials $P_{L,n}[f](x)$ and $\overline{P}_{L,n}[f](x)$ become

$$P_{L,n}[f](x) = \int_0^1 f(x)dx + \sum_{i=1}^{n} \frac{f^{(i-1)}(1) - f^{(i-1)}(0)}{i!} B_i(x), \tag{133}$$

$$\overline{P}_{L,n}[f](x) = f(0) + \sum_{i=1}^{n} \frac{f^{(i-1)}(1) - f^{(i-1)}(0)}{i!} \left(B_i(x) - B_i(0)\right), \tag{134}$$

where $B_i(x)$ are the classical Bernoulli polynomials ([17, 23]);

- Bernoulli interpolation of second kind ([19]):
  Assuming

$$L(f) = \left[D\Delta^{-1}f\right]_{x=0}, \tag{135}$$

where Δ^{-1} denote the indefinite summation operator and is defined as the linear operator inverse of the finite difference operator Δ, the interpolation polynomials $P^*_{L,n}[f](x)$ and $\overline{P}^*_{L,n}[f](x)$ become

$$P^*_{L,n}[f](x) = [\Delta^{-1}Df]_{x=0} + \sum_{i=0}^{n-1} f'(i)\mathcal{B}^{II}_{n,i}(x), \tag{136}$$

$$\overline{P}^*_{L,n}[f](x) = f(0) + \sum_{i=0}^{n-1} f'(i)\left(\mathcal{B}^{II}_{n,i}(x) - \mathcal{B}^{II}_{n,i}(0)\right), \tag{137}$$

where

$$\mathcal{B}^{II}_{n,i}(x) = \sum_{j=i}^{n-1} \binom{j}{i} \frac{(-1)^{j-i}}{(j+1)!} B^{II}_{j+1}(x), \tag{138}$$

and $B^{II}_j(x)$ are the Bernoulli polynomials of second kind ([19]);

c) Euler and Boole interpolation:

- Euler interpolation ([21]):
  Assuming

$$L(f) = \frac{f(0) + f(1)}{2}, \tag{139}$$

the interpolation polynomials $P_{L,n}[f](x)$ and $\overline{P}_{L,n}[f](x)$ become

$$P_{L,n}[f](x) = \frac{f(0) + f(1)}{2} + \sum_{i=1}^{n} \frac{f^{(i)}(0) + f^{(i)}(1)}{2i!} E_i(x), \tag{140}$$

$$\overline{P}_{L,n}[f](x) = f(0) + \sum_{i=1}^{n} \frac{f^{(i)}(0) + f^{(i)}(1)}{2i!} \left(E_i(x) - E_i(0)\right); \tag{141}$$

- Boole interpolation ([19]):
  Assuming

$$L(f) = [Mf]_{x=0}, \tag{142}$$

where Mf is defined by

$$Mf(x) = \frac{f(x) + f(x+1)}{2}, \tag{143}$$

the interpolation polynomials $P^*_{L,n}[f](x)$ and $\overline{P}^*_{L,n}[f](x)$ become

$$P^*_{L,n}[f](x) = \frac{f(0) + f(1)}{2} \mathcal{E}^{II}_{n,0}(x) + \sum_{i=1}^{n} \frac{f(i) + f(i+1)}{2} \mathcal{E}^{II}_{n,i}(x), \tag{144}$$

$$\overline{P}^*_{L,n}[f](x) = f(0) + \sum_{i=1}^{n} \frac{f(i)+f(i+1)}{2}\left(\mathcal{E}^{II}_{n,i}(x) - \mathcal{E}^{II}_{n,i}(0)\right), \tag{145}$$

where

$$\mathcal{E}^{II}_{n,i}(x) = \sum_{j=i}^{n} \binom{j}{i} \frac{(-1)^{j-i}}{j!} E^{II}_j(x), \tag{146}$$

and $E^{II}_j(x)$ are the Boole polynomials ([19]).

10. The algebraic approach of Yang and Youn

Yang and Youn ([18]) also proposed an algebraic approach to Appell polynomial sequences but with different methods. In fact, they referred the Appell sequence, $s_n(x)$, to an invertible analytic function g(t):

$$s_n(x) = \left[\frac{d^n}{dt}\left(\frac{1}{g(t)}e^{xt}\right)\right]_{t=0}, \tag{147}$$

and called Appell vector for $g(t)$ the vector

$$\overline{S}_n(x) = [s_0(x), ..., s_n(x)]^T. \tag{148}$$

Then, they proved that

$$\overline{S}_n(x) = P_n\left[\frac{1}{g(t)}\right]_{t=0} W_n\left[e^{xt}\right]_{t=0} = W_n\left[\frac{1}{g(t)}e^{xt}\right]_{t=0}, \tag{149}$$

being $W_n[f(t)] = \left[f(t), f'(t), ..., f^{(n)}(t)\right]^T$ and $P_n[f(t)]$ the generalized Pascal functional matrix of $f(t)$ ([28]) defined by

$$(P_n[f(t)])_{i,j} = \begin{cases} \binom{i}{j} f^{(i-j)}(t) & i \geq j \\ 0 & otherwise \end{cases}, \qquad i,j = 0, ..., n. \tag{150}$$

Expressing the (149) in matrix form we have

$$\overline{S}_n(x) = SX(x), \tag{151}$$

with

$$S = \begin{bmatrix} s_{00} & 0 & 0 & \cdots & 0 \\ s_{10} & s_{11} & 0 & \cdots & 0 \\ s_{20} & s_{21} & s_{22} & \cdots & 0 \\ \vdots & \vdots & \vdots & \ddots & \vdots \\ s_{n0} & s_{n1} & s_{n2} & \cdots & s_{nn} \end{bmatrix}, \quad X(x) = [1, x, ..., x^n]^T, \tag{152}$$

where

$$s_{i,j} = \binom{i}{j}\left[\left(\frac{1}{g(t)}\right)^{(i-j)}\right]_{t=0}, \qquad i = 0, ..., n, \quad j = 0, ..., i. \tag{153}$$

It is easy to see that the matrix S coincides with the matrix M^{-1} introduced in Section 5, Theorem 7.

11. Conclusions

We have presented an elementary algebraic approach to the theory of Appell polynomials. Given a sequence of real numbers β_i, $i = 0, 1, ...,$ $\beta_0 \neq 0$, a polynomial sequence on determinantal form, called of Appell, has been built. The equivalence of this approach with others existing was proven and, almost always using elementary tools of linear algebra, most important properties od Appell polynomials were proven too. A dual theory referred to the finite difference operator Δ has been proposed. This theory has provided a class of polynomials called Appell polynomials of second kind. Finally, given a linear functional L, with $L(1) \neq 0$, and defined

$$L(x^i) = \beta_i, \quad (L((x)_i) = \beta_i), \tag{154}$$

the linear interpolation problem

$$L(P_n^{(i)}) = i!\omega_i, \qquad \left(L(\Delta^i P_n) = i!\omega_i\right), \quad P_n \in \mathcal{P}_n, \quad \omega_i \in \mathbb{R}, \tag{155}$$

has been considered and its solution has been expressed by the basis of Appell polynomials related to the functional L by (154). This problem can be extended to appropriate real functions, providing a new approximating polynomial, the remainder of which can be estimated too. This theory is susceptible of extension to the more general class of Sheffer polynomials and to the bi-dimensional case.

Author details

Costabile Francesco Aldo and Longo Elisabetta
Department of Mathematics, University of Calabria, Rende, CS, Italy.

12. References

[1] Appell P.E (1880) Sur une classe de polynomes. Annales scientifique de l'E.N.S. 2 (9): 119-144.

[2] Avram F, Taqqu M.S (1987) Noncentral Limit Theorems and Appell Polynomial. Ann. Probab. 15 (2): 767-775.

[3] Tempesta P (2007) Formal groups, Bernoulli-type polynomials and L-series. C. R. Acad. Sci. Paris, Ser I, 345: 303-306

[4] Levi D, Tempesta P, Winternitz P (2004) Umbral calculus, differential equations and the discrete Schrodinger equation. J. Math. Phys. 45 (11): 4077-4105.

[5] Levi D, Tempesta P, Winternitz P (2004) Lorentz and Galilei invariance on lattices. Phys. Rev. D. 69 (10): 105011,1-6.

[6] Davis H.T (1936) The Theory of Linear Operator. Principia Press, Bloomington, Indiana.

[7] Sheffer I.M (1939) Some properties of polynomial sets of type zero. Duke Math. J. 5: 590-622.

[8] Bell E.T (1938) The history of Blissard's symbolic calculus, with a sketch of the inventor's life. Amer. Math. Monthly 45: 414-421.

[9] Steffensen J.F (1941) The poweroid, an extension of the mathematical notion of power. Acta Math. 73: 333-366.

[10] Mullin R, Rota G.C (1970) On the foundations of combinatorial theory III. Theory of binomial enumeration. B. Harris (Ed.) Graph Theory and its Applications, Academic Press: 167-213.

[11] Roman S (1984) The Umbral Calculus. Academic Press. New York.

[12] Roman S, Rota G.C (1978) The Umbral Calculus. Adv. Math. 27: 95-188.

[13] Di Bucchianico A, Loeb D (2000) A Selected Survey of Umbral Calculus, Electronic J. Combinatorics Dynamical Survey DS3: 1-34.

[14] Lehmer D.H (1988) A New Approach to Bernoulli Polynomials. Amer. Math. Monthly 95: 905-911.

[15] Costabile F.A (1999) On expansion of a real function in Bernoulli polynomials and application. Conferenze del Seminario Matem. - Univ. Bari 273.

[16] Costabile F.A, Dell'Accio F, Gualtieri M.I (2006) A new approach to Bernoulli polynomials. Rendiconti di matematica e delle sue applicazioni 26: 1-12.

[17] Costabile F.A, Longo E (2010) A determinantal approach to Appell polynomials. Journal of Computational and Applied Mathematics 234 (5): 1528-1542.

[18] Yang Y, Youn H (2009) Appell polynomial sequences: a linear algebra approach. JP Journal of Algebra, Number Theory and Applications 13 (1): 65-98.

[19] Costabile F.A, Longo E. Appell polynomials sequences of second kind and related interpolation problem. Under submission.

[20] Fort T (1942) Generalizations of the Bernoulli polynomials and numbers and corresponding summation formulas. Bull. Amer. Math. Soc. 48: 567-574.

[21] Costabile F.A, Longo E (2011) The Appell interpolation problem. Journal of Computational and Applied Mathematics 236 (6): 1024-1032.

[22] Highman N.H (1996) Accuracy and stability of numerical Algorithms. SIAM. Philadelphia.

[23] Jordan C (1965) Calculus of finite difference. Chealsea Pub. Co. New York.

[24] Boas R.P, Buck R.C (1964) Polynomial Expansions of Analytic Functions. Springer-Verlag, New York.

[25] Tempesta P (2008) On Appell sequences of polynomials of Bernoulli and Euler type. Journal of Mathematical Analysis and Applications 341: 1295-1310.

[26] He M.X, Ricci P.E (2002) Differential equation of Appell polynomials via the factorization method. J. Comput. Appl. Math. 139: 231-237.

[27] Davis P.J (1975) Interpolation & Approximation. Dover Publication, Inc. New York.

[28] Yang Y, Micek C (2007) Generalized Pascal functional matrix and its applications. Linear Algebra Appl. 423: 230-245.

Chapter 3

An Interpretation of Rosenbrock's Theorem via Local Rings

A. Amparan, S. Marcaida and I. Zaballa

Additional information is available at the end of the chapter

http://dx.doi.org/10.5772/81051

1. Introduction

Consider a linear time invariant system

$$\dot{x}(t) = Ax(t) + Bu(t) \tag{1}$$

to be identified with the pair of matrices (A, B) where $A \in \mathbb{F}^{n\times n}$, $B \in \mathbb{F}^{n\times m}$ and $\mathbb{F} = \mathbb{R}$ or $\mathbb{C}$ the fields of the real or complex numbers. If state-feedback $u(t) = Fx(t) + v(t)$ is applied to system (1), Rosenbrock's Theorem on pole assignment (see [14]) characterizes for the closed-loop system

$$\dot{x}(t) = (A + BF)x(t) + Bv(t), \tag{2}$$

the invariant factors of its state-space matrix $A + BF$. This result can be seen as the solution of an inverse problem; that of finding a non-singular polynomial matrix with prescribed invariant factors and left Wiener–Hopf factorization indices at infinity. To see this we recall that the invariant factors form a complete system of invariants for the finite equivalence of polynomial matrices (this equivalence relation will be revisited in Section 2) and it will be seen in Section 4 that any polynomial matrix is left Wiener–Hopf equivalent at infinity to a diagonal matrix $\text{Diag}(s^{k_1}, \ldots, s^{k_m})$, where the non-negative integers $k_1, \ldots, k_m$ (that can be assumed in non-increasing order) form a complete system of invariants for the left Wiener–Hopf equivalence at infinity. Consider now the transfer function matrix $G(s) = (sI - (A + BF))^{-1}B$ of (2). This is a rational matrix that can be written as an irreducible matrix fraction description $G(s) = N(s)P(s)^{-1}$, where $N(s)$ and $P(s)$ are right coprime polynomial matrices. In the terminology of [18], $P(s)$ is a polynomial matrix representation of (2), concept that is closely related to that of polynomial model introduced by Fuhrmann (see for example [8] and the references therein). It turns out that all polynomial matrix representations of a system are right equivalent (see [8, 18]), that is, if $P_1(s)$ and $P_2(s)$ are polynomial matrix representations of the same system there exists a unimodular matrix $U(s)$ such that $P_2(s) = P_1(s)U(s)$. Therefore all polynomial matrix representations of (2) have the same invariant factors, which are the invariant factors of $sI_n - (A + BF)$ except for some trivial ones. Furthermore, all polynomial matrix representations also have the same left Wiener– Hopf factorization indices at infinity, which are equal to the controllability indices of (2) and (1), because the controllability indices are invariant under feedback. With all this in mind it is not hard to see that Rosenbrock's Theorem on pole assignment is equivalent to finding necessary and sufficient conditions for the existence of a non-singular polynomial matrix with prescribed invariant factors and left Wiener–Hopf factorization indices at infinity. This result will be precisely stated in Section 5 once all the elements that appear are properly defined. In addition, there is a similar result to Rosenbrock's Theorem on pole assignment but involving the infinite structure (see [1]).

Our goal is to generalize both results (the finite and infinite versions of Rosenbrock's Theorem) for rational matrices defined on arbitrary fields via local rings. This will be done in Section 5 and an extension to arbitrary fields of the concept of Wiener–Hopf equivalence will be needed. This concept is very well established for complex valued rational matrix functions (see for example [6, 10]). Originally it requires a closed contour, γ, that divides the extended complex plane ($\mathbb{C} \cup \{\infty\}$) into two parts: the inner domain (Ω_+) and the region outside γ (Ω_-), which contains the point at infinity. Then two non-singular $m \times m$ complex rational matrices $T_1(s)$ and $T_2(s)$, with no poles and no zeros in γ, are said to be left Wiener–Hopf equivalent with respect to γ if there are $m \times m$ matrices $U_-(s)$ and $U_+(s)$ with no poles and no zeros in $\Omega_- \cup \gamma$ and $\Omega_+ \cup \gamma$, respectively, such that

$$T_2(s) = U_-(s)T_1(s)U_+(s). \tag{3}$$

It can be seen, then, that any non-singular $m \times m$ complex rational matrix $T(s)$ is left Wiener–Hopf equivalent with respect to γ to a diagonal matrix

$$\text{Diag}\left((s - z_0)^{k_1}, \ldots, (s - z_0)^{k_m}\right) \tag{4}$$

where z_0 is any complex number in Ω_+ and $k_1 \geq \cdots \geq k_m$ are integers uniquely determined by $T(s)$. They are called the left Wiener–Hopf factorization indices of $T(s)$ with respect to γ (see again [6, 10]). The generalization to arbitrary fields relies on the following idea: We can identify $\Omega_+ \cup \gamma$ and $(\Omega_- \cup \gamma) \setminus \{\infty\}$ with two sets M and M', respectively, of maximal ideals of $\mathbb{C}[s]$. In fact, to each $z_0 \in \mathbb{C}$ we associate the ideal generated by $s - z_0$, which is a maximal ideal of $\mathbb{C}[s]$. Notice that $s - z_0$ is also a prime polynomial of $\mathbb{C}[s]$ but M and M', as defined, cannot contain the zero ideal, which is prime. Thus we are led to consider the set $\text{Specm}(\mathbb{C}[s])$ of maximal ideals of $\mathbb{C}[s]$. By using this identification we define the left Wiener–Hopf equivalence of rational matrices over an arbitrary field $\mathbb{F}$ with respect to a subset M of $\text{Specm}(\mathbb{F}[s])$, the set of all maximal ideals of $\mathbb{F}[s]$. In this study local rings play a fundamental role. They will be introduced in Section 2. Localization techniques have been used previously in the algebraic theory of linear systems (see, for example, [7]). In Section 3 the algebraic structure of the rings of proper rational functions with prescribed finite poles is studied (i.e., for a fixed $M \subseteq \text{Specm}(\mathbb{F}[s])$ the ring of proper rational functions $\frac{p(s)}{q(s)}$ with $\gcd(g(s), \pi(s)) = 1$ for all $(\pi(s)) \in M$). It will be shown that if there is an ideal generated by a linear polynomial outside M then the set of proper rational functions with no poles in M is an Euclidean domain and all rational matrices can be classified according to their Smith–McMillan invariants. In this case, two types of invariants live together for any non-singular rational matrix and any set $M \subseteq \text{Specm}(\mathbb{F}[s])$: its Smith–McMillan and left Wiener–Hopf invariants. In Section 5 we show that a Rosenbrock-like Theorem holds true that completely characterizes the relationship between these two types of invariants.

2. Preliminaries

In the sequel $\mathbb{F}[s]$ will denote the ring of polynomials with coefficients in an arbitrary field $\mathbb{F}$ and $\mathrm{Specm}(\mathbb{F}[s])$ the set of all maximal ideals of $\mathbb{F}[s]$, that is,

$$\mathrm{Specm}(\mathbb{F}[s]) = \{(\pi(s)) : \pi(s) \in \mathbb{F}[s], \text{ irreducible, monic, different from } 1\}. \tag{5}$$

Let $\pi(s) \in \mathbb{F}[s]$ be a monic irreducible non-constant polynomial. Let $S = \mathbb{F}[s] \setminus (\pi(s))$ be the multiplicative subset of $\mathbb{F}[s]$ whose elements are coprime with $\pi(s)$. We denote by $\mathbb{F}_\pi(s)$ the quotient ring of $\mathbb{F}[s]$ by S; i.e., $S^{-1}\mathbb{F}[s]$:

$$\mathbb{F}_\pi(s) = \left\{ \frac{p(s)}{q(s)} : p(s), q(s) \in \mathbb{F}[s],\ \gcd(q(s), \pi(s)) = 1 \right\}. \tag{6}$$

This is the localization of $\mathbb{F}[s]$ at $(\pi(s))$ (see [5]). The units of $\mathbb{F}_\pi(s)$ are the rational functions $u(s) = \frac{p(s)}{q(s)}$ such that $\gcd(p(s), \pi(s)) = 1$ and $\gcd(q(s), \pi(s)) = 1$. Consequentially,

$$\mathbb{F}_\pi(s) = \left\{ u(s)\pi(s)^d : u(s) \text{ is a unit and } d \geq 0 \right\} \cup \{0\}. \tag{7}$$

For any $M \subseteq \mathrm{Specm}(\mathbb{F}[s])$, let

$$\begin{aligned} \mathbb{F}_M(s) &= \textstyle\bigcap_{(\pi(s)) \in M} \mathbb{F}_\pi(s) \\ &= \left\{ \frac{p(s)}{q(s)} : p(s), q(s) \in \mathbb{F}[s],\ \gcd(q(s), \pi(s)) = 1\ \forall\, (\pi(s)) \in M \right\}. \end{aligned} \tag{8}$$

This is a ring whose units are the rational functions $u(s) = \frac{p(s)}{q(s)}$ such that for all ideals $(\pi(s)) \in M$, $\gcd(p(s), \pi(s)) = 1$ and $\gcd(q(s), \pi(s)) = 1$. Notice that, in particular, if $M = \mathrm{Specm}(\mathbb{F}[s])$ then $\mathbb{F}_M(s) = \mathbb{F}[s]$ and if $M = \emptyset$ then $\mathbb{F}_M(s) = \mathbb{F}(s)$, the field of rational functions.

Moreover, if $\alpha(s) \in \mathbb{F}[s]$ is a non-constant polynomial whose prime factorization, $\alpha(s) = k\alpha_1(s)^{d_1} \cdots \alpha_m(s)^{d_m}$, satisfies the condition that $(\alpha_i(s)) \in M$ for all i, we will say that $\alpha(s)$ factorizes in M or $\alpha(s)$ has all its zeros in M. We will consider that the only polynomials that factorize in $M = \emptyset$ are the constants. We say that a non-zero rational function factorizes in M if both its numerator and denominator factorize in M. In this case we will say that the rational function has all its zeros and poles in M. Similarly, we will say that $\frac{p(s)}{q(s)}$ has no poles in M if $p(s) \neq 0$ and $\gcd(q(s), \pi(s)) = 1$ for all ideals $(\pi(s)) \in M$. And it has no zeros in M if $\gcd(p(s), \pi(s)) = 1$ for all ideals $(\pi(s)) \in M$. In other words, it is equivalent that $\frac{p(s)}{q(s)}$ has no poles and no zeros in M and that $\frac{p(s)}{q(s)}$ is a unit of $\mathbb{F}_M(s)$. So, a non-zero rational function factorizes in M if and only if it is a unit in $\mathbb{F}_{\mathrm{Specm}(\mathbb{F}[s]) \setminus M}(s)$.

Let $\mathbb{F}_M(s)^{m \times m}$ denote the set of $m \times m$ matrices with elements in $\mathbb{F}_M(s)$. A matrix is invertible in $\mathbb{F}_M(s)^{m \times m}$ if all its elements are in $\mathbb{F}_M(s)$ and its determinant is a unit in $\mathbb{F}_M(s)$. We denote by $\mathrm{Gl}_m(\mathbb{F}_M(s))$ the group of units of $\mathbb{F}_M(s)^{m \times m}$.

Remark 1. Let $M_1, M_2 \subseteq \mathrm{Specm}(\mathbb{F}[s])$. Notice that

1. If $M_1 \subseteq M_2$ then $\mathbb{F}_{M_1}(s) \supseteq \mathbb{F}_{M_2}(s)$ and $\mathrm{Gl}_m(\mathbb{F}_{M_1}(s)) \supseteq \mathrm{Gl}_m(\mathbb{F}_{M_2}(s))$.
2. $\mathbb{F}_{M_1 \cup M_2}(s) = \mathbb{F}_{M_1}(s) \cap \mathbb{F}_{M_2}(s)$ and $\mathrm{Gl}_m(\mathbb{F}_{M_1 \cup M_2}(s)) = \mathrm{Gl}_m(\mathbb{F}_{M_1}(s)) \cap \mathrm{Gl}_m(\mathbb{F}_{M_2}(s))$.

For any $M \subseteq \text{Specm}(\mathbb{F}[s])$ the ring $\mathbb{F}_M(s)$ is a principal ideal domain (see [3]) and its field of fractions is $\mathbb{F}(s)$. Two matrices $T_1(s), T_2(s) \in \mathbb{F}(s)^{m\times m}$ are equivalent with respect to M if there exist matrices $U(s), V(s) \in \text{Gl}_m(\mathbb{F}_M(s))$ such that $T_2(s) = U(s)T_1(s)V(s)$. Since $\mathbb{F}_M(s)$ is a principal ideal domain, for all non-singular $G(s) \in \mathbb{F}_M(s)^{m\times m}$ (see [13]) there exist matrices $U(s), V(s) \in \text{Gl}_m(\mathbb{F}_M(s))$ such that

$$G(s) = U(s)\,\text{Diag}(\alpha_1(s), \dots, \alpha_m(s))V(s) \tag{9}$$

with $\alpha_1(s) \mid \cdots \mid \alpha_m(s)$ ("$|$" stands for divisibility) monic polynomials factorizing in M, unique up to multiplication by units of $\mathbb{F}_M(s)$. The diagonal matrix is the Smith normal form of $G(s)$ with respect to M and $\alpha_1(s), \dots, \alpha_m(s)$ are called the invariant factors of $G(s)$ with respect to M. Now we introduce the Smith–McMillan form with respect to M. Assume that $T(s) \in \mathbb{F}(s)^{m\times m}$ is a non-singular rational matrix. Then $T(s) = \frac{G(s)}{d(s)}$ with $G(s) \in \mathbb{F}_M(s)^{m\times m}$ and $d(s) \in \mathbb{F}[s]$ monic, factorizing in M. Let $G(s) = U(s)\,\text{Diag}(\alpha_1(s), \dots, \alpha_m(s))V(s)$ be the Smith normal form with respect to M of $G(s)$, i.e., $U(s), V(s)$ invertible in $\mathbb{F}_M(s)^{m\times m}$ and $\alpha_1(s) \mid \cdots \mid \alpha_m(s)$ monic polynomials factorizing in M. Then

$$T(s) = U(s)\,\text{Diag}\left(\frac{\epsilon_1(s)}{\psi_1(s)}, \dots, \frac{\epsilon_m(s)}{\psi_m(s)}\right)V(s) \tag{10}$$

where $\frac{\epsilon_i(s)}{\psi_i(s)}$ are irreducible rational functions, which are the result of dividing $\alpha_i(s)$ by $d(s)$ and canceling the common factors. They satisfy that $\epsilon_1(s) \mid \cdots \mid \epsilon_m(s)$, $\psi_m(s) \mid \cdots \mid \psi_1(s)$ are monic polynomials factorizing in M. The diagonal matrix in (10) is the Smith–McMillan form with respect to M. The rational functions $\frac{\epsilon_i(s)}{\psi_i(s)}$, $i = 1, \dots, m$, are called the invariant rational functions of $T(s)$ with respect to M and constitute a complete system of invariants of the equivalence with respect to M for rational matrices.

In particular, if $M = \text{Specm}(\mathbb{F}[s])$ then $\mathbb{F}_{\text{Specm}(\mathbb{F}[s])}(s) = \mathbb{F}[s]$, the matrices $U(s), V(s) \in \text{Gl}_m(\mathbb{F}[s])$ are unimodular matrices, (10) is the global Smith–McMillan form of a rational matrix (see [15] or [14] when $\mathbb{F} = \mathbb{R}$ or $\mathbb{C}$) and $\frac{\epsilon_i(s)}{\psi_i(s)}$ are the global invariant rational functions of $T(s)$.

From now on rational matrices will be assumed to be non-singular unless the opposite is specified. Given any $M \subseteq \text{Specm}(\mathbb{F}[s])$ we say that an $m \times m$ non-singular rational matrix has no zeros and no poles in M if its global invariant rational functions are units of $\mathbb{F}_M(s)$. If its global invariant rational functions factorize in M, the matrix has its global finite structure localized in M and we say that the matrix has all zeros and poles in M. The former means that $T(s) \in \text{Gl}_m(\mathbb{F}_M(s))$ and the latter that $T(s) \in \text{Gl}_m(\mathbb{F}_{\text{Specm}(\mathbb{F}[s])\setminus M}(s))$ because $\det T(s) = \det U(s) \det V(s) \frac{\epsilon_1(s)\cdots\epsilon_m(s)}{\psi_1(s)\cdots\psi_m(s)}$ and $\det U(s), \det V(s)$ are non-zero constants. The following result clarifies the relationship between the global finite structure of any rational matrix and its local structure with respect to any $M \subseteq \text{Specm}(\mathbb{F}[s])$.

Proposition 2. *Let $M \subseteq \text{Specm}(\mathbb{F}[s])$. Let $T(s) \in \mathbb{F}(s)^{m\times m}$ be non-singular with $\frac{\alpha_1(s)}{\beta_1(s)}, \dots, \frac{\alpha_m(s)}{\beta_m(s)}$ its global invariant rational functions and let $\frac{\epsilon_1(s)}{\psi_1(s)}, \dots, \frac{\epsilon_m(s)}{\psi_m(s)}$ be irreducible rational functions such that $\epsilon_1(s) \mid \cdots \mid \epsilon_m(s)$, $\psi_m(s) \mid \cdots \mid \psi_1(s)$ are monic polynomials factorizing in M. The following properties are equivalent:*

1. *There exist* $T_L(s), T_R(s) \in \mathbb{F}(s)^{m\times m}$ *such that the global invariant rational functions of* $T_L(s)$ *are* $\frac{\epsilon_1(s)}{\psi_1(s)}, \ldots, \frac{\epsilon_m(s)}{\psi_m(s)}$, $T_R(s) \in \mathrm{Gl}_m(\mathbb{F}_M(s))$ *and* $T(s) = T_L(s)T_R(s)$.
2. *There exist matrices* $U_1(s), U_2(s)$ *invertible in* $\mathbb{F}_M(s)^{m\times m}$ *such that*

$$T(s) = U_1(s)\,\mathrm{Diag}\left(\frac{\epsilon_1(s)}{\psi_1(s)}, \ldots, \frac{\epsilon_m(s)}{\psi_m(s)}\right) U_2(s), \tag{11}$$

i.e., $\frac{\epsilon_1(s)}{\psi_1(s)}, \ldots, \frac{\epsilon_m(s)}{\psi_m(s)}$ *are the invariant rational functions of* $T(s)$ *with respect to* M.
3. $\alpha_i(s) = \epsilon_i(s)\epsilon_i'(s)$ *and* $\beta_i(s) = \psi_i(s)\psi_i'(s)$ *with* $\epsilon_i'(s), \psi_i'(s) \in \mathbb{F}[s]$ *units of* $\mathbb{F}_M(s)$, *for* $i = 1, \ldots, m$.

Proof.- $1 \Rightarrow 2$. Since the global invariant rational functions of $T_L(s)$ are $\frac{\epsilon_1(s)}{\psi_1(s)}, \ldots, \frac{\epsilon_m(s)}{\psi_m(s)}$, there exist $W_1(s), W_2(s) \in \mathrm{Gl}_m(\mathbb{F}[s])$ such that $T_L(s) = W_1(s)\,\mathrm{Diag}\left(\frac{\epsilon_1(s)}{\psi_1(s)}, \ldots, \frac{\epsilon_m(s)}{\psi_m(s)}\right) W_2(s)$. As $\mathbb{F}_{\mathrm{Specm}(\mathbb{F}[s])}(s) = \mathbb{F}[s]$, by Remark 1.1, $W_1(s), W_2(s) \in \mathrm{Gl}_m(\mathbb{F}_M(s))$. Therefore, putting $U_1(s) = W_1(s)$ and $U_2(s) = W_2(s)T_R(s)$ it follows that $U_1(s)$ and $U_2(s)$ are invertible in $\mathbb{F}_M(s)^{m\times m}$ and $T(s) = U_1(s)\,\mathrm{Diag}\left(\frac{\epsilon_1(s)}{\psi_1(s)}, \ldots, \frac{\epsilon_m(s)}{\psi_m(s)}\right) U_2(s)$.

$2 \Rightarrow 3$. There exist unimodular matrices $V_1(s), V_2(s) \in \mathbb{F}[s]^{m\times m}$ such that

$$T(s) = V_1(s)\,\mathrm{Diag}\left(\frac{\alpha_1(s)}{\beta_1(s)}, \ldots, \frac{\alpha_m(s)}{\beta_m(s)}\right) V_2(s) \tag{12}$$

with $\frac{\alpha_i(s)}{\beta_i(s)}$ irreducible rational functions such that $\alpha_1(s) \mid \cdots \mid \alpha_m(s)$ and $\beta_m(s) \mid \cdots \mid \beta_1(s)$ are monic polynomials. Write $\frac{\alpha_i(s)}{\beta_i(s)} = \frac{p_i(s)p_i'(s)}{q_i(s)q_i'(s)}$ such that $p_i(s), q_i(s)$ factorize in M and $p_i'(s), q_i'(s)$ factorize in $\mathrm{Specm}(\mathbb{F}[s]) \setminus M$. Then

$$T(s) = V_1(s)\,\mathrm{Diag}\left(\frac{p_1(s)}{q_1(s)}, \ldots, \frac{p_m(s)}{q_m(s)}\right) \mathrm{Diag}\left(\frac{p_1'(s)}{q_1'(s)}, \ldots, \frac{p_m'(s)}{q_m'(s)}\right) V_2(s) \tag{13}$$

with $V_1(s)$ and $\mathrm{Diag}\left(\frac{p_1'(s)}{q_1'(s)}, \ldots, \frac{p_m'(s)}{q_m'(s)}\right) V_2(s)$ invertible in $\mathbb{F}_M(s)^{m\times m}$. Since the Smith–McMillan form with respect to M is unique we get that $\frac{p_i(s)}{q_i(s)} = \frac{\epsilon_i(s)}{\psi_i(s)}$.

$3 \Rightarrow 1$. Write (12) as

$$T(s) = V_1(s)\,\mathrm{Diag}\left(\frac{\epsilon_1(s)}{\psi_1(s)}, \ldots, \frac{\epsilon_m(s)}{\psi_m(s)}\right) \mathrm{Diag}\left(\frac{\epsilon_1'(s)}{\psi_1'(s)}, \ldots, \frac{\epsilon_m'(s)}{\psi_m'(s)}\right) V_2(s). \tag{14}$$

It follows that $T(s) = T_L(s)T_R(s)$ with $T_L(s) = V_1(s)\,\mathrm{Diag}\left(\frac{\epsilon_1(s)}{\psi_1(s)}, \ldots, \frac{\epsilon_m(s)}{\psi_m(s)}\right)$ and $T_R(s) = \mathrm{Diag}\left(\frac{\epsilon_1'(s)}{\psi_1'(s)}, \ldots, \frac{\epsilon_m'(s)}{\psi_m'(s)}\right) V_2(s) \in \mathrm{Gl}_m(\mathbb{F}_M(s))$. ∎

Corollary 3. *Let* $T(s) \in \mathbb{F}(s)^{m\times m}$ *be non-singular and* $M_1, M_2 \subseteq \mathrm{Specm}(\mathbb{F}[s])$ *such that* $M_1 \cap M_2 = \emptyset$. *If* $\frac{\epsilon_1^i(s)}{\psi_1^i(s)}, \ldots, \frac{\epsilon_m^i(s)}{\psi_m^i(s)}$ *are the invariant rational functions of* $T(s)$ *with respect to* M_i, $i = 1, 2$, *then* $\frac{\epsilon_1^1(s)\epsilon_1^2(s)}{\psi_1^1(s)\psi_1^2(s)}, \ldots, \frac{\epsilon_m^1(s)\epsilon_m^2(s)}{\psi_m^1(s)\psi_m^2(s)}$ *are the invariant rational functions of* $T(s)$ *with respect to* $M_1 \cup M_2$.

Proof.- Let $\frac{\alpha_1(s)}{\beta_1(s)}, \ldots, \frac{\alpha_m(s)}{\beta_m(s)}$ be the global invariant rational functions of $T(s)$. By Proposition 2, $\alpha_i(s) = \epsilon_i^1(s)n_i^1(s), \beta_i(s) = \psi_i^1(s)d_i^1(s)$, with $n_i^1(s), d_i^1(s) \in \mathbb{F}[s]$ units of $\mathbb{F}_{M_1}(s)$. On the other hand $\alpha_i(s) = \epsilon_i^2(s)n_i^2(s), \beta_i(s) = \psi_i^2(s)d_i^2(s)$, with $n_i^2(s), d_i^2(s) \in \mathbb{F}[s]$ units of $\mathbb{F}_{M_2}(s)$. So, $\epsilon_i^1(s)n_i^1(s) = \epsilon_i^2(s)n_i^2(s)$ or equivalently $n_i^1(s) = \frac{\epsilon_i^2(s)n_i^2(s)}{\epsilon_i^1(s)}, n_i^2(s) = \frac{\epsilon_i^1(s)n_i^1(s)}{\epsilon_i^2(s)}$. The polynomials $\epsilon_i^1(s), \epsilon_i^2(s)$ are coprime because $\epsilon_i^1(s)$ factorizes in M_1, $\epsilon_i^2(s)$ factorizes in M_2 and $M_1 \cap M_2 = \emptyset$. In consequence $\epsilon_i^1(s) \mid n_i^2(s)$ and $\epsilon_i^2(s) \mid n_i^1(s)$. Therefore, there exist polynomials $a(s)$, unit of $\mathbb{F}_{M_2}(s)$, and $a'(s)$, unit of $\mathbb{F}_{M_1}(s)$, such that $n_i^2(s) = \epsilon_i^1(s)a(s), n_i^1(s) = \epsilon_i^2(s)a'(s)$. Since $\alpha_i(s) = \epsilon_i^1(s)n_i^1(s) = \epsilon_i^1(s)\epsilon_i^2(s)a'(s)$ and $\alpha_i(s) = \epsilon_i^2(s)n_i^2(s) = \epsilon_i^2(s)\epsilon_i^1(s)a(s)$. This implies that $a(s) = a'(s)$ unit of $\mathbb{F}_{M_1}(s) \cap \mathbb{F}_{M_2}(s) = \mathbb{F}_{M_1 \cup M_2}(s)$. Following the same ideas we can prove that $\beta_i(s) = \psi_i^1(s)\psi_i^2(s)b(s)$ with $b(s)$ a unit of $\mathbb{F}_{M_1 \cup M_2}(s)$. By Proposition 2 $\frac{\epsilon_1^1(s)\epsilon_1^2(s)}{\psi_1^1(s)\psi_1^2(s)}, \ldots, \frac{\epsilon_m^1(s)\epsilon_m^2(s)}{\psi_m^1(s)\psi_m^2(s)}$ are the invariant rational functions of $T(s)$ with respect to $M_1 \cup M_2$.■

Corollary 4. *Let $M_1, M_2 \subseteq \mathrm{Specm}(\mathbb{F}[s])$. Two non-singular matrices are equivalent with respect to $M_1 \cup M_2$ if and only if they are equivalent with respect to M_1 and with respect to M_2.*

Proof.- Notice that by Remark 1.2 two matrices $T_1(s), T_2(s) \in \mathbb{F}(s)^{m \times m}$ are equivalent with respect to $M_1 \cup M_2$ if and only if there exist $U_1(s), U_2(s)$ invertible in $\mathbb{F}_{M_1}(s)^{m \times m} \cap \mathbb{F}_{M_2}(s)^{m \times m}$ such that $T_2(s) = U_1(s)T_1(s)U_2(s)$. Since $U_1(s)$ and $U_2(s)$ are invertible in both $\mathbb{F}_{M_1}(s)^{m \times m}$ and $\mathbb{F}_{M_2}(s)^{m \times m}$ then $T_1(s)$ and $T_2(s)$ are equivalent with respect to M_1 and with respect to M_2.

Conversely, if $T_1(s)$ and $T_2(s)$ are equivalent with respect to M_1 and with respect to M_2 then, by the necessity of this result, they are equivalent with respect to $M_1 \setminus (M_1 \cap M_2)$, with respect to $M_2 \setminus (M_1 \cap M_2)$ and with respect to $M_1 \cap M_2$. Let $\frac{\epsilon_1^1(s)}{\psi_1^1(s)}, \ldots, \frac{\epsilon_m^1(s)}{\psi_m^1(s)}$ be the invariant rational functions of $T_1(s)$ and $T_2(s)$ with respect to $M_1 \setminus (M_1 \cap M_2)$, $\frac{\epsilon_1^2(s)}{\psi_1^2(s)}, \ldots, \frac{\epsilon_m^2(s)}{\psi_m^2(s)}$ be the invariant rational functions of $T_1(s)$ and $T_2(s)$ with respect to $M_2 \setminus (M_1 \cap M_2)$ and $\frac{\epsilon_1^3(s)}{\psi_1^3(s)}, \ldots, \frac{\epsilon_m^3(s)}{\psi_m^3(s)}$ be the invariant rational functions of $T_1(s)$ and $T_2(s)$ with respect to $M_1 \cap M_2$. By Corollary 3 $\frac{\epsilon_1^1(s)}{\psi_1^1(s)}\frac{\epsilon_1^2(s)}{\psi_1^2(s)}\frac{\epsilon_1^3(s)}{\psi_1^3(s)}, \ldots, \frac{\epsilon_m^1(s)}{\psi_m^1(s)}\frac{\epsilon_m^2(s)}{\psi_m^2(s)}\frac{\epsilon_m^3(s)}{\psi_m^3(s)}$ must be the invariant rational functions of $T_1(s)$ and $T_2(s)$ with respect to $M_1 \cup M_2$. Therefore, $T_1(s)$ and $T_2(s)$ are equivalent with respect to $M_1 \cup M_2$.■

Let $\mathbb{F}_{pr}(s)$ be the ring of proper rational functions, that is, rational functions with the degree of the numerator at most the degree of the denominator. The units in this ring are the rational functions whose numerators and denominators have the same degree. They are called biproper rational functions. A matrix $B(s) \in \mathbb{F}_{pr}(s)^{m \times m}$ is said to be biproper if it is a unit in $\mathbb{F}_{pr}(s)^{m \times m}$ or, what is the same, if its determinant is a biproper rational function.

Recall that a rational function $t(s)$ has a pole (zero) at ∞ if $t\left(\frac{1}{s}\right)$ has a pole (zero) at 0. Following this idea, we can define the local ring at ∞ as the set of rational functions, $t(s)$, such that $t\left(\frac{1}{s}\right)$ does not have 0 as a pole, that is, $\mathbb{F}_\infty(s) = \left\{t(s) \in \mathbb{F}(s) : t\left(\frac{1}{s}\right) \in \mathbb{F}_s(s)\right\}$. If $t(s) = \frac{p(s)}{q(s)}$ with $p(s) = a_t s^t + a_{t+1}s^{t+1} + \cdots + a_p s^p$, $a_p \neq 0$, $q(s) = b_r s^r + b_{r+1}s^{r+1} + \cdots +$

$b_q s^q$, $b_q \neq 0$, $p = d(p(s))$, $q = d(q(s))$, where $d(\cdot)$ stands for "degree of", then

$$t\left(\frac{1}{s}\right) = \frac{\frac{a_t}{s^t} + \frac{a_{t+1}}{s^{t+1}} + \cdots + \frac{a_p}{s^p}}{\frac{b_r}{s^r} + \frac{b_{r+1}}{s^{r+1}} + \cdots + \frac{b_q}{s^q}} = \frac{a_t s^{p-t} + a_{t+1}s^{p-t-1} + \cdots + a_p}{b_r s^{q-r} + b_{r+1}s^{q-r-1} + \cdots + b_q} s^{q-p} = \frac{f(s)}{g(s)} s^{q-p}. \tag{15}$$

As $\mathbb{F}_s(s) = \left\{\frac{f(s)}{g(s)} s^d : f(0) \neq 0, g(0) \neq 0 \text{ and } d \geq 0\right\} \cup \{0\}$, then

$$\mathbb{F}_\infty(s) = \left\{\frac{p(s)}{q(s)} \in \mathbb{F}(s) : d(q(s)) \geq d(p(s))\right\}. \tag{16}$$

Thus, this set is the ring of proper rational functions, $\mathbb{F}_{pr}(s)$.

Two rational matrices $T_1(s), T_2(s) \in \mathbb{F}(s)^{m\times m}$ are equivalent at infinity if there exist biproper matrices $B_1(s), B_2(s) \in \mathrm{Gl}_m(\mathbb{F}_{pr}(s))$ such that $T_2(s) = B_1(s)T_1(s)B_2(s)$. Given a non-singular rational matrix $T(s) \in \mathbb{F}(s)^{m\times m}$ (see [15]) there always exist $B_1(s), B_2(s) \in \mathrm{Gl}_m(\mathbb{F}_{pr}(s))$ such that

$$T(s) = B_1(s)\,\mathrm{Diag}(s^{q_1}, \ldots, s^{q_m})B_2(s) \tag{17}$$

where $q_1 \geq \cdots \geq q_m$ are integers. They are called the invariant orders of $T(s)$ at infinity and the rational functions $s^{q_1}, \ldots, s^{q_m}$ are called the invariant rational functions of $T(s)$ at infinity.

3. Structure of the ring of proper rational functions with prescribed finite poles

Let $M' \subseteq \mathrm{Specm}(\mathbb{F}[s])$. Any non-zero rational function $t(s)$ can be uniquely written as $t(s) = \frac{n(s)}{d(s)}\frac{n'(s)}{d'(s)}$ where $\frac{n(s)}{d(s)}$ is an irreducible rational function factorizing in M' and $\frac{n'(s)}{d'(s)}$ is a unit of $\mathbb{F}_{M'}(s)$. Define the following function over $\mathbb{F}(s) \setminus \{0\}$ (see [15], [16]):

$$\begin{array}{rcl} \delta : \mathbb{F}(s) \setminus \{0\} & \to & \mathbb{Z} \\ t(s) & \mapsto & d(d'(s)) - d(n'(s)). \end{array} \tag{18}$$

This mapping is not a discrete valuation of $\mathbb{F}(s)$ if $M' \neq \emptyset$: Given two non-zero elements $t_1(s), t_2(s) \in \mathbb{F}(s)$ it is clear that $\delta(t_1(s)t_2(s)) = \delta(t_1(s)) + \delta(t_2(s))$; but it may not satisfy that $\delta(t_1(s) + t_2(s)) \geq \min(\delta(t_1(s)), \delta(t_2(s)))$. For example, let $M' = \{(s-a) \in \mathrm{Specm}(\mathbb{R}[s]) : a \notin [-2,-1]\}$. Put $t_1(s) = \frac{s+0.5}{s+1.5}$ and $t_2(s) = \frac{s+2.5}{s+1.5}$. We have that $\delta(t_1(s)) = d(s+1.5) - d(1) = 1$, $\delta(t_2(s)) = d(s+1.5) - d(1) = 1$ but $\delta(t_1(s) + t_2(s)) = \delta(2) = 0$.

However, if $M' = \emptyset$ and $t(s) = \frac{n(s)}{d(s)} \in \mathbb{F}(s)$ where $n(s), d(s) \in \mathbb{F}[s]$, $d(s) \neq 0$, the map

$$\delta_\infty : \mathbb{F}(s) \to \mathbb{Z} \cup \{+\infty\} \tag{19}$$

defined via $\delta_\infty(t(s)) = d(d(s)) - d(n(s))$ if $t(s) \neq 0$ and $\delta_\infty(t(s)) = +\infty$ if $t(s) = 0$ is a discrete valuation of $\mathbb{F}(s)$.

Consider the subset of $\mathbb{F}(s)$, $\mathbb{F}_{M'}(s) \cap \mathbb{F}_{pr}(s)$, consisting of all proper rational functions with poles in $\mathrm{Specm}(\mathbb{F}[s]) \setminus M'$, that is, the elements of $\mathbb{F}_{M'}(s) \cap \mathbb{F}_{pr}(s)$ are proper rational functions whose denominators are coprime with all the polynomials $\pi(s)$ such that $(\pi(s)) \in M'$. Notice that $g(s) \in \mathbb{F}_{M'}(s) \cap \mathbb{F}_{pr}(s)$ if and only if $g(s) = n(s)\frac{n'(s)}{d'(s)}$ where:

(a) $n(s) \in \mathbb{F}[s]$ is a polynomial factorizing in M',

(b) $\frac{n'(s)}{d'(s)}$ is an irreducible rational function and a unit of $\mathbb{F}_{M'}(s)$,

(c) $\delta(g(s)) - d(n(s)) \geq 0$ or equivalently $\delta_\infty(g(s)) \geq 0$.

In particular (c) implies that $\frac{n'(s)}{d'(s)} \in \mathbb{F}_{pr}(s)$. The units in $\mathbb{F}_{M'}(s) \cap \mathbb{F}_{pr}(s)$ are biproper rational functions $\frac{n'(s)}{d'(s)}$, that is $d(n'(s)) = d(d'(s))$, with $n'(s), d'(s)$ factorizing in $\text{Specm}(\mathbb{F}[s]) \setminus M'$. Furthermore, $\mathbb{F}_{M'}(s) \cap \mathbb{F}_{pr}(s)$ is an integral domain whose field of fractions is $\mathbb{F}(s)$ provided that $M' \neq \text{Specm}(\mathbb{F}[s])$(see, for example, [15, Prop.5.22]). Notice that for $M' = \text{Specm}(\mathbb{F}[s])$, $\mathbb{F}_{M'}(s) \cap \mathbb{F}_{pr}(s) = \mathbb{F}[s] \cap \mathbb{F}_{pr}(s) = \mathbb{F}$.

Assume that there are ideals in $\text{Specm}(\mathbb{F}[s]) \setminus M'$ generated by linear polynomials and let $(s - a)$ be any of them. The elements of $\mathbb{F}_{M'}(s) \cap \mathbb{F}_{pr}(s)$ can be written as $g(s) = n(s)u(s)\frac{1}{(s-a)^d}$ where $n(s) \in \mathbb{F}[s]$ factorizes in M', $u(s)$ is a unit in $\mathbb{F}_{M'}(s) \cap \mathbb{F}_{pr}(s)$ and $d = \delta(g(s)) \geq d(n(s))$. If $\mathbb{F}$ is algebraically closed, for example $\mathbb{F} = \mathbb{C}$, and $M' \neq \text{Specm}(\mathbb{F}[s])$ the previous condition is always fulfilled.

The divisibility in $\mathbb{F}_{M'}(s) \cap \mathbb{F}_{pr}(s)$ is characterized in the following lemma.

Lemma 5. *Let $M' \subseteq \text{Specm}(\mathbb{F}[s])$. Let $g_1(s), g_2(s) \in \mathbb{F}_{M'}(s) \cap \mathbb{F}_{pr}(s)$ be such that $g_1(s) = n_1(s)\frac{n_1'(s)}{d_1'(s)}$ and $g_2(s) = n_2(s)\frac{n_2'(s)}{d_2'(s)}$ with $n_1(s), n_2(s) \in \mathbb{F}[s]$ factorizing in M' and $\frac{n_1'(s)}{d_1'(s)}, \frac{n_2'(s)}{d_2'(s)}$ irreducible rational functions, units of $\mathbb{F}_{M'}(s)$. Then $g_1(s)$ divides $g_2(s)$ in $\mathbb{F}_{M'}(s) \cap \mathbb{F}_{pr}(s)$ if and only if*

$$n_1(s) \mid n_2(s) \text{ in } \mathbb{F}[s] \tag{20}$$

$$\delta(g_1(s)) - d(n_1(s)) \leq \delta(g_2(s)) - d(n_2(s)). \tag{21}$$

Proof.- If $g_1(s) \mid g_2(s)$ then there exists $g(s) = n(s)\frac{n'(s)}{d'(s)} \in \mathbb{F}_{M'}(s) \cap \mathbb{F}_{pr}(s)$, with $n(s) \in \mathbb{F}[s]$ factorizing in M' and $n'(s), d'(s) \in \mathbb{F}[s]$ coprime, factorizing in $\text{Specm}(\mathbb{F}[s]) \setminus M'$, such that $g_2(s) = g(s)g_1(s)$. Equivalently, $n_2(s)\frac{n_2'(s)}{d_2'(s)} = n(s)\frac{n'(s)}{d'(s)}n_1(s)\frac{n_1'(s)}{d_1'(s)} = n(s)n_1(s)\frac{n'(s)n_1'(s)}{d'(s)d_1'(s)}$. So $n_2(s) = n(s)n_1(s)$ and $\delta(g_2(s)) - d(n_2(s)) = \delta(g(s)) - d(n(s)) + \delta(g_1(s)) - d(n_1(s))$. Moreover, as $g(s)$ is a proper rational function, $\delta(g(s)) - d(n(s)) \geq 0$ and $\delta(g_2(s)) - d(n_2(s)) \geq \delta(g_1(s)) - d(n_1(s))$.

Conversely, if $n_1(s) \mid n_2(s)$ then there is $n(s) \in \mathbb{F}[s]$, factorizing in M', such that $n_2(s) = n(s)n_1(s)$. Write $g(s) = n(s)\frac{n'(s)}{d'(s)}$ where $\frac{n'(s)}{d'(s)}$ is an irreducible fraction representation of $\frac{n_2'(s)d_1'(s)}{d_2'(s)n_1'(s)}$, i.e., $\frac{n'(s)}{d'(s)} = \frac{n_2'(s)d_1'(s)}{d_2'(s)n_1'(s)}$ after canceling possible common factors. Thus $\frac{n_2'(s)}{d_2'(s)} = \frac{n'(s)}{d'(s)}\frac{n_1'(s)}{d_1'(s)}$ and

$$\begin{aligned}\delta(g(s)) - d(n(s)) &= d(d'(s)) - d(n'(s)) - d(n(s)) \\ &= d(d_2'(s)) + d(n_1'(s)) - d(n_2'(s)) - d(d_1'(s)) - d(n_2(s)) + d(n_1(s)) \\ &= \delta(g_2(s)) - d(n_2(s)) - (\delta(g_1(s)) - d(n_1(s))) \geq 0.\end{aligned} \tag{22}$$

Then $g(s) \in \mathbb{F}_{M'}(s) \cap \mathbb{F}_{pr}(s)$ and $g_2(s) = g(s)g_1(s)$. ■

Notice that condition (20) means that $g_1(s) \mid g_2(s)$ in $\mathbb{F}_{M'}(s)$ and condition (21) means that $g_1(s) \mid g_2(s)$ in $\mathbb{F}_{pr}(s)$. So, $g_1(s) \mid g_2(s)$ in $\mathbb{F}_{M'}(s) \cap \mathbb{F}_{pr}(s)$ if and only if $g_1(s) \mid g_2(s)$ simultaneously in $\mathbb{F}_{M'}(s)$ and $\mathbb{F}_{pr}(s)$.

Lemma 6. *Let $M' \subseteq$ Specm($\mathbb{F}[s]$). Let $g_1(s), g_2(s) \in \mathbb{F}_{M'}(s) \cap \mathbb{F}_{pr}(s)$ be such that $g_1(s) = n_1(s)\frac{n_1'(s)}{d_1'(s)}$ and $g_2(s) = n_2(s)\frac{n_2'(s)}{d_2'(s)}$ as in Lemma 5. If $n_1(s)$ and $n_2(s)$ are coprime in $\mathbb{F}[s]$ and either $\delta(g_1(s)) = d(n_1(s))$ or $\delta(g_2(s)) = d(n_2(s))$ then $g_1(s)$ and $g_2(s)$ are coprime in $\mathbb{F}_{M'}(s) \cap \mathbb{F}_{pr}(s)$.*

Proof.- Suppose that $g_1(s)$ and $g_2(s)$ are not coprime. Then there exists a non-unit $g(s) = n(s)\frac{n'(s)}{d'(s)} \in \mathbb{F}_{M'}(s) \cap \mathbb{F}_{pr}(s)$ such that $g(s) \mid g_1(s)$ and $g(s) \mid g_2(s)$. As $g(s)$ is not a unit, $n(s)$ is not a constant or $\delta(g(s)) > 0$. If $n(s)$ is not a constant then $n(s) \mid n_1(s)$ and $n(s) \mid n_2(s)$ which is impossible because $n_1(s)$ and $n_2(s)$ are coprime. Otherwise, if $n(s)$ is a constant then $\delta(g(s)) > 0$ and we have that $\delta(g(s)) \leq \delta(g_1(s)) - d(n_1(s))$ and $\delta(g(s)) \leq \delta(g_2(s)) - d(n_2(s))$. But this is again impossible. ■

It follows from this Lemma that if $g_1(s), g_2(s)$ are coprime in both rings $\mathbb{F}_{M'}(s)$ and $\mathbb{F}_{pr}(s)$ then $g_1(s), g_2(s)$ are coprime in $\mathbb{F}_{M'}(s) \cap \mathbb{F}_{pr}(s)$. The following example shows that the converse is not true in general.

Example 7. Suppose that $\mathbb{F} = \mathbb{R}$ and $M' =$ Specm$(\mathbb{R}[s]) \setminus \{(s^2+1)\}$. It is not difficult to prove that $g_1(s) = \frac{s^2}{s^2+1}$ and $g_2(s) = \frac{s}{s^2+1}$ are coprime elements in $\mathbb{R}_{M'}(s) \cap \mathbb{R}_{pr}(s)$. Assume that there exists a non-unit $g(s) = n(s)\frac{n'(s)}{d'(s)} \in \mathbb{R}_{M'}(s) \cap \mathbb{R}_{pr}(s)$ such that $g(s) \mid g_1(s)$ and $g(s) \mid g_2(s)$. Then $n(s) \mid s^2$, $n(s) \mid s$ and $\delta(g(s)) - d(n(s)) = 0$. Since $g(s)$ is not a unit, $n(s)$ cannot be a constant. Hence, $n(s) = cs$, $c \neq 0$, and $\delta(g(s)) = 1$, but this is impossible because $d'(s)$ and $n'(s)$ are powers of s^2+1. Therefore $g_1(s)$ and $g_2(s)$ must be coprime. However $n_1(s) = s^2$ and $n_2(s) = s$ are not coprime.

Now, we have the following property when there are ideals in Specm$(\mathbb{F}[s]) \setminus M'$, $M' \subseteq$ Specm$(\mathbb{F}[s])$, generated by linear polynomials.

Lemma 8. *Let $M' \subseteq$ Specm$(\mathbb{F}[s])$. Assume that there are ideals in Specm$(\mathbb{F}[s]) \setminus M'$ generated by linear polynomials and let $(s-a)$ be any of them. Let $g_1(s), g_2(s) \in \mathbb{F}_{M'}(s) \cap \mathbb{F}_{pr}(s)$ be such that $g_1(s) = n_1(s)u_1(s)\frac{1}{(s-a)^{d_1}}$ and $g_2(s) = n_2(s)u_2(s)\frac{1}{(s-a)^{d_2}}$. If $g_1(s)$ and $g_2(s)$ are coprime in $\mathbb{F}_{M'}(s) \cap \mathbb{F}_{pr}(s)$ then $n_1(s)$ and $n_2(s)$ are coprime in $\mathbb{F}[s]$ and either $d_1 = d(n_1(s))$ or $d_2 = d(n_2(s))$.*

Proof.- Suppose that $n_1(s)$ and $n_2(s)$ are not coprime in $\mathbb{F}[s]$. Then there exists a non-constant $n(s) \in \mathbb{F}[s]$ such that $n(s) \mid n_1(s)$ and $n(s) \mid n_2(s)$. Let $d = d(n(s))$. Then $g(s) = n(s)\frac{1}{(s-a)^d}$ is not a unit in $\mathbb{F}_{M'}(s) \cap \mathbb{F}_{pr}(s)$ and divides $g_1(s)$ and $g_2(s)$ because $0 = d - d(n(s)) \leq d_1 - d(n_1(s))$ and $0 = d - d(n(s)) \leq d_2 - d(n_2(s))$. This is impossible, so $n_1(s)$ and $n_2(s)$ must be coprime.

Now suppose that $d_1 > d(n_1(s))$ and $d_2 > d(n_2(s))$. Let $d = \min\{d_1 - d(n_1(s)), d_2 - d(n_2(s))\}$. We have that $d > 0$. Thus $g(s) = \frac{1}{(s-a)^d}$ is not a unit in $\mathbb{F}_{M'}(s) \cap \mathbb{F}_{pr}(s)$ and divides $g_1(s)$ and $g_2(s)$ because $d \leq d_1 - d(n_1(s))$ and $d \leq d_2 - d(n_2(s))$. This is again impossible and either $d_1 = d(n_1(s))$ or $d_2 = d(n_2(s))$. ■

The above lemmas yield a characterization of coprimeness of elements in $\mathbb{F}_{M'}(s) \cap \mathbb{F}_{pr}(s)$ when M' excludes at least one ideal generated by a linear polynomial.

Following the same steps as in [16, p. 11] and [15, p. 271] we get the following result.

Lemma 9. *Let $M' \subseteq \text{Specm}(\mathbb{F}[s])$ and assume that there is at least an ideal in $\text{Specm}(\mathbb{F}[s]) \setminus M'$ generated by a linear polynomial. Then $\mathbb{F}_{M'}(s) \cap \mathbb{F}_{pr}(s)$ is a Euclidean domain.*

The following examples show that if all ideals generated by polynomials of degree one are in M', the ring $\mathbb{F}_{M'}(s) \cap \mathbb{F}_{pr}(s)$ may not be a Bezout domain. Thus, it may not be a Euclidean domain. Even more, it may not be a greatest common divisor domain.

Example 10. Let $\mathbb{F} = \mathbb{R}$ and $M' = \text{Specm}(\mathbb{R}[s]) \setminus \{(s^2+1)\}$. Let $g_1(s) = \frac{s^2}{s^2+1}, g_2(s) = \frac{s}{s^2+1} \in \mathbb{R}_{M'}(s) \cap \mathbb{R}_{pr}(s)$. We have seen, in the previous example, that $g_1(s), g_2(s)$ are coprime. We show now that the Bezout identity is not fulfilled, that is, there are not $a(s), b(s) \in \mathbb{R}_{M'}(s) \cap \mathbb{R}_{pr}(s)$ such that $a(s)g_1(s) + b(s)g_2(s) = u(s)$, with $u(s)$ a unit in $\mathbb{R}_{M'}(s) \cap \mathbb{R}_{pr}(s)$. Elements in $\mathbb{R}_{M'}(s) \cap \mathbb{R}_{pr}(s)$ are of the form $\frac{n(s)}{(s^2+1)^d}$ with $n(s)$ relatively prime with s^2+1 and $2d \geq d(n(s))$ and the units in $\mathbb{R}_{M'}(s) \cap \mathbb{R}_{pr}(s)$ are non-zero constants. We will see that there are not elements $a(s) = \frac{n(s)}{(s^2+1)^d}, b(s) = \frac{n'(s)}{(s^2+1)^{d'}}$ with $n(s)$ and $n'(s)$ coprime with s^2+1, $2d \geq d(n(s))$ and $2d' \geq d(n'(s))$ such that $a(s)g_1(s) + b(s)g_2(s) = c$, with c non-zero constant. Assume that $\frac{n(s)}{(s^2+1)^d}\frac{s^2}{s^2+1} + \frac{n'(s)}{(s^2+1)^{d'}}\frac{s}{s^2+1} = c$. We conclude that $c(s^2+1)^{d+1}$ or $c(s^2+1)^{d'+1}$ is a multiple of s, which is impossible.

Example 11. Let $\mathbb{F} = \mathbb{R}$ and $M' = \text{Specm}(\mathbb{R}[s]) \setminus \{(s^2+1)\}$. A fraction $g(s) = \frac{n(s)}{(s^2+1)^d} \in \mathbb{R}_{M'}(s) \cap \mathbb{R}_{pr}(s)$ if and only if $2d - d(n(s)) \geq 0$. Let $g_1(s) = \frac{s^2}{(s^2+1)^3}, g_2(s) = \frac{s(s+1)}{(s^2+1)^4} \in \mathbb{R}_{M'}(s) \cap \mathbb{R}_{pr}(s)$. By Lemma 5:

- $g(s) \mid g_1(s) \Leftrightarrow n(s) \mid s^2$ and $0 \leq 2d - d(n(s)) \leq 6 - 2 = 4$
- $g(s) \mid g_2(s) \Leftrightarrow n(s) \mid s(s+1)$ and $0 \leq 2d - d(n(s)) \leq 8 - 2 = 6$.

If $n(s) \mid s^2$ and $n(s) \mid s(s+1)$ then $n(s) = c$ or $n(s) = cs$ with c a non-zero constant. Then $g(s) \mid g_1(s)$ and $g(s) \mid g_2(s)$ if and only if $n(s) = c$ and $d \leq 2$ or $n(s) = cs$ and $2d \leq 5$. So, the list of common divisors of $g_1(s)$ and $g_2(s)$ is:

$$\left\{c, \frac{c}{s^2+1}, \frac{c}{(s^2+1)^2}, \frac{cs}{s^2+1}, \frac{cs}{(s^2+1)^2} : c \in \mathbb{F}, c \neq 0\right\}. \tag{23}$$

If there would be a greatest common divisor, say $\frac{n(s)}{(s^2+1)^d}$, then $n(s) = cs$ because $n(s)$ must be a multiple of c and cs. Thus such a greatest common divisor should be either $\frac{cs}{s^2+1}$ or $\frac{cs}{(s^2+1)^2}$, but $\frac{c}{(s^2+1)^2}$ does not divide neither of them because

$$4 = \delta\left(\frac{c}{(s^2+1)^2}\right) - d(c) > max\left\{\delta\left(\frac{cs}{s^2+1}\right) - d(cs), \delta\left(\frac{cs}{(s^2+1)^2}\right) - d(cs)\right\} = 3. \tag{24}$$

Thus, $g_1(s)$ and $g_2(s)$ do not have greatest common divisor.

3.1. Smith–McMillan form

A matrix $U(s)$ is invertible in $\mathbb{F}_{M'}(s)^{m\times m} \cap \mathbb{F}_{pr}(s)^{m\times m}$ if $U(s) \in \mathbb{F}_{M'}(s)^{m\times m} \cap \mathbb{F}_{pr}(s)^{m\times m}$ and its determinant is a unit in both rings, $\mathbb{F}_{M'}(s)$ and $\mathbb{F}_{pr}(s)$, i.e., $U(s) \in \mathrm{Gl}_m(\mathbb{F}_{M'}(s) \cap \mathbb{F}_{pr}(s))$ if and only if $U(s) \in \mathrm{Gl}_m(\mathbb{F}_{M'}(s)) \cap \mathrm{Gl}_m(\mathbb{F}_{pr}(s))$.

Two matrices $G_1(s), G_2(s) \in \mathbb{F}_{M'}(s)^{m\times m} \cap \mathbb{F}_{pr}(s)^{m\times m}$ are equivalent in $\mathbb{F}_{M'}(s) \cap \mathbb{F}_{pr}(s)$ if there exist $U_1(s), U_2(s)$ invertible in $\mathbb{F}_{M'}(s)^{m\times m} \cap \mathbb{F}_{pr}(s)^{m\times m}$ such that

$$G_2(s) = U_1(s) G_1(s) U_2(s). \tag{25}$$

If there are ideals in $\mathrm{Specm}(\mathbb{F}[s]) \setminus M'$ generated by linear polynomials then $\mathbb{F}_{M'}(s) \cap \mathbb{F}_{pr}(s)$ is an Euclidean ring and any matrix with elements in $\mathbb{F}_{M'}(s) \cap \mathbb{F}_{pr}(s)$ admits a Smith normal form (see [13], [15] or [16]). Bearing in mind the characterization of divisibility in $\mathbb{F}_{M'}(s) \cap \mathbb{F}_{pr}(s)$ given in Lemma 5 we have

Theorem 12. *(Smith normal form in $\mathbb{F}_{M'}(s) \cap \mathbb{F}_{pr}(s)$) Let $M' \subseteq \mathrm{Specm}(\mathbb{F}[s])$. Assume that there are ideals in $\mathrm{Specm}(\mathbb{F}[s]) \setminus M'$ generated by linear polynomials and let $(s-a)$ be one of them. Let $G(s) \in \mathbb{F}_{M'}(s)^{m\times m} \cap \mathbb{F}_{pr}(s)^{m\times m}$ be non-singular. Then there exist $U_1(s), U_2(s)$ invertible in $\mathbb{F}_{M'}(s)^{m\times m} \cap \mathbb{F}_{pr}(s)^{m\times m}$ such that*

$$G(s) = U_1(s)\,\mathrm{Diag}\left(n_1(s)\frac{1}{(s-a)^{d_1}}, \ldots, n_m(s)\frac{1}{(s-a)^{d_m}}\right) U_2(s) \tag{26}$$

with $n_1(s)|\cdots|n_m(s)$ monic polynomials factorizing in M' and $d_1, \ldots, d_m$ integers such that $0 \leq d_1 - d(n_1(s)) \leq \cdots \leq d_m - d(n_m(s))$.

Under the hypothesis of the last theorem $n_1(s)\frac{1}{(s-a)^{d_1}}, \ldots, n_m(s)\frac{1}{(s-a)^{d_m}}$ form a complete system of invariants for the equivalence in $\mathbb{F}_{M'}(s) \cap \mathbb{F}_{pr}(s)$ and are called the invariant rational functions of $G(s)$ in $\mathbb{F}_{M'}(s) \cap \mathbb{F}_{pr}(s)$. Notice that $0 \leq d_1 \leq \cdots \leq d_m$ because $n_i(s)$ divides $n_{i+1}(s)$.

Recall that the field of fractions of $\mathbb{F}_{M'}(s) \cap \mathbb{F}_{pr}(s)$ is $\mathbb{F}(s)$ when $M' \neq \mathrm{Specm}(\mathbb{F}[s])$. Thus we can talk about equivalence of matrix rational functions. Two rational matrices $T_1(s), T_2(s) \in \mathbb{F}(s)^{m\times m}$ are equivalent in $\mathbb{F}_{M'}(s) \cap \mathbb{F}_{pr}(s)$ if there are $U_1(s), U_2(s)$ invertible in $\mathbb{F}_{M'}(s)^{m\times m} \cap \mathbb{F}_{pr}(s)^{m\times m}$ such that

$$T_2(s) = U_1(s) T_1(s) U_2(s). \tag{27}$$

When all ideals generated by linear polynomials are not in M', each rational matrix admits a reduction to Smith–McMillan form with respect to $\mathbb{F}_{M'}(s) \cap \mathbb{F}_{pr}(s)$.

Theorem 13. *(Smith–McMillan form in $\mathbb{F}_{M'}(s) \cap \mathbb{F}_{pr}(s)$) Let $M' \subseteq \mathrm{Specm}(\mathbb{F}[s])$. Assume that there are ideals in $\mathrm{Specm}(\mathbb{F}[s]) \setminus M'$ generated by linear polynomials and let $(s-a)$ be any of them. Let $T(s) \in \mathbb{F}(s)^{m\times m}$ be a non-singular matrix. Then there exist $U_1(s), U_2(s)$ invertible in $\mathbb{F}_{M'}(s)^{m\times m} \cap \mathbb{F}_{pr}(s)^{m\times m}$ such that*

$$T(s) = U_1(s)\,\mathrm{Diag}\left(\frac{\frac{\epsilon_1(s)}{(s-a)^{n_1}}}{\frac{\psi_1(s)}{(s-a)^{d_1}}}, \ldots, \frac{\frac{\epsilon_m(s)}{(s-a)^{n_m}}}{\frac{\psi_m(s)}{(s-a)^{d_m}}}\right) U_2(s) \tag{28}$$

with $\frac{\epsilon_i(s)}{(s-a)^{n_i}}, \frac{\psi_i(s)}{(s-a)^{d_i}} \in \mathbb{F}_{M'}(s) \cap \mathbb{F}_{pr}(s)$ *coprime for all i such that* $\epsilon_i(s), \psi_i(s)$ *are monic polynomials factorizing in* M', $\frac{\epsilon_i(s)}{(s-a)^{n_i}}$ *divides* $\frac{\epsilon_{i+1}(s)}{(s-a)^{n_{i+1}}}$ *for* $i = 1, \ldots, m-1$ *while* $\frac{\psi_i(s)}{(s-a)^{d_i}}$ *divides* $\frac{\psi_{i-1}(s)}{(s-a)^{d_{i-1}}}$ *for* $i = 2, \ldots, m$.

The elements $\frac{\frac{\epsilon_i(s)}{(s-a)^{n_i}}}{\frac{\psi_i(s)}{(s-a)^{d_i}}}$ of the diagonal matrix, satisfying the conditions of the previous theorem, constitute a complete system of invariant for the equivalence in $\mathbb{F}_{M'}(s) \cap \mathbb{F}_{pr}(s)$ of rational matrices. However, this system of invariants is not minimal. A smaller one can be obtained by substituting each pair of positive integers (n_i, d_i) by its difference $l_i = n_i - d_i$.

Theorem 14. *Under the conditions of Theorem 13,* $\frac{\epsilon_i(s)}{\psi_i(s)} \frac{1}{(s-a)^{l_i}}$ *with* $\epsilon_i(s), \psi_i(s)$ *monic and coprime polynomials factorizing in* M', $\epsilon_i(s) \mid \epsilon_{i+1}(s)$ *while* $\psi_i(s) \mid \psi_{i-1}(s)$ *and* $l_1, \ldots, l_m$ *integers such that* $l_1 + d(\psi_1(s)) - d(\epsilon_1(s)) \leq \cdots \leq l_m + d(\psi_m(s)) - d(\epsilon_m(s))$ *also constitute a complete system of invariants for the equivalence in* $\mathbb{F}_{M'}(s) \cap \mathbb{F}_{pr}(s)$.

Proof.- We only have to show that from the system $\frac{\epsilon_i(s)}{\psi_i(s)} \frac{1}{(s-a)^{l_i}}$, $i = 1, \ldots, m$, satisfying the conditions of Theorem 14, the system $\frac{\frac{\epsilon_i(s)}{(s-a)^{n_i}}}{\frac{\psi_i(s)}{(s-a)^{d_i}}}$, $i = 1, \ldots, n$, can be constructed satisfying the conditions of Theorem 13.

Suppose that $\epsilon_i(s)$, $\psi_i(s)$ are monic and coprime polynomials factorizing in M' such that $\epsilon_i(s) \mid \epsilon_{i+1}(s)$ and $\psi_i(s) \mid \psi_{i-1}(s)$. And suppose also that $l_1, \ldots, l_m$ are integers such that $l_1 + d(\psi_1(s)) - d(\epsilon_1(s)) \leq \cdots \leq l_m + d(\psi_m(s)) - d(\epsilon_m(s))$. If $l_i + d(\psi_i(s)) - d(\epsilon_i(s)) \leq 0$ for all i, we define non-negative integers $n_i = d(\epsilon_i(s))$ and $d_i = d(\epsilon_i(s)) - l_i$ for $i = 1, \ldots, m$. If $l_i + d(\psi_i(s)) - d(\epsilon_i(s)) > 0$ for all i, we define $n_i = l_i + d(\psi_i(s))$ and $d_i = d(\psi_i(s))$. Otherwise there is an index $k \in \{2, \ldots, m\}$ such that

$$l_{k-1} + d(\psi_{k-1}(s)) - d(\epsilon_{k-1}(s)) \leq 0 < l_k + d(\psi_k(s)) - d(\epsilon_k(s)). \tag{29}$$

Define now the non-negative integers n_i, d_i as follows:

$$n_i = \begin{cases} d(\epsilon_i(s)) & \text{if } i < k \\ l_i + d(\psi_i(s)) & \text{if } i \geq k \end{cases} \quad d_i = \begin{cases} d(\epsilon_i(s)) - l_i & \text{if } i < k \\ d(\psi_i(s)) & \text{if } i \geq k \end{cases} \tag{30}$$

Notice that $l_i = n_i - d_i$. Moreover,

$$n_i - d(\epsilon_i(s)) = \begin{cases} 0 & \text{if } i < k \\ l_i + d(\psi_i(s)) - d(\epsilon_i(s)) & \text{if } i \geq k \end{cases} \tag{31}$$

$$d_i - d(\psi_i(s)) = \begin{cases} -l_i - d(\psi_i(s)) + d(\epsilon_i(s)) & \text{if } i < k \\ 0 & \text{if } i \geq k \end{cases} \tag{32}$$

and using (29), (30)

$$n_1 - d(\epsilon_1(s)) = \cdots = n_{k-1} - d(\epsilon_{k-1}(s)) = 0 < n_k - d(\epsilon_k(s)) \leq \cdots \leq n_m - d(\epsilon_m(s)) \tag{33}$$

$$d_1 - d(\psi_1(s)) \geq \cdots \geq d_{k-1} - d(\psi_{k-1}(s)) \geq 0 = d_k - d(\psi_k(s)) = \cdots = d_m - d(\psi_m(s)). \tag{34}$$

In any case $\frac{\epsilon_i(s)}{(s-a)^{n_i}}$ and $\frac{\psi_i(s)}{(s-a)^{d_i}}$ are elements of $\mathbb{F}_{M'}(s) \cap \mathbb{F}_{pr}(s)$. Now, on the one hand $\epsilon_i(s), \psi_i(s)$ are coprime and $n_i - d(\epsilon_i(s)) = 0$ or $d_i - d(\psi_i(s)) = 0$. This means (Lemma 6) that $\frac{\epsilon_i(s)}{(s-a)^{n_i}}, \frac{\psi_i(s)}{(s-a)^{d_i}}$ are coprime for all i. On the other hand $\epsilon_i(s) \mid \epsilon_{i+1}(s)$ and $0 \leq n_i - d(\epsilon_i(s)) \leq n_{i+1} - d(\epsilon_{i+1}(s))$. Then (Lemma 5) $\frac{\epsilon_i(s)}{(s-a)^{n_i}}$ divides $\frac{\epsilon_{i+1}(s)}{(s-a)^{n_{i+1}}}$. Similarly, since $\psi_i(s) \mid \psi_{i-1}(s)$ and $0 \leq d_i - d(\psi_i(s)) \leq d_{i-1} - d(\psi_{i-1}(s))$, it follows that $\frac{\psi_i(s)}{(s-a)^{d_i}}$ divides $\frac{\psi_{i-1}(s)}{(s-a)^{d_{i-1}}}$. ∎

We call $\frac{\epsilon_i(s)}{\psi_i(s)} \frac{1}{(s-a)^{l_i}}$, $i = 1, \dots, m$, the invariant rational functions of $T(s)$ in $\mathbb{F}_{M'}(s) \cap \mathbb{F}_{pr}(s)$.

There is a particular case worth considering: If $M' = \emptyset$ then $\mathbb{F}_{\emptyset}(s) \cap \mathbb{F}_{pr}(s) = \mathbb{F}_{pr}(s)$ and $(s) \in \text{Specm}(\mathbb{F}[s]) \setminus M' = \text{Specm}(\mathbb{F}[s])$. In this case, we obtain the invariant rational functions of $T(s)$ at infinity (recall (17)).

4. Wiener–Hopf equivalence

The left Wiener–Hopf equivalence of rational matrices with respect to a closed contour in the complex plane has been extensively studied ([6] or [10]). Now we present the generalization to arbitrary fields ([4]).

Definition 15. *Let M and M' be subsets of* $\text{Specm}(\mathbb{F}[s])$ *such that* $M \cup M' = \text{Specm}(\mathbb{F}[s])$. *Let $T_1(s), T_2(s) \in \mathbb{F}(s)^{m \times m}$ be two non-singular rational matrices with no zeros and no poles in $M \cap M'$. The matrices $T_1(s), T_2(s)$ are said to be left Wiener–Hopf equivalent with respect to (M, M') if there exist both $U_1(s)$ invertible in $\mathbb{F}_{M'}(s)^{m \times m} \cap \mathbb{F}_{pr}(s)^{m \times m}$ and $U_2(s)$ invertible in $\mathbb{F}_M(s)^{m \times m}$ such that*

$$T_2(s) = U_1(s) T_1(s) U_2(s). \tag{35}$$

This is, in fact, an equivalence relation as it is easily seen. It would be an equivalence relation even if no condition about the union and intersection of M and M' were imposed. It will be seen later on that these conditions are natural assumptions for the existence of unique diagonal representatives in each class.

The right Wiener–Hopf equivalence with respect to (M, M') is defined in a similar manner: There are invertible matrices $U_1(s)$ in $\mathbb{F}_{M'}(s)^{m \times m} \cap \mathbb{F}_{pr}(s)^{m \times m}$ and $U_2(s)$ in $\mathbb{F}_M(s)^{m \times m}$ such that

$$T_2(s) = U_2(s) T_1(s) U_1(s). \tag{36}$$

In the following only the left Wiener–Hopf equivalence will be considered, but, by transposition, all results hold for the right Wiener–Hopf equivalence as well.

The aim of this section is to obtain a complete system of invariants for the Wiener–Hopf equivalence with respect to (M, M') of rational matrices, and to obtain, if possible, a canonical form.

There is a particular case that is worth-considering: If $M = \text{Specm}(\mathbb{F}[s])$ and $M' = \emptyset$, the invertible matrices in $\mathbb{F}_{\emptyset}(s)^{m \times m} \cap \mathbb{F}_{pr}(s)^{m \times m}$ are the biproper matrices and the invertible matrices in $\mathbb{F}_{\text{Specm}(\mathbb{F}[s])}(s)^{m \times m}$ are the unimodular matrices. In this case, the left Wiener–Hopf equivalence with respect to $(M, M') = (\text{Specm}(\mathbb{F}[s]), \emptyset)$ is the so-called left Wiener–Hopf equivalence at infinity (see [9]). It is known that any non-singular rational matrix is left Wiener–Hopf equivalent at infinity to a diagonal matrix $\text{Diag}(s^{g_1}, \dots, s^{g_m})$ where $g_1, \dots, g_m$

are integers, that is, for any non-singular $T(s) \in \mathbb{F}(s)^{m\times m}$ there exist both a biproper matrix $B(s) \in \mathrm{Gl}_m(\mathbb{F}_{pr}(s))$ and a unimodular matrix $U(s) \in \mathrm{Gl}_m(\mathbb{F}[s])$ such that

$$T(s) = B(s)\,\mathrm{Diag}(s^{g_1},\ldots,s^{g_m})U(s) \tag{37}$$

where $g_1 \geq \cdots \geq g_m$ are integers uniquely determined by $T(s)$. They are called the left Wiener–Hopf factorization indices at infinity and form a complete system of invariants for the left Wiener–Hopf equivalence at infinity. These are the basic objects that will produce the complete system of invariants for the left Wiener–Hopf equivalence with respect to (M, M').

For polynomial matrices, their left Wiener–Hopf factorization indices at infinity are the column degrees of any right equivalent (by a unimodular matrix) column proper matrix. Namely, a polynomial matrix is column proper if it can be written as $P_c\,\mathrm{Diag}(s^{g_1},\ldots,s^{g_m}) + L(s)$ with $P_c \in \mathbb{F}^{m\times m}$ non-singular, $g_1,\ldots,g_m$ non-negative integers and $L(s)$ a polynomial matrix such that the degree of the ith column of $L(s)$ smaller than g_i, $1 \leq i \leq m$. Let $P(s) \in \mathbb{F}[s]^{m\times m}$ be non-singular polynomial. There exists a unimodular matrix $V(s) \in \mathbb{F}[s]^{m\times m}$ such that $P(s)V(s)$ is column proper. The column degrees of $P(s)V(s)$ are uniquely determined by $P(s)$, although $V(s)$ is not (see [9], [12, p. 388], [17]). Since $P(s)V(s)$ is column proper, it can be written as $P(s)V(s) = P_cD(s) + L(s)$ with P_c non-singular, $D(s) = \mathrm{Diag}(s^{g_1},\ldots,s^{g_m})$ and the degree of the ith column of $L(s)$ smaller than g_i, $1 \leq i \leq m$. Then $P(s)V(s) = (P_c + L(s)D(s)^{-1})D(s)$. Put $B(s) = P_c + L(s)D(s)^{-1}$. Since P_c is non-singular and $L(s)D(s)^{-1}$ is a strictly proper matrix, $B(s)$ is biproper, and $P(s) = B(s)D(s)U(s)$ where $U(s) = V(s)^{-1}$.

The left Wiener–Hopf factorization indices at infinity can be used to associate a sequence of integers with every non-singular rational matrix and every $M \subseteq \mathrm{Specm}(\mathbb{F}[s])$. This is done as follows: If $T(s) \in \mathbb{F}(s)^{m\times m}$ then it can always be written as $T(s) = T_L(s)T_R(s)$ such that the global invariant rational functions of $T_L(s)$ factorize in M and $T_R(s) \in \mathrm{Gl}_m(\mathbb{F}_M(s))$ or, equivalently, the global invariant rational functions of $T_R(s)$ factorize in $\mathrm{Specm}(\mathbb{F}[s]) \setminus M$ (see Proposition 2). There may be many factorizations of this type, but it turns out (see [1, Proposition 3.2] for the polynomial case) that the left factors in all of them are right equivalent. This means that if $T(s) = T_{L1}(s)T_{R1}(s) = T_{L2}(s)T_{R2}(s)$ with the global invariant rational functions of $T_{L1}(s)$ and $T_{L2}(s)$ factorizing in M and the global invariant rational functions of $T_{R1}(s)$ and $T_{R2}(s)$ factorizing in $\mathrm{Specm}(\mathbb{F}[s]) \setminus M$ then there is a unimodular matrix $U(s)$ such that $T_{L1}(s) = T_{L2}(s)U(s)$. In particular, $T_{L1}(s)$ and $T_{L2}(s)$ have the same left Wiener–Hopf factorization indices at infinity. Thus the following definition makes sense:

Definition 16. *Let $T(s) \in \mathbb{F}(s)^{m\times m}$ be a non-singular rational matrix and $M \subseteq \mathrm{Specm}(\mathbb{F}[s])$. Let $T_L(s), T_R(s) \in \mathbb{F}(s)^{m\times m}$ such that*

i) $T(s) = T_L(s)T_R(s)$,

ii) the global invariant rational functions of $T_L(s)$ factorize in M, and

iii) the global invariant rational functions of $T_R(s)$ factorize in $\mathrm{Specm}(\mathbb{F}[s]) \setminus M$.

Then the left Wiener–Hopf factorization indices of $T(s)$ with respect to M are defined to be the left Wiener–Hopf factorization indices of $T_L(s)$ at infinity.

In the particular case that $M = \mathrm{Specm}(\mathbb{F}[s])$, we can put $T_L(s) = T(s)$ and $T_R(s) = I_m$. Therefore, the left Wiener–Hopf factorization indices of $T(s)$ with respect to $\mathrm{Specm}(\mathbb{F}[s])$ are the left Wiener–Hopf factorization indices of $T(s)$ at infinity.

We prove now that the left Wiener–Hopf equivalence with respect to (M, M') can be characterized through the left Wiener–Hopf factorization indices with respect to M.

Theorem 17. *Let $M, M' \subseteq \text{Specm}(\mathbb{F}[s])$ be such that $M \cup M' = \text{Specm}(\mathbb{F}[s])$. Let $T_1(s), T_2(s) \in \mathbb{F}(s)^{m\times m}$ be two non-singular rational matrices with no zeros and no poles in $M \cap M'$. The matrices $T_1(s)$ and $T_2(s)$ are left Wiener–Hopf equivalent with respect to (M, M') if and only if $T_1(s)$ and $T_2(s)$ have the same left Wiener–Hopf factorization indices with respect to M.*

Proof.- By Proposition 2 we can write $T_1(s) = T_{L1}(s)T_{R1}(s)$, $T_2(s) = T_{L2}(s)T_{R2}(s)$ with the global invariant rational functions of $T_{L1}(s)$ and of $T_{L2}(s)$ factorizing in $M \setminus M'$ (recall that $T_1(s)$ and $T_2(s)$ have no zeros and no poles in $M \cap M'$) and the global invariant rational functions of $T_{R1}(s)$ and of $T_{R2}(s)$ factorizing in $M' \setminus M$.

Assume that $T_1(s)$, $T_2(s)$ have the same left Wiener–Hopf factorization indices with respect to M. By definition, $T_1(s)$ and $T_2(s)$ have the same left Wiener–Hopf factorization indices with respect to M if $T_{L1}(s)$ and $T_{L2}(s)$ have the same left Wiener–Hopf factorization indices at infinity. This means that there exist matrices $B(s) \in \text{Gl}_m(\mathbb{F}_{pr}(s))$ and $U(s) \in \text{Gl}_m(\mathbb{F}[s])$ such that $T_{L2}(s) = B(s)T_{L1}(s)U(s)$. We have that $T_2(s) = T_{L2}(s)T_{R2}(s) = B(s)T_{L1}(s)U(s)T_{R2}(s) = B(s)T_1(s)(T_{R1}(s)^{-1}U(s)T_{R2}(s))$. We aim to prove that $B(s) = T_{L2}(s)U(s)^{-1}T_{L1}(s)^{-1}$ is invertible in $\mathbb{F}_{M'}(s)^{m\times m}$ and $T_{R1}(s)^{-1}U(s)T_{R2}(s) \in Gl_m(\mathbb{F}_M(s))$. Since the global invariant rational functions of $T_{L2}(s)$ and $T_{L1}(s)$ factorize in $M \setminus M'$, $T_{L2}(s), T_{L1}(s) \in \mathbb{F}_{M'}(s)^{m\times m}$ and $B(s) \in \mathbb{F}_{M'}(s)^{m\times m}$. Moreover, $\det B(s)$ is a unit in $\mathbb{F}_{M'}(s)^{m\times m}$ as desired. Now, $T_{R1}(s)^{-1}U(s)T_{R2}(s) \in Gl_m(\mathbb{F}_M(s))$ because $T_{R1}(s), T_{R2}(s) \in \mathbb{F}_M(s)^{m\times m}$ and $\det T_{R1}(s)$ and $\det T_{R2}(s)$ factorize in $M' \setminus M$. Therefore $T_1(s)$ and $T_2(s)$ are left Wiener–Hopf equivalent with respect to (M, M').

Conversely, let $U_1(s) \in \text{Gl}_m(\mathbb{F}_{M'}(s)) \cap \text{Gl}_m(\mathbb{F}_{pr}(s))$ and $U_2(s) \in \text{Gl}_m(\mathbb{F}_M(s))$ such that $T_1(s) = U_1(s)T_2(s)U_2(s)$. Hence, $T_1(s) = T_{L1}(s)T_{R1}(s) = U_1(s)T_{L2}(s)T_{R2}(s)U_2(s)$. Put $\overline{T}_{L2}(s) = U_1(s)T_{L2}(s)$ and $\overline{T}_{R2}(s) = T_{R2}(s)U_2(s)$. Therefore,

(i) $T_1(s) = T_{L1}(s)T_{R1}(s) = \overline{T}_{L2}(s)\overline{T}_{R2}(s)$,

(ii) the global invariant rational functions of $T_{L1}(s)$ and of $\overline{T}_{L2}(s)$ factorize in M, and

(iii)the global invariant rational functions of $T_{R1}(s)$ and of $\overline{T}_{R2}(s)$ factorize in $\text{Specm}(\mathbb{F}[s]) \setminus M$.

Then $T_{L1}(s)$ and $\overline{T}_{L2}(s)$ are right equivalent (see the remark previous to Definition 16). So, there exists $U(s) \in \text{Gl}_m(\mathbb{F}[s])$ such that $T_{L1}(s) = \overline{T}_{L2}(s)U(s)$. Thus, $T_{L1}(s) = U_1(s)T_{L2}(s)U(s)$. Since $U_1(s)$ is biproper and $U(s)$ is unimodular $T_{L1}(s)$, $T_{L2}(s)$ have the same left Wiener–Hopf factorization indices at infinity. Consequentially, $T_1(s)$ and $T_2(s)$ have the same left Wiener–Hopf factorization indices with respect to M. ∎

In conclusion, for non-singular rational matrices with no zeros and no poles in $M \cap M'$ the left Wiener–Hopf factorization indices with respect to M form a complete system of invariants for the left Wiener–Hopf equivalence with respect to (M, M') with $M \cup M' = \text{Specm}(\mathbb{F}[s])$.

A straightforward consequence of the above theorem is the following Corollary

Corollary 18. *Let $M, M' \subseteq \text{Specm}(\mathbb{F}[s])$ be such that $M \cup M' = \text{Specm}(\mathbb{F}[s])$. Let $T_1(s)$, $T_2(s) \in \mathbb{F}(s)^{m\times m}$ be non-singular with no zeros and no poles in $M \cap M'$. Then $T_1(s)$ and $T_2(s)$ are left Wiener–Hopf equivalent with respect to (M, M') if and only if for any factorizations $T_1(s) = T_{L1}(s)T_{R1}(s)$ and $T_2(s) = T_{L2}(s)T_{R2}(s)$ satisfying the conditions (i)–(iii) of Definition 16, $T_{L1}(s)$ and $T_{L2}(s)$ are left Wiener–Hopf equivalent at infinity.*

Next we deal with the problem of factorizing or reducing a rational matrix to diagonal form by Wiener–Hopf equivalence. It will be shown that if there exists in M an ideal generated by a monic irreducible polynomial of degree equal to 1 which is not in M', then any non-singular rational matrix, with no zeros and no poles in $M \cap M'$ admits a factorization with respect to (M, M'). Afterwards, some examples will be given in which these conditions on M and M' are removed and factorization fails to exist.

Theorem 19. *Let $M, M' \subseteq \text{Specm}(\mathbb{F}[s])$ be such that $M \cup M' = \text{Specm}(\mathbb{F}[s])$. Assume that there are ideals in $M \setminus M'$ generated by linear polynomials. Let $(s-a)$ be any of them and $T(s) \in \mathbb{F}(s)^{m \times m}$ a non-singular matrix with no zeros and no poles in $M \cap M'$. There exist both $U_1(s)$ invertible in $\mathbb{F}_{M'}(s)^{m\times m} \cap \mathbb{F}_{pr}(s)^{m\times m}$ and $U_2(s)$ invertible in $\mathbb{F}_M(s)^{m\times m}$ such that*

$$T(s) = U_1(s)\,\text{Diag}((s-a)^{k_1}, \ldots, (s-a)^{k_m})U_2(s), \tag{38}$$

where $k_1 \geq \cdots \geq k_m$ are integers uniquely determined by $T(s)$. Moreover, they are the left Wiener–Hopf factorization indices of $T(s)$ with respect to M.

Proof.- The matrix $T(s)$ can be written (see Proposition 2) as $T(s) = T_L(s)T_R(s)$ with the global invariant rational functions of $T_L(s)$ factorizing in $M \setminus M'$ and the global invariant rational functions of $T_R(s)$ factorizing in $\text{Specm}(\mathbb{F}[s]) \setminus M = M' \setminus M$. As $k_1, \ldots, k_m$ are the left Wiener–Hopf factorization indices of $T_L(s)$ at infinity, there exist matrices $U(s) \in \text{Gl}_m(\mathbb{F}[s])$ and $B(s) \in \text{Gl}_m(\mathbb{F}_{pr}(s))$ such that $T_L(s) = B(s)D_1(s)U(s)$ with $D_1(s) = \text{Diag}(s^{k_1}, \ldots, s^{k_m})$. Put $D(s) = \text{Diag}((s-a)^{k_1}, \ldots, (s-a)^{k_m})$ and $U_1(s) = B(s)\,\text{Diag}\left(\frac{s^{k_1}}{(s-a)^{k_1}}, \ldots, \frac{s^{k_m}}{(s-a)^{k_m}}\right)$. Then $T_L(s) = U_1(s)D(s)U(s)$. If $U_2(s) = U(s)T_R(s)$ then this matrix is invertible in $\mathbb{F}_M(s)^{m\times m}$ and $T(s) = U_1(s)\,\text{Diag}((s-a)^{k_1}, \ldots, (s-a)^{k_m})U_2(s)$. We only have to prove that $U_1(s)$ is invertible in $\mathbb{F}_{M'}(s)^{m\times m} \cap \mathbb{F}_{pr}(s)^{m\times m}$. It is clear that $U_1(s)$ is in $\mathbb{F}_{pr}(s)^{m\times m}$ and biproper. Moreover, the global invariant rational functions of $T_L(s)$ $U_1(s) = T_L(s)(D(s)U(s))^{-1}$ factorize in $M \setminus M'$. Therefore, $U_1(s)$ is invertible in $\mathbb{F}_{M'}(s)^{m\times m}$.

We prove now the uniqueness of the factorization. Assume that $T(s)$ also factorizes as

$$T(s) = \tilde{U}_1(s)\,\text{Diag}((s-a)^{\tilde{k}_1}, \ldots, (s-a)^{\tilde{k}_m})\tilde{U}_2(s), \tag{39}$$

with $\tilde{k}_1 \geq \cdots \geq \tilde{k}_m$ integers. Then,

$$\text{Diag}((s-a)^{\tilde{k}_1}, \ldots, (s-a)^{\tilde{k}_m}) = \tilde{U}_1(s)^{-1}U_1(s)\,\text{Diag}((s-a)^{k_1}, \ldots, (s-a)^{k_m})U_2(s)\tilde{U}_2(s)^{-1}. \tag{40}$$

The diagonal matrices have no zeros and no poles in $M \cap M'$ (because $(s-a) \in M \setminus M'$) and they are left Wiener–Hopf equivalent with respect to (M, M'). By Theorem 17, they have the same left Wiener–Hopf factorization indices with respect to M. Thus, $\tilde{k}_i = k_i$ for all $i = 1, \ldots, m$. ∎

Following [6] we could call left Wiener–Hopf factorization indices with respect to (M, M') the exponents $k_1 \geq \cdots \geq k_m$ appearing in the diagonal matrix of Theorem 19. They are, actually, the left Wiener–Hopf factorization indices with respect to M.

Several examples follow that exhibit some remarkable features about the results that have been proved so far. The first two examples show that if no assumption is made on the intersection and/or union of M and M' then existence and/or uniqueness of diagonal factorization may fail to exist.

Example 20. If $P(s)$ is a polynomial matrix with zeros in $M \cap M'$ then the existence of invertible matrices $U_1(s) \in \mathrm{Gl}_m(\mathbb{F}_{M'}(s)) \cap \mathrm{Gl}_m(\mathbb{F}_{pr}(s))$ and $U_2(s) \in \mathrm{Gl}_m(\mathbb{F}_M(s))$ such that $P(s) = U_1(s)\,\mathrm{Diag}((s-a)^{k_1}, \ldots, (s-a)^{k_m})U_2(s)$ with $(s-a) \in M \setminus M'$ may fail. In fact, suppose that $M = \{(s), (s+1)\}$, $M' = \mathrm{Specm}\,\mathbb{F}[s] \setminus \{(s)\}$. Therefore, $M \cap M' = \{(s+1)\}$ and $(s) \in M \setminus M'$. Consider $p_1(s) = s+1$. Assume that $s+1 = u_1(s)s^k u_2(s)$ with $u_1(s)$ a unit in $\mathbb{F}_{M'}(s) \cap \mathbb{F}_{pr}(s)$ and $u_2(s)$ a unit in $\mathbb{F}_M(s)$. Thus, $u_1(s) = c$ a nonzero constant and $u_2(s) = \frac{1}{c}\frac{s+1}{s^k}$ which is not a unit in $\mathbb{F}_M(s)$.

Example 21. If $M \cup M' \neq \mathrm{Specm}\,\mathbb{F}[s]$ then the factorization indices with respect to (M, M') may be not unique. Suppose that $(\beta(s)) \notin M \cup M'$, $(\pi(s)) \in M \setminus M'$ with $d(\pi(s)) = 1$ and $p(s) = u_1(s)\pi(s)^k u_2(s)$, with $u_1(s)$ a unit in $\mathbb{F}_{M'}(s) \cap \mathbb{F}_{pr}(s)$ and $u_2(s)$ a unit in $\mathbb{F}_M(s)$. Then $p(s)$ can also be factorized as $p(s) = \tilde{u}_1(s)\pi(s)^{k-d(\beta(s))}\tilde{u}_2(s)$ with $\tilde{u}_1(s) = u_1(s)\frac{\pi(s)^{d(\beta(s))}}{\beta(s)}$ a unit in $\mathbb{F}_{M'}(s) \cap \mathbb{F}_{pr}(s)$ and $\tilde{u}_2(s) = \beta(s)u_2(s)$ a unit in $\mathbb{F}_M(s)$.

The following example shows that if all ideals generated by polynomials of degree equal to one are in $M' \setminus M$ then a factorization as in Theorem 19 may not exist.

Example 22. Suppose that $\mathbb{F} = \mathbb{R}$. Consider $M = \{(s^2+1)\} \subseteq \mathrm{Specm}(\mathbb{R}[s])$ and $M' = \mathrm{Specm}(\mathbb{R}[s]) \setminus \{(s^2+1)\}$. Let

$$P(s) = \begin{bmatrix} s & 0 \\ -s^2 & (s^2+1)^2 \end{bmatrix}. \tag{41}$$

Notice that $P(s)$ has no zeros and no poles in $M \cap M' = \emptyset$. We will see that it is not possible to find invertible matrices $U_1(s) \in \mathbb{R}_{M'}(s)^{2\times 2} \cap \mathbb{R}_{pr}(s)^{2\times 2}$ and $U_2(s) \in \mathbb{R}_M(s)^{2\times 2}$ such that

$$U_1(s)P(s)U_2(s) = \mathrm{Diag}((p(s)/q(s))^{c_1}, (p(s)/q(s))^{c_2}). \tag{42}$$

We can write $\frac{p(s)}{q(s)} = u(s)(s^2+1)^a$ with $u(s)$ a unit in $\mathbb{R}_M(s)$ and $a \in \mathbb{Z}$. Therefore,

$$\mathrm{Diag}((p(s)/q(s))^{c_1}, (p(s)/q(s))^{c_2}) = \mathrm{Diag}((s^2+1)^{ac_1}, (s^2+1)^{ac_2})\,\mathrm{Diag}(u(s)^{c_1}, u(s)^{c_2}). \tag{43}$$

$\mathrm{Diag}(u(s)^{c_1}, u(s)^{c_2})$ is invertible in $\mathbb{R}_M(s)^{2\times 2}$ and $P(s)$ is also left Wiener–Hopf equivalent with respect to (M, M') to the diagonal matrix $\mathrm{Diag}((s^2+1)^{ac_1}, (s^2+1)^{ac_2})$.

Assume that there exist invertible matrices $U_1(s) \in \mathbb{R}_{M'}(s)^{2\times 2} \cap \mathbb{R}_{pr}(s)^{2\times 2}$ and $U_2(s) \in \mathbb{R}_M(s)^{2\times 2}$ such that $U_1(s)P(s)U_2(s) = \mathrm{Diag}((s^2+1)^{d_1}, (s^2+1)^{d_2})$, with $d_1 \geq d_2$ integers. Notice first that $\det U_1(s)$ is a nonzero constant and since $\det P(s) = s(s^2+1)^2$ and $\det U_2(s)$ is a rational function with numerator and denominator relatively prime with s^2+1, it follows that $cs(s^2+1)^2 \det U_2(s) = (s^2+1)^{d_1+d_2}$. Thus, $d_1 + d_2 = 2$. Let

$$U_1(s)^{-1} = \begin{bmatrix} b_{11}(s) & b_{12}(s) \\ b_{21}(s) & b_{22}(s) \end{bmatrix}, \quad U_2(s) = \begin{bmatrix} u_{11}(s) & u_{12}(s) \\ u_{21}(s) & u_{22}(s) \end{bmatrix}. \tag{44}$$

From $P(s)U_2(s) = U_1(s)^{-1}\,\mathrm{Diag}((s^2+1)^{d_1}, (s^2+1)^{d_2})$ we get

$$su_{11}(s) = b_{11}(s)(s^2+1)^{d_1}, \tag{45}$$

$$-s^2u_{11}(s) + (s^2+1)^2u_{21}(s) = b_{21}(s)(s^2+1)^{d_1}, \tag{46}$$

$$su_{12}(s) = b_{12}(s)(s^2+1)^{d_2}, \tag{47}$$

$$-s^2u_{12}(s) + (s^2+1)^2u_{22}(s) = b_{22}(s)(s^2+1)^{d_2}. \tag{48}$$

As $u_{11}(s) \in \mathbb{R}_M(s)$ and $b_{11}(s) \in \mathbb{R}_{M'}(s) \cap \mathbb{R}_{pr}(s)$, we can write $u_{11}(s) = \frac{f_1(s)}{g_1(s)}$ and $b_{11}(s) = \frac{h_1(s)}{(s^2+1)^{q_1}}$ with $f_1(s), g_1(s), h_1(s) \in \mathbb{R}[s]$, $\gcd(g_1(s), s^2+1) = 1$ and $d(h_1(s)) \leq 2q_1$. Therefore, by (45), $s\frac{f_1(s)}{g_1(s)} = \frac{h_1(s)}{(s^2+1)^{q_1}}(s^2+1)^{d_1}$. Hence, $u_{11}(s) = f_1(s)$ or $u_{11}(s) = \frac{f_1(s)}{s}$. In the same way and using (47), $u_{12}(s) = f_2(s)$ or $u_{12}(s) = \frac{f_2(s)}{s}$ with $f_2(s)$ a polynomial. Moreover, by (47), d_2 must be non-negative. Hence, $d_1 \geq d_2 \geq 0$. Using now (46) and (48) and bearing in mind again that $u_{21}(s), u_{22}(s) \in \mathbb{R}_M(s)$ and $b_{21}(s), b_{22}(s) \in \mathbb{R}_{M'}(s) \cap \mathbb{R}_{pr}(s)$, we conclude that $u_{21}(s)$ and $u_{22}(s)$ are polynomials.

We can distinguish two cases: $d_1 = 2, d_2 = 0$ and $d_1 = d_2 = 1$. If $d_1 = 2$ and $d_2 = 0$, by (47), $b_{12}(s)$ is a polynomial and since $b_{12}(s)$ is proper, it is constant: $b_{12}(s) = c_1$. Thus $u_{12}(s) = \frac{c_1}{s}$. By (48), $b_{22}(s) = -c_1s + (s^2+1)^2u_{22}(s)$. Since $u_{22}(s)$ is polynomial and $b_{22}(s)$ is proper, $b_{22}(s)$ is also constant and then $u_{22}(s) = 0$ and $c_1 = 0$. Consequentially, $b_{22}(s) = 0$, and $b_{12}(s) = 0$. This is impossible because $U_1(s)$ is invertible.

If $d_1 = d_2 = 1$ then , using (46),

$$\begin{aligned} b_{21}(s) &= \frac{-s^2u_{11}(s)+(s^2+1)^2u_{21}(s)}{s^2+1} = \frac{-s^2\frac{b_{11}(s)}{s}(s^2+1)+(s^2+1)^2u_{21}(s)}{s^2+1} \\ &= -sb_{11}(s) + (s^2+1)u_{21}(s) = -s\frac{h_1(s)}{(s^2+1)^{q_1}} + (s^2+1)u_{21}(s) \\ &= \frac{-sh_1(s)+(s^2+1)^{q_1+1}u_{21}(s)}{(s^2+1)^{q_1}}. \end{aligned} \tag{49}$$

Notice that $d(-sh_1(s)) \leq 1 + 2q_1$ and $d((s^2+1)^{q_1+1}u_{21}(s)) = 2(q_1+1) + d(u_{21}(s)) \geq 2q_1+2$ unless $u_{21}(s) = 0$. Hence, if $u_{21}(s) \neq 0$, $d(-sh_1(s) + (s^2+1)^{q_1+1}u_{21}(s)) \geq 2q_1+2$ which is greater than $d((s^2+1)^{q_1}) = 2q_1$. This cannot happen because $b_{21}(s)$ is proper. Thus, $u_{21}(s) = 0$. In the same way and reasoning with (48) we get that $u_{22}(s)$ is also zero. This is again impossible because $U_2(s)$ is invertible. Therefore no left Wiener–Hopf factorization of $P(s)$ with respect to (M, M') exits.

We end this section with an example where the left Wiener–Hopf factorization indices of the matrix polynomial in the previous example are computed. Then an ideal generated by a polynomial of degree 1 is added to M and the Wiener–Hopf factorization indices of the same matrix are obtained in two different cases.

Example 23. Let $\mathbb{F} = \mathbb{R}$ and $M = \{(s^2+1)\}$. Consider the matrix

$$P(s) = \begin{bmatrix} s & 0 \\ -s^2 & (s^2+1)^2 \end{bmatrix}, \tag{50}$$

which has a zero at 0. It can be written as $P(s) = P_1(s)P_2(s)$ with

$$P_1(s) = \begin{bmatrix} 1 & 0 \\ -s & (s^2+1)^2 \end{bmatrix}, \quad P_2(s) = \begin{bmatrix} s & 0 \\ 0 & 1 \end{bmatrix}, \tag{51}$$

where the global invariant factors of $P_1(s)$ are powers of s^2+1 and the global invariant factors of $P_2(s)$ are relatively prime with s^2+1. Moreover, the left Wiener–Hopf factorization indices of $P_1(s)$ at infinity are 3, 1 (add the first column multiplied by s^3+2s to the second column; the result is a column proper matrix with column degrees 1 and 3). Therefore, the left Wiener–Hopf factorization indices of $P(s)$ with respect to M are 3, 1.

Consider now $\tilde{M} = \{(s^2+1),(s)\}$ and $\tilde{M}' = \mathrm{Specm}(\mathbb{R}[s]) \setminus \tilde{M}$. There is a unimodular matrix $U(s) = \begin{bmatrix} 1 & s^2+2 \\ 0 & 1 \end{bmatrix}$, invertible in $\mathbb{R}_{\tilde{M}}(s)^{2\times 2}$, such that $P(s)U(s) = \begin{bmatrix} s & s^3+2s \\ -s^2 & 1 \end{bmatrix}$ is column proper with column degrees 3 and 2. We can write

$$P(s)U(s) = \begin{bmatrix} 0 & 1 \\ -1 & 0 \end{bmatrix} \begin{bmatrix} s^2 & 0 \\ 0 & s^3 \end{bmatrix} + \begin{bmatrix} s & 2s \\ 0 & 1 \end{bmatrix} = B(s) \begin{bmatrix} s^2 & 0 \\ 0 & s^3 \end{bmatrix}, \tag{52}$$

where $B(s)$ is the following biproper matrix

$$B(s) = \begin{bmatrix} 0 & 1 \\ -1 & 0 \end{bmatrix} + \begin{bmatrix} s & 2s \\ 0 & 1 \end{bmatrix} \begin{bmatrix} s^{-2} & 0 \\ 0 & s^{-3} \end{bmatrix} = \begin{bmatrix} \frac{1}{s} & \frac{s^2+2}{s^2} \\ -1 & \frac{1}{s^3} \end{bmatrix}. \tag{53}$$

Moreover, the denominators of its entries are powers of s and $\det B(s) = \frac{(s^2+1)^2}{s^4}$. Therefore, $B(s)$ is invertible in $\mathbb{R}_{\tilde{M}'}(s)^{2\times 2} \cap \mathbb{R}_{pr}(s)^{2\times 2}$. Since $B(s)^{-1}P(s)U(s) = \mathrm{Diag}(s^2, s^3)$, the left Wiener–Hopf factorization indices of $P(s)$ with respect to $\tilde{M}$ are 3, 2.

If $\tilde{M} = \{(s^2+1),(s-1)\}$, for example, a similar procedure shows that $P(s)$ has 3, 1 as left Wiener–Hopf factorization indices with respect to $\tilde{M}$; the same indices as with respect to M. The reason is that $s-1$ is not a divisor of $\det P(s)$ and so $P(s) = P_1(s)P_2(s)$ with $P_1(s)$ and $P_2(s)$ as in (51) and $P_1(s)$ factorizing in $\tilde{M}$.

Remark 24. It must be noticed that a procedure has been given to compute, at least theoretically, the left Wiener–Hopf factorization indices of any rational matrix with respect to any subset M of $\mathrm{Specm}(\mathbb{F}[s])$. In fact, given a rational matrix $T(s)$ and M, write $T(s) = T_L(s)T_R(s)$ with the global invariant rational functions of $T_L(s)$ factorizing in M, and the global invariant rational functions of $T_R(s)$ factorizing in $\mathrm{Specm}(\mathbb{F}[s]) \setminus M$ (for example, using the global Smith–McMillan form of $T(s)$). We need to compute the left Wiener–Hopf factorization indices at infinity of the rational matrix $T_L(s)$. The idea is as follows: Let $d(s)$ be the monic least common denominator of all the elements of $T_L(s)$. The matrix $T_L(s)$ can be written as $T_L(s) = \frac{P(s)}{d(s)}$, with $P(s)$ polynomial. The left Wiener–Hopf factorization indices of $P(s)$ at infinity are the column degrees of any column proper matrix right equivalent to $P(s)$. If $k_1, \ldots, k_m$ are the left Wiener–Hopf factorization indices at infinity of $P(s)$ then $k_1+d, \ldots, k_m+d$ are the left Wiener–Hopf factorization indices of $T_L(s)$, where $d = d(d(s))$ (see [1]). Free and commercial software exists that compute such column degrees.

5. Rosenbrock's Theorem via local rings

As said in the Introduction, Rosenbrock's Theorem ([14]) on pole assignment by state feedback provides, in its polynomial formulation, a complete characterization of the relationship between the invariant factors and the left Wiener–Hopf factorization indices at infinity of any non-singular matrix polynomial. The precise statement of this result is the following theorem:

Theorem 25. *Let $g_1 \geq \cdots \geq g_m$ and $\alpha_1(s) \mid \cdots \mid \alpha_m(s)$ be non-negative integers and monic polynomials, respectively. Then there exists a non-singular matrix $P(s) \in \mathbb{F}[s]^{m\times m}$ with $\alpha_1(s),\ldots,\alpha_m(s)$ as invariant factors and $g_1,\ldots,g_m$ as left Wiener–Hopf factorization indices at infinity if and only if the following relation holds:*

$$(g_1,\ldots,g_m) \prec (d(\alpha_m(s)),\ldots,d(\alpha_1(s))). \tag{54}$$

Symbol $\prec$ appearing in (54) is the majorization symbol (see [11]) and it is defined as follows: If $(a_1,\ldots,a_m)$ and $(b_1,\ldots,b_m)$ are two finite sequences of real numbers and $a_{[1]} \geq \cdots \geq a_{[m]}$ and $b_{[1]} \geq \cdots \geq b_{[m]}$ are the given sequences arranged in non-increasing order then $(a_1,\ldots,a_m) \prec (b_1,\ldots,b_m)$ if

$$\sum_{i=1}^{j} a_{[i]} \leq \sum_{i=1}^{j} b_{[i]}, \quad 1 \leq j \leq m-1 \tag{55}$$

with equality for $j = m$.

The above Theorem 25 can be extended to cover rational matrix functions. Any rational matrix $T(s)$ can be written as $\frac{N(s)}{d(s)}$ where $d(s)$ is the monic least common denominator of all the elements of $T(s)$ and $N(s)$ is polynomial. It turns out that the invariant rational functions of $T(s)$ are the invariant factors of $N(s)$ divided by $d(s)$ after canceling common factors. We also have the following characterization of the left Wiener– Hopf factorization indices at infinity of $T(s)$: these are those of $N(s)$ plus the degree of $d(s)$ (see [1]). Bearing all this in mind one can easily prove (see [1])

Theorem 26. *Let $g_1 \geq \cdots \geq g_m$ be integers and $\frac{\alpha_1(s)}{\beta_1(s)},\ldots,\frac{\alpha_m(s)}{\beta_m(s)}$ irreducible rational functions, where $\alpha_i(s),\beta_i(s) \in \mathbb{F}[s]$ are monic such that $\alpha_1(s) \mid \cdots \mid \alpha_m(s)$ while $\beta_m(s) \mid \cdots \mid \beta_1(s)$. Then there exists a non-singular rational matrix $T(s) \in \mathbb{F}(s)^{m\times m}$ with $g_1,\ldots,g_m$ as left Wiener–Hopf factorization indices at infinity and $\frac{\alpha_1(s)}{\beta_1(s)},\ldots,\frac{\alpha_m(s)}{\beta_m(s)}$ as global invariant rational functions if and only if*

$$(g_1,\ldots,g_m) \prec (d(\alpha_m(s)) - d(\beta_m(s)),\ldots,d(\alpha_1(s)) - d(\beta_1(s))). \tag{56}$$

Recall that for $M \subseteq \text{Specm}(\mathbb{F}[s])$ any rational matrix $T(s)$ can be factorized into two matrices (see Proposition 2) such that the global invariant rational functions and the left Wiener–Hopf factorization indices at infinity of the left factor of $T(s)$ give the invariant rational functions and the left Wiener–Hopf factorization indices of $T(s)$ with respect to M. Using Theorem 26 on the left factor of $T(s)$ we get:

Theorem 27. *Let $M \subseteq \text{Specm}(\mathbb{F}[s])$. Let $k_1 \geq \cdots \geq k_m$ be integers and $\frac{\epsilon_1(s)}{\psi_1(s)},\ldots,\frac{\epsilon_m(s)}{\psi_m(s)}$ be irreducible rational functions such that $\epsilon_1(s) \mid \cdots \mid \epsilon_m(s)$, $\psi_m(s) \mid \cdots \mid \psi_1(s)$ are monic polynomials factorizing in M. Then there exists a non-singular matrix $T(s) \in \mathbb{F}(s)^{m\times m}$ with $\frac{\epsilon_1(s)}{\psi_1(s)},\ldots,\frac{\epsilon_m(s)}{\psi_m(s)}$ as invariant rational functions with respect to M and $k_1,\ldots,k_m$ as left Wiener–Hopf factorization indices with respect to M if and only if*

$$(k_1,\ldots,k_m) \prec (d(\epsilon_m(s)) - d(\psi_m(s)),\ldots,d(\epsilon_1(s)) - d(\psi_1(s))). \tag{57}$$

Theorem 27 relates the left Wiener–Hopf factorization indices with respect to M and the finite structure inside M. Our last result will relate the left Wiener–Hopf factorization indices with respect to M and the structure outside M, including that at infinity. The next Theorem is an extension of Rosenbrock's Theorem to the point at infinity, which was proved in [1]:

Theorem 28. *Let $g_1 \geq \cdots \geq g_m$ and $q_1 \geq \cdots \geq q_m$ be integers. Then there exists a non-singular matrix $T(s) \in \mathbb{F}(s)^{m\times m}$ with $g_1, \ldots, g_m$ as left Wiener–Hopf factorization indices at infinity and $s^{q_1}, \ldots, s^{q_m}$ as invariant rational functions at infinity if and only if*

$$(g_1, \ldots, g_m) \prec (q_1, \ldots, q_m). \tag{58}$$

Notice that Theorem 26 can be obtained from Theorem 27 when $M = \text{Specm}(\mathbb{F}[s])$. In the same way, taking into account that the equivalence at infinity is a particular case of the equivalence in $\mathbb{F}_{M'}(s) \cap \mathbb{F}_{pr}(s)$ when $M' = \emptyset$, we can give a more general result than that of Theorem 28. Specifically, necessary and sufficient conditions can be provided for the existence of a non-singular rational matrix with prescribed left Wiener–Hopf factorization indices with respect to M and invariant rational functions in $\mathbb{F}_{M'}(s) \cap \mathbb{F}_{pr}(s)$.

Theorem 29. *Let $M, M' \subseteq \text{Specm}(\mathbb{F}[s])$ be such that $M \cup M' = \text{Specm}(\mathbb{F}[s])$. Assume that there are ideals in $M \setminus M'$ generated by linear polynomials and let $(s-a)$ be any of them. Let $k_1 \geq \cdots \geq k_m$ be integers, $\frac{\epsilon_1(s)}{\psi_1(s)}, \ldots, \frac{\epsilon_m(s)}{\psi_m(s)}$ irreducible rational functions such that $\epsilon_1(s)|\cdots|\epsilon_m(s)$, $\psi_m(s)|\cdots|\psi_1(s)$ are monic polynomials factorizing in $M' \setminus M$ and $l_1, \ldots, l_m$ integers such that $l_1 + d(\psi_1(s)) - d(\epsilon_1(s)) \leq \cdots \leq l_m + d(\psi_m(s)) - d(\epsilon_m(s))$. Then there exists a non-singular matrix $T(s) \in \mathbb{F}(s)^{m\times m}$ with no zeros and no poles in $M \cap M'$ with $k_1, \ldots, k_m$ as left Wiener–Hopf factorization indices with respect to M and $\frac{\epsilon_1(s)}{\psi_1(s)} \frac{1}{(s-a)^{l_1}}, \ldots, \frac{\epsilon_m(s)}{\psi_m(s)} \frac{1}{(s-a)^{l_m}}$ as invariant rational functions in $\mathbb{F}_{M'}(s) \cap \mathbb{F}_{pr}(s)$ if and only if the following condition holds:*

$$(k_1, \ldots, k_m) \prec (-l_1, \ldots, -l_m). \tag{59}$$

The proof of this theorem will be given along the following two subsections. We will use several auxiliary results that will be stated and proved when needed.

5.1. Necessity

We can give the following result for rational matrices using a similar result given in Lemma 4.2 in [2] for matrix polynomials.

Lemma 30. *Let $M, M' \subseteq \text{Specm}(\mathbb{F}[s])$ be such that $M \cup M' = \text{Specm}(\mathbb{F}[s])$. Let $T(s) \in \mathbb{F}(s)^{m\times m}$ be a non-singular matrix with no zeros and no poles in $M \cap M'$ with $g_1 \geq \cdots \geq g_m$ as left Wiener–Hopf factorization indices at infinity and $k_1 \geq \cdots \geq k_m$ as left Wiener–Hopf factorization indices with respect to M. If $\frac{\epsilon_1(s)}{\psi_1(s)}, \ldots, \frac{\epsilon_m(s)}{\psi_m(s)}$ are the invariant rational functions of $T(s)$ with respect to M' then*

$$(g_1 - k_1, \ldots, g_m - k_m) \prec (d(\epsilon_m(s)) - d(\psi_m(s)), \ldots, d(\epsilon_1(s)) - d(\psi_1(s))). \tag{60}$$

It must be pointed out that $(g_1 - k_1, \ldots, g_m - k_m)$ may be an unordered m-tuple.

Proof.- By Proposition 2 there exist unimodular matrices $U(s), V(s) \in \mathbb{F}[s]^{m\times m}$ such that

$$T(s) = U(s)\,\mathrm{Diag}\left(\frac{\alpha_1(s)}{\beta_1(s)}, \ldots, \frac{\alpha_m(s)}{\beta_m(s)}\right)\mathrm{Diag}\left(\frac{\epsilon_1(s)}{\psi_1(s)}, \ldots, \frac{\epsilon_m(s)}{\psi_m(s)}\right)V(s) \tag{61}$$

with $\alpha_i(s) \mid \alpha_{i+1}(s)$, $\beta_i(s) \mid \beta_{i-1}(s)$, $\epsilon_i(s) \mid \epsilon_{i+1}(s)$, $\psi_i(s) \mid \psi_{i-1}(s)$, $\alpha_i(s), \beta_i(s)$ units in $\mathbb{F}_{M'\setminus M}(s)$ and $\epsilon_i(s), \psi_i(s)$ factorizing in $M' \setminus M$ because $T(s)$ has no poles and no zeros in $M \cap M'$. Therefore $T(s) = T_L(s)T_R(s)$, where $T_L(s) = U(s)\,\mathrm{Diag}\left(\frac{\alpha_1(s)}{\beta_1(s)}, \ldots, \frac{\alpha_m(s)}{\beta_m(s)}\right)$ has $k_1, \ldots, k_m$ as left Wiener–Hopf factorization indices at infinity and $T_R(s) = \mathrm{Diag}\left(\frac{\epsilon_1(s)}{\psi_1(s)}, \ldots, \frac{\epsilon_m(s)}{\psi_m(s)}\right)V(s)$ has $\frac{\epsilon_1(s)}{\psi_1(s)}, \ldots, \frac{\epsilon_m(s)}{\psi_m(s)}$ as global invariant rational functions. Let $d(s) = \beta_1(s)\psi_1(s)$. Hence,

$$d(s)T(s) = U(s)\,\mathrm{Diag}(\overline{\alpha}_1(s), \ldots, \overline{\alpha}_m(s))\,\mathrm{Diag}(\overline{\epsilon}_1(s), \ldots, \overline{\epsilon}_m(s))V(s) \tag{62}$$

with $\overline{\alpha}_i(s) = \frac{\alpha_i(s)}{\beta_i(s)}\beta_1(s)$ units in $\mathbb{F}_{M'\setminus M}(s)$ and $\overline{\epsilon}_i(s) = \frac{\epsilon_i(s)}{\psi_i(s)}\psi_1(s)$ factorizing in $M' \setminus M$. Put $P(s) = d(s)T(s)$. Its left Wiener–Hopf factorization indices at infinity are $g_1 + d(d(s)), \ldots, g_m + d(d(s))$ [1, Lemma 2.3]. The matrix $P_1(s) = U(s)\,\mathrm{Diag}(\overline{\alpha}_1(s), \ldots, \overline{\alpha}_m(s)) = \beta_1(s)T_L(s)$ has $k_1 + d(\beta_1(s)), \ldots, k_m + d(\beta_1(s))$ as left Wiener–Hopf factorization indices at infinity. Now if $P_2(s) = \mathrm{Diag}(\overline{\epsilon}_1(s), \ldots, \overline{\epsilon}_m(s))V(s) = \psi_1(s)T_R(s)$ then its invariant factors are $\overline{\epsilon}_1(s), \ldots, \overline{\epsilon}_m(s)$, $P(s) = P_1(s)P_2(s)$ and, by [2, Lemma 4.2],

$$(g_1 + d(d(s)) - k_1 - d(\beta_1(s)), \ldots, g_m + d(d(s)) - k_m - d(\beta_1(s))) \prec (d(\overline{\epsilon}_m(s)), \ldots, d(\overline{\epsilon}_1(s))). \tag{63}$$

Therefore, (60) follows. ∎

5.1.1. *Proof of Theorem 29: Necessity*

If $\frac{\epsilon_1(s)}{\psi_1(s)}\frac{1}{(s-a)^{l_1}}, \ldots, \frac{\epsilon_m(s)}{\psi_m(s)}\frac{1}{(s-a)^{l_m}}$ are the invariant rational functions of $T(s)$ in $\mathbb{F}_{M'}(s) \cap \mathbb{F}_{pr}(s)$ then there exist matrices $U_1(s), U_2(s)$ invertible in $\mathbb{F}_{M'}(s)^{m\times m} \cap \mathbb{F}_{pr}(s)^{m\times m}$ such that

$$T(s) = U_1(s)\,\mathrm{Diag}\left(\frac{\epsilon_1(s)}{\psi_1(s)}\frac{1}{(s-a)^{l_1}}, \ldots, \frac{\epsilon_m(s)}{\psi_m(s)}\frac{1}{(s-a)^{l_m}}\right)U_2(s). \tag{64}$$

We analyze first the finite structure of $T(s)$ with respect to M'. If $D_1(s) = \mathrm{Diag}((s-a)^{-l_1}, \ldots, (s-a)^{-l_m}) \in \mathbb{F}_{M'}(s)^{m\times m}$, we can write $T(s)$ as follows:

$$T(s) = U_1(s)\,\mathrm{Diag}\left(\frac{\epsilon_1(s)}{\psi_1(s)}, \ldots, \frac{\epsilon_m(s)}{\psi_m(s)}\right)D_1(s)U_2(s), \tag{65}$$

with $U_1(s)$ and $D_1(s)U_2(s)$ invertible matrices in $\mathbb{F}_{M'}(s)^{m\times m}$. Thus $\frac{\epsilon_1(s)}{\psi_1(s)}, \ldots, \frac{\epsilon_m(s)}{\psi_m(s)}$ are the invariant rational functions of $T(s)$ with respect to M'. Let $g_1 \geq \cdots \geq g_m$ be the left Wiener–Hopf factorization indices of $T(s)$ at infinity. By Lemma 30 we have

$$(g_1 - k_1, \ldots, g_m - k_m) \prec (d(\epsilon_m(s)) - d(\psi_m(s)), \ldots, d(\epsilon_1(s)) - d(\psi_1(s))). \tag{66}$$

As far as the structure of $T(s)$ at infinity is concerned, let

$$D_2(s) = \mathrm{Diag}\left(\frac{\epsilon_1(s)}{\psi_1(s)}\frac{s^{l_1+d(\psi_1(s))-d(\epsilon_1(s))}}{(s-a)^{l_1}}, \ldots, \frac{\epsilon_m(s)}{\psi_m(s)}\frac{s^{l_m+d(\psi_m(s))-d(\epsilon_m(s))}}{(s-a)^{l_m}}\right). \tag{67}$$

Then $D_2(s) \in Gl(\mathbb{F}_{pr}(s))$ and

$$T(s) = U_1(s)\,\mathrm{Diag}\left(s^{-l_1-d(\psi_1(s))+d(\epsilon_1(s))}, \ldots, s^{-l_m-d(\psi_m(s))+d(\epsilon_m(s))}\right) D_2(s)U_2(s) \qquad (68)$$

where $U_1(s) \in \mathbb{F}_{pr}(s)^{m\times m}$ and $D_2(s)U_2(s) \in \mathbb{F}_{pr}(s)^{m\times m}$ are biproper matrices. Therefore $s^{-l_1-d(\psi_1(s))+d(\epsilon_1(s))}, \ldots, s^{-l_m-d(\psi_m(s))+d(\epsilon_m(s))}$ are the invariant rational functions of $T(s)$ at infinity. By Theorem 28

$$(g_1, \ldots, g_m) \prec (-l_1 - d(\psi_1(s)) + d(\epsilon_1(s)), \ldots, -l_m - d(\psi_m(s)) + d(\epsilon_m(s))). \qquad (69)$$

Let $\sigma \in \Sigma_m$ (the symmetric group of order m) be a permutation such that $g_{\sigma(1)} - k_{\sigma(1)} \geq \cdots \geq g_{\sigma(m)} - k_{\sigma(m)}$ and define $c_i = g_{\sigma(i)} - k_{\sigma(i)}$, $i = 1, \ldots, m$. Using (66) and (69) we obtain

$$\begin{aligned} \sum_{j=1}^{r} k_j + \sum_{j=1}^{r} (d(\epsilon_j(s)) - d(\psi_j(s))) &\leq \sum_{j=1}^{r} k_j + \sum_{j=m-r+1}^{m} c_j \\ &\leq \sum_{j=1}^{r} k_j + \sum_{j=1}^{r} (g_j - k_j) = \sum_{j=1}^{r} g_j \\ &\leq \sum_{j=1}^{r} -l_j + \sum_{j=1}^{r} (d(\epsilon_j(s)) - d(\psi_j(s))) \end{aligned} \qquad (70)$$

for $r = 1, \ldots, m-1$. When $r = m$ the previous inequalities are all equalities and condition (59) is satisfied. ∎

Remark 31. It has been seen in the above proof that if a matrix has $\frac{\epsilon_1(s)}{\psi_1(s)} \frac{1}{(s-a)^{l_1}}, \ldots, \frac{\epsilon_m(s)}{\psi_m(s)} \frac{1}{(s-a)^{l_m}}$ as invariant rational functions in $\mathbb{F}_{M'}(s) \cap \mathbb{F}_{pr}(s)$ then $\frac{\epsilon_1(s)}{\psi_1(s)}, \ldots, \frac{\epsilon_m(s)}{\psi_m(s)}$ are its invariant rational functions with respect to M' and $s^{-l_1-d(\psi_1(s))+d(\epsilon_1(s))}$, $\ldots, s^{-l_m-d(\psi_m(s))+d(\epsilon_m(s))}$ are its invariant rational functions at infinity.

5.2. Sufficiency

Let $a, b \in \mathbb{F}$ be arbitrary elements such that $ab \neq 1$. Consider the changes of indeterminate

$$f(s) = a + \frac{1}{s-b}, \quad \tilde{f}(s) = b + \frac{1}{s-a} \qquad (71)$$

and notice that $f(\tilde{f}(s)) = \tilde{f}(f(s)) = s$. For $\alpha(s) \in \mathbb{F}[s]$, let $\mathbb{F}[s] \setminus (\alpha(s))$ denote the multiplicative subset of $\mathbb{F}[s]$ whose elements are coprime with $\alpha(s)$. For $a, b \in \mathbb{F}$ as above define

$$\begin{aligned} t_{a,b} : \mathbb{F}[s] &\rightarrow \mathbb{F}[s] \setminus (s-b) \\ \pi(s) &\mapsto (s-b)^{d(\pi(s))}\, \pi\left(a + \tfrac{1}{s-b}\right) = (s-b)^{d(\pi(s))}\, \pi(f(s)). \end{aligned} \qquad (72)$$

In words, if $\pi(s) = p_d(s-a)^d + p_{d-1}(s-a)^{d-1} + \cdots + p_1(s-a) + p_0$ $(p_d \neq 0)$ then

$$t_{a,b}(\pi(s)) = p_0(s-b)^d + p_1(s-b)^{d-1} + \cdots + p_{d-1}(s-b) + p_d. \qquad (73)$$

In general $d(t_{a,b}(\pi(s))) \leq d(\pi(s))$ with equality if and only if $\pi(s) \in \mathbb{F}[s] \setminus (s-a)$. This shows that the restriction $h_{a,b} : \mathbb{F}[s] \setminus (s-a) \rightarrow \mathbb{F}[s] \setminus (s-b)$ of $t_{a,b}$ to $\mathbb{F}[s] \setminus (s-a)$ is a bijection. In addition $h_{a,b}^{-1}$ is the restriction of $t_{b,a}$ to $\mathbb{F}[s] \setminus (s-b)$; i.e.,

$$\begin{aligned} h_{a,b}^{-1} : \mathbb{F}[s] \setminus (s-b) &\rightarrow \mathbb{F}[s] \setminus (s-a) \\ \alpha(s) &\mapsto (s-a)^{d(\alpha(s))}\, \alpha\left(b + \tfrac{1}{s-a}\right) = (s-a)^{d(\alpha(s))}\, \alpha(\tilde{f}(s)) \end{aligned} \qquad (74)$$

or $h_{a,b}^{-1} = h_{b,a}$.

In what follows we will think of a, b as given elements of $\mathbb{F}$ and the subindices of $t_{a,b}$, $h_{a,b}$ and $h_{a,b}^{-1}$ will be removed. The following are properties of h (and h^{-1}) that can be easily proved.

Lemma 32. *Let $\pi_1(s), \pi_2(s) \in \mathbb{F}[s] \setminus (s-a)$. The following properties hold:*

1. $h(\pi_1(s)\pi_2(s)) = h(\pi_1(s))h(\pi_2(s))$.
2. *If $\pi_1(s) \mid \pi_2(s)$ then $h(\pi_1(s)) \mid h(\pi_2(s))$.*
3. *If $\pi_1(s)$ is an irreducible polynomial then $h(\pi_1(s))$ is an irreducible polynomial.*
4. *If $\pi_1(s), \pi_2(s)$ are coprime polynomials then $h(\pi_1(s)), h(\pi_2(s))$ are coprime polynomials.*

As a consequence the map

$$\begin{array}{rcl} H : \text{Specm}\,(\mathbb{F}[s]) \setminus \{(s-a)\} & \rightarrow & \text{Specm}\,(\mathbb{F}[s]) \setminus \{(s-b)\} \\ (\pi(s)) & \mapsto & (\frac{1}{p_0} h(\pi(s))) \end{array} \tag{75}$$

with $p_0 = \pi(a)$, is a bijection whose inverse is

$$\begin{array}{rcl} H^{-1} : \text{Specm}\,(\mathbb{F}[s]) \setminus \{(s-b)\} & \rightarrow & \text{Specm}\,(\mathbb{F}[s]) \setminus \{(s-a)\} \\ (\alpha(s)) & \mapsto & (\frac{1}{a_0} h^{-1}(\alpha(s))) \end{array} \tag{76}$$

where $a_0 = \alpha(b)$. In particular, if $M' \subseteq \text{Specm}(\mathbb{F}[s]) \setminus \{(s-a)\}$ and $\tilde{M} = \text{Specm}(\mathbb{F}[s]) \setminus (M' \cup \{(s-a)\})$ (i.e. the complementary subset of M' in Specm $(\mathbb{F}[s]) \setminus \{(s-a)\}$) then

$$H(\tilde{M}) = \text{Specm}\,(\mathbb{F}[s]) \setminus (H(M') \cup \{(s-b)\}). \tag{77}$$

In what follows and for notational simplicity we will assume $b = 0$.

Lemma 33. *Let $M' \subseteq \text{Specm}\,(\mathbb{F}[s]) \setminus \{(s-a)\}$ where $a \in \mathbb{F}$ is an arbitrary element of $\mathbb{F}$.*

1. *If $\pi(s) \in \mathbb{F}[s]$ factorizes in M' then $h(\pi(s))$ factorizes in $H(M')$.*
2. *If $\pi(s) \in \mathbb{F}[s]$ is a unit of $\mathbb{F}_{M'}(s)$ then $t(\pi(s))$ is a unit of $\mathbb{F}_{H(M')}(s)$.*

Proof.- 1. Let $\pi(s) = c\pi_1(s)^{g_1} \cdots \pi_m(s)^{g_m}$ with $c \neq 0$ constant, $(\pi_i(s)) \in M'$ and $g_i \geq 1$. Then $h(\pi(s)) = c(h(\pi_1(s)))^{g_1} \cdots (h(\pi_m(s)))^{g_m}$. By Lemma 32 $h(\pi_i(s))$ is an irreducible polynomial (that may not be monic). If c_i is the leading coefficient of $h(\pi_i(s))$ then $\frac{1}{c_i} h(\pi_i(s))$ is monic, irreducible and $(\frac{1}{c_i} h(\pi_i(s))) \in H(M')$. Hence $h(\pi(s))$ factorizes in $H(M')$.

2. If $\pi(s) \in \mathbb{F}[s]$ is a unit of $\mathbb{F}_{M'}(s)$ then it can be written as $\pi(s) = (s-a)^g \pi_1(s)$ where $g \geq 0$ and $\pi_1(s)$ is a unit of $\mathbb{F}_{M' \cup \{(s-a)\}}(s)$. Therefore $\pi_1(s)$ factorizes in $\text{Specm}(\mathbb{F}[s]) \setminus (M' \cup \{(s-a)\})$. Since $t(\pi(s)) = h(\pi_1(s))$, it factorizes in (recall that we are assuming $b = 0$) $H(\text{Specm}(\mathbb{F}[s]) \setminus (M' \cup \{(s-a)\}) = \text{Specm}(\mathbb{F}[s]) \setminus (H(M') \cup \{(s)\})$. So, $t(\pi(s))$ is a unit of $\mathbb{F}_{H(M')}(s)$. ■

Lemma 34. *Let $a \in \mathbb{F}$ be an arbitrary element. Then*

1. *If $M' \subseteq \text{Specm}\,(\mathbb{F}[s]) \setminus \{(s-a)\}$ and $U(s) \in \text{Gl}_m(\mathbb{F}_{M'}(s))$ then $U(f(s)) \in \text{Gl}_m(\mathbb{F}_{H(M')}(s))$.*

2. *If* $U(s) \in \mathrm{Gl}_m(\mathbb{F}_{s-a}(s))$ *then* $U(f(s)) \in \mathrm{Gl}_m(\mathbb{F}_{pr}(s))$.
3. *If* $U(s) \in \mathrm{Gl}_m(\mathbb{F}_{pr}(s))$ *then* $U(f(s)) \in \mathrm{Gl}_m(\mathbb{F}_s(s))$.
4. *If* $(s-a) \in M' \subseteq \mathrm{Specm}(\mathbb{F}[s])$ *and* $U(s) \in \mathrm{Gl}_m(\mathbb{F}_{M'}(s))$ *then the matrix* $U(f(s)) \in \mathrm{Gl}_m(\mathbb{F}_{H(M'\setminus\{(s-a)\})}(s)) \cap \mathrm{Gl}_m(\mathbb{F}_{pr}(s))$

Proof.- Let $\frac{p(s)}{q(s)}$ with $p(s), q(s) \in \mathbb{F}[s]$.

$$\frac{p(f(s))}{q(f(s))} = \frac{s^{d(p(s))}p(f(s))}{s^{d(q(s))}q(f(s))}s^{d(q(s))-d(p(s))} = \frac{t(p(s))}{t(q(s))}s^{d(q(s))-d(p(s))}. \tag{78}$$

1. Assume that $U(s) \in \mathrm{Gl}_m(\mathbb{F}_{M'}(s))$ and let $\frac{p(s)}{q(s)}$ be any element of $U(s)$. Therefore $q(s)$ is a unit of $\mathbb{F}_{M'}(s)$ and, by Lemma 33.2, $t(q(s))$ is a unit of $\mathbb{F}_{H(M')}(s)$. Moreover, s is also a unit of $\mathbb{F}_{H(M')}(s)$. Hence, $\frac{p(f(s))}{q(f(s))} \in \mathbb{F}_{H(M')}(s)$. Furthermore, if $\det U(s) = \frac{\tilde{p}(s)}{\tilde{q}(s)}$, it is a unit of $\mathbb{F}_{M'}(s)$ and $\det U(f(s)) = \frac{\tilde{p}(f(s))}{\tilde{q}(f(s))}$ is a unit of $\mathbb{F}_{H(M')}(s)$.

2. If $\frac{p(s)}{q(s)}$ is any element of $U(s) \in \mathrm{Gl}_m(\mathbb{F}_{s-a}(s))$ then $q(s) \in \mathbb{F}[s] \setminus (s-a)$ and so $d(h(q(s))) = d(q(s))$. Since $s-a$ may divide $p(s)$ we have that $d(t(p(s))) \leq d(p(s))$. Hence, $d(h(q(s))) - d(q(s)) \geq d(t(p(s)) - d(p(s))$ and $\frac{p(f(s))}{q(f(s))} = \frac{t(p(s))}{h(q(s))}s^{d(q(s))-d(p(s))} \in \mathbb{F}_{pr}(s)$. Moreover if $\det U(s) = \frac{\tilde{p}(s)}{\tilde{q}(s)}$ then $\tilde{p}(s), \tilde{q}(s) \in \mathbb{F}[s] \setminus (s-a)$, $d(h(\tilde{p}(s))) = d(\tilde{p}(s))$ and $d(h(\tilde{q}(s))) = d(\tilde{q}(s))$. Thus, $\det U(f(s)) = \frac{h(\tilde{p}(s))}{h(\tilde{q}(s))}s^{d(\tilde{q}(s))-d(\tilde{p}(s))}$ is a biproper rational function, i.e., a unit of $\mathbb{F}_{pr}(s)$.

3. If $U(s) \in \mathrm{Gl}_m(\mathbb{F}_{pr}(s))$ and $\frac{p(s)}{q(s)}$ is any element of $U(s)$ then $d(q(s)) \geq d(p(s))$. Since $\frac{p(f(s))}{q(f(s))} = \frac{t(p(s))}{t(q(s))}s^{d(q(s))-d(p(s))}$ and $t(p(s)), t(q(s)) \in \mathbb{F}[s] \setminus (s)$ we obtain that $U(f(s)) \in \mathbb{F}_s(s)^{m\times m}$. In addition, if $\det U(s) = \frac{\tilde{p}(s)}{\tilde{q}(s)}$, which is a unit of $\mathbb{F}_{pr}(s)$, then $d(\tilde{q}(s)) = d(\tilde{p}(s))$ and since $t(\tilde{p}(s)), t(\tilde{q}(s)) \in \mathbb{F}[s] \setminus (s)$ we conclude that $\det U(f(s)) = \frac{t(\tilde{p}(s))}{t(\tilde{q}(s))}$ is a unit of $\mathbb{F}_s(s)$.

4. It is a consequence of 1., 2. and Remark 1.2. ∎

Proposition 35. *Let* $M \subseteq \mathrm{Specm}(\mathbb{F}[s])$ *and* $(s-a) \in M$. *If* $T(s) \in \mathbb{F}(s)^{m\times m}$ *is non-singular with* $\frac{n_i(s)}{d_i(s)} = (s-a)^{g_i}\frac{\epsilon_i(s)}{\psi_i(s)}$ $(\epsilon_i(s), \psi_i(s) \in \mathbb{F}[s] \setminus (s-a))$ *as invariant rational functions with respect to* M *then* $T(f(s))^T \in \mathbb{F}(s)^{m\times m}$ *is a non-singular matrix with* $\frac{1}{c_i}\frac{h(\epsilon_i(s))}{h(\psi_i(s))}s^{-g_i+d(\psi_i(s))-d(\epsilon_i(s))}$ *as invariant rational functions in* $\mathbb{F}_{H(M\setminus\{(s-a)\})}(s)^{m\times m} \cap \mathbb{F}_{pr}(s)^{m\times m}$ *where* $c_i = \frac{\epsilon_i(a)}{\psi_i(a)}$.

Proof.- Since $(s-a)^{g_i}\frac{\epsilon_i(s)}{\psi(s)}$ are the invariant rational functions of $T(s)$ with respect to M, there are $U_1(s), U_2(s) \in Gl_m(\mathbb{F}_M(s))$ such that

$$T(s) = U_1(s)\,\mathrm{Diag}\left((s-a)^{g_1}\frac{\epsilon_1(s)}{\psi_1(s)}, \ldots, (s-a)^{g_m}\frac{\epsilon_m(s)}{\psi_m(s)}\right)U_2(s). \tag{79}$$

Notice that $(f(s)-a)^{g_i}\frac{\epsilon_i(f(s))}{\psi_i(f(s))} = \frac{h(\epsilon_i(s))}{h(\psi_i(s))}s^{-g_i+d(\psi_i(s))-d(\epsilon_i(s))}$. Let $c_i = \frac{\epsilon_i(a)}{\psi_i(a)}$, which is a non-zero constant, and put $D = \mathrm{Diag}(c_1,\ldots,c_m)$. Hence,

$$T(f(s))^T = U_2(f(s))^T DL(s)U_1(f(s))^T \tag{80}$$

with

$$L(s) = \mathrm{Diag}\left(\frac{1}{c_1}\frac{h(\epsilon_1(s))}{h(\psi_1(s))}s^{-g_1+d(\psi_1(s))-d(\epsilon_1(s))},\ldots,\frac{1}{c_m}\frac{h(\epsilon_m(s))}{h(\psi_m(s))}s^{-g_m+d(\psi_m(s))-d(\epsilon_m(s))}\right). \tag{81}$$

By 4 of Lemma 34 matrices $U_1(f(s))^T, U_2(f(s))^T \in \mathrm{Gl}_m(\mathbb{F}_{H(M\setminus\{(s-a)\})}(s)) \cap \mathrm{Gl}_m(\mathbb{F}_{pr}(s))$ and the Proposition follows. ∎

Proposition 36. *Let $M, M' \subseteq \mathrm{Specm}(\mathbb{F}[s])$ such that $M \cup M' = \mathrm{Specm}(\mathbb{F}[s])$. Assume that there are ideals in $M \setminus M'$ generated by linear polynomials and let $(s-a)$ be any of them. If $T(s) \in \mathbb{F}(s)^{m\times m}$ is a non-singular rational matrix with no poles and no zeros in $M \cap M'$ and $k_1,\ldots,k_m$ as left Wiener–Hopf factorization indices with respect to M then $T(f(s))^T \in \mathbb{F}(s)^{m\times m}$ is a non-singular rational matrix with no poles and no zeros in $H(M \cap M')$ and $-k_m,\ldots,-k_1$ as left Wiener–Hopf factorization indices with respect to $H(M') \cup \{(s)\}$.*

Proof.- By Theorem 19 there are matrices $U_1(s)$ invertible in $\mathbb{F}_{M'}(s)^{m\times m} \cap \mathbb{F}_{pr}(s)^{m\times m}$ and $U_2(s)$ invertible in $\mathbb{F}_M(s)^{m\times m}$ such that $T(s) = U_1(s)\,\mathrm{Diag}\left((s-a)^{k_1},\ldots,(s-a)^{k_m}\right)U_2(s)$. By Lemma 34 $U_2(f(s))^T$ is invertible in $\mathbb{F}_{H(M\setminus\{(s-a)\})}(s)^{m\times m} \cap \mathbb{F}_{pr}(s)^{m\times m}$ and $U_1(f(s))^T$ is invertible in $\mathbb{F}_{H(M')}(s)^{m\times m} \cap \mathbb{F}_s(s)^{m\times m} = \mathbb{F}_{H(M')\cup\{(s)\}}(s)^{m\times m}$. Moreover, $H(M\setminus\{(s-a)\}) \cup H(M') \cup \{(s)\} = \mathrm{Specm}(\mathbb{F}[s])$ and $H(M\setminus\{(s-a)\}) \cap (H(M') \cup \{(s)\}) = H(M \cap M')$. Thus, $T(f(s))^T = U_2(f(s))^T\,\mathrm{Diag}\left(s^{-k_1},\ldots,s^{-k_m}\right)U_1(f(s))^T$ has no poles and no zeros in $H(M \cap M')$ and $-k_m,\ldots,-k_1$ are its left Wiener–Hopf factorization indices with respect to $H(M') \cup \{(s)\}$. ∎

5.2.1. *Proof of Theorem 29: Sufficiency*

Let $k_1 \geq \cdots \geq k_m$ be integers, $\frac{\epsilon_1(s)}{\psi_1(s)},\ldots,\frac{\epsilon_m(s)}{\psi_m(s)}$ irreducible rational functions such that $\epsilon_1(s) \mid \cdots \mid \epsilon_m(s)$, $\psi_m(s) \mid \cdots \mid \psi_1(s)$ are monic polynomials factorizing in $M' \setminus M$ and $l_1,\ldots,l_m$ integers such that $l_1 + d(\psi_1(s)) - d(\epsilon_1(s)) \leq \cdots \leq l_m + d(\psi_m(s)) - d(\epsilon_m(s))$ and satisfying (59).

Since $\epsilon_i(s)$ and $\psi_i(s)$ are coprime polynomials that factorize in $M' \setminus M$ and $(s-a) \in M \setminus M'$, by Lemmas 32 and 33, $\frac{h(\epsilon_1(s))}{h(\psi_1(s))}s^{l_1+d(\psi_1(s))-d(\epsilon_1(s))},\ldots,\frac{h(\epsilon_m(s))}{h(\psi_m(s))}s^{l_m+d(\psi_m(s))-d(\epsilon_m(s))}$ are irreducible rational functions with numerators and denominators polynomials factorizing in $H(M') \cup \{(s)\}$ (actually, in $H(M' \setminus M) \cup \{(s)\}$) and such that each numerator divides the next one and each denominator divides the previous one.

By (59) and Theorem 27 there is a matrix $G(s) \in \mathbb{F}(s)^{m\times m}$ with $-k_m,\ldots,-k_1$ as left Wiener–Hopf factorization indices with respect to $H(M') \cup \{(s)\}$ and $\frac{1}{c_1}\frac{h(\epsilon_1(s))}{h(\psi_1(s))}s^{l_1+d(\psi_1(s))-d(\epsilon_1(s))},\ldots,\frac{1}{c_m}\frac{h(\epsilon_m(s))}{h(\psi_m(s))}s^{l_m+d(\psi_m(s))-d(\epsilon_m(s))}$ as invariant rational functions with respect to $H(M') \cup \{(s)\}$ where $c_i = \frac{\epsilon_i(a)}{\psi_i(a)}$, $i = 1,\ldots,m$. Notice that $G(s)$ has no zeros and poles in $H(M \cap M')$ because the numerator and denominator of each rational function $\frac{h(\epsilon_i(s))}{h(\psi_i(s))} s^{l_i + d(\psi_i(s)) - d(\epsilon_i(s))}$ factorizes in $H(M' \setminus M) \cup \{(s)\}$ and so it is a unit of $\mathbb{F}_{H(M \cap M')}(s)$.

Put $\widehat{M} = H(M') \cup \{(s)\}$ and $\widehat{M}' = H(M \setminus \{(s-a)\})$. As remarked in the proof of Proposition 36, $\widehat{M} \cup \widehat{M}' = \text{Specm}(\mathbb{F}[s])$ and $\widehat{M} \cap \widehat{M}' = H(M \cap M')$. Now $(s) \in \widehat{M}$ so that we can apply Proposition 35 to $G(s)$ with the change of indeterminate $\widetilde{f}(s) = \frac{1}{s-a}$. Thus the invariant rational functions of $G(\widetilde{f}(s))^T$ in $\mathbb{F}_{M'}(s) \cap \mathbb{F}_{pr}(s)$ are $\frac{\epsilon_1(s)}{\psi_1(s)} \frac{1}{(s-a)^{l_1}}, \ldots, \frac{\epsilon_m(s)}{\psi_m(s)} \frac{1}{(s-a)^{l_m}}$.

On the other hand $\widehat{M}' = H(M \setminus \{(s-a)\}) \subseteq \text{Specm}(\mathbb{F}[s]) \setminus \{(s)\}$ and so $(s) \in \widehat{M} \setminus \widehat{M}'$. Then we can apply Proposition 36 to $G(s)$ with $\widetilde{f}(s) = \frac{1}{s-a}$ so that $G(\widetilde{f}(s))^T$ is a non-singular matrix with no poles and no zeros in $H^{-1}(\widehat{M} \cap \widehat{M}') = H^{-1}(H(M \cap M')) = M \cap M'$ and $k_1, \ldots, k_m$ as left Wiener–Hopf factorization indices with respect to $H^{-1}(\widehat{M}') \cup \{(s-a)\} = (M \setminus \{(s-a)\}) \cup \{(s-a)\} = M$. The theorem follows by letting $T(s) = G(\widetilde{f}(s))^T$. ∎

Remark 37. Notice that when $M' = \emptyset$ and $M = \text{Specm}(\mathbb{F}[s])$ in Theorem 29 we obtain Theorem 28 ($q_i = -l_i$).

Author details

A. Amparan, S. Marcaida, I. Zaballa
Universidad del País Vasco/Euskal Herriko Unibertsitatea UPV/EHU, Spain

6. References

[1] Amparan, A., Marcaida, S. & Zaballa, I. [2004]. Wiener–hopf factorization indices and infinite structure of rational matrices, *SIAM J. Control Optim.* Vol. 42(No. 6): 2130–2144.

[2] Amparan, A., Marcaida, S. & Zaballa, I. [2006]. On the existence of linear systems with prescribed invariants for system similarity, *Linear Algebra and its Applications* Vol. 413: 510–533.

[3] Amparan, A., Marcaida, S. & Zaballa, I. [2007]. Local realizations and local polynomial matrix representations of systems, *Linear Algebra and its Applications* Vol. 425: 757–775.

[4] Amparan, A., Marcaida, S. & Zaballa, I. [2009]. Local wiener–hopf factorization and indices over arbitrary fields, *Linear Algebra and its Applications* Vol. 430: 1700–1722.

[5] Atiyah, M. F. & MacDonald, I. G. [1969]. *Introduction to commutative algebra*, Addison-Wesley.

[6] Clancey, K. & Gohberg, I. [1981]. *Factorization of matrix functions and singular integral operators*, Birkhäuser Verlag, Basel, Boston, Stuttgart.

[7] Cullen, D. J. [1986]. Local system equivalence, *Mathematical Systems Theory* 19: 67–78.

[8] Fuhrmann, P. & Helmke, U. [2001]. On the parametrization of conditioned invariant subspaces and observer theory, *Linear Algebra and its Applications* 332–334: 265–353.

[9] Fuhrmann, P. & Willems, J. C. [1979]. Factorization indices at infinity for rational matrix functions, *Integral Equations Operator Theory* 2/3: 287–301.

[10] Gohberg, I., Kaashoek, M. A. & van Schagen F. [1995]. *Partially specified matrices and operators: classification, completion, applications*, Birkhäuser Verlag, Basel.

[11] Hardy, G. H., Littlewood, J. E. & G., P. [1967]. *Inequalities*, Cambridge Univ. Press, Cambridge.

[12] Kailath, T. [1980]. *Linear systems*, Prentice Hall, New Jersey.
[13] Newman, M. [1972]. *Integral matrices*, Academic Press, New York and London.
[14] Rosenbrock, H. H. [1970]. *State-space and multivariable theory*, Thomas Nelson and Sons, London.
[15] Vardulakis, A. I. G. [1991]. *Linear multivariable control*, John Wiley and Sons, New York.
[16] Vidyasagar, M. [1985]. *Control system synthesis. A factorization approach*, The MIT Press, New York.
[17] Wolovich, W. A. [1974]. *Linear multivariable systems*, Springer-Verlag, New York.
[18] Zaballa, I. [1997]. Controllability and hermite indices of matrix pairs, *Int. J. Control* 68(1): 61–86.

Chapter 4

Gauge Theory, Combinatorics, and Matrix Models

Taro Kimura

Additional information is available at the end of the chapter

http://dx.doi.org/10.5772/46481

1. Introduction

Quantum field theory is the most universal method in physics, applied to all the area from condensed-matter physics to high-energy physics. The standard tool to deal with quantum field theory is the perturbation method, which is quite useful if we know the vacuum of the system, namely the starting point of our analysis. On the other hand, sometimes the vacuum itself is not obvious due to the quantum nature of the system. In that case, since the perturbative method is not available any longer, we have to treat the theory in a non-perturbative way.

Supersymmetric gauge theory plays an important role in study on the non-perturbative aspects of quantum field theory. The milestone paper by Seiberg and Witten proposed a *solution* to $\mathcal{N} = 2$ supersymmetric gauge theory [48, 49], which completely describes the low energy effective behavior of the theory. Their solution can be written down by an auxiliary complex curve, called *Seiberg-Witten curve*, but its meaning was not yet clear and the origin was still mysterious. Since the establishment of Seiberg-Witten theory, tremendous number of works are devoted to understand the Seiberg-Witten's solution, not only by physicists but also mathematicians. In this sense the solution was not a *solution* at that time, but just a *starting point* of the exploration.

One of the most remarkable progress in $\mathcal{N} = 2$ theories referring to Seiberg-Witten theory is then the exact derivation of the gauge theory partition function by performing the integral over the instanton moduli space [43]. The partition function is written down by multiple partitions, thus we can discuss it in a combinatorial way. It was mathematically proved that the partition function correctly reproduces the Seiberg-Witten solution. This means Seiberg-Witten theory was mathematically established at that time.

The recent progress on the four dimensional $\mathcal{N} = 2$ supersymmetric gauge theory has revealed a remarkable relation to the two dimensional conformal field theory [1]. This relation provides the explicit interpretation for the partition function of the four dimensional gauge theory as the conformal block of the two dimensional Liouville field theory. It is naturally regarded as a consequence of the M-brane compactifications [23, 60], and also reproduces the results of Seiberg-Witten theory. It shows how Seiberg-Witten curve characterizes the corresponding four dimensional gauge theory, and thus we can obtain a novel viewpoint of Seiberg-Witten theory.

Based on the connection between the two and four dimensional theories, established results on the two dimensional side can be reconsidered from the viewpoint of the four dimensional theory, and vice versa. One of the useful applications is the matrix model description of the supersymmetric gauge theory [12, 16, 17, 47]. This is based on the fact that the conformal block on the sphere can be also regarded as the matrix integral, which is called Dotsenko-Fateev integral representation [14, 15]. In this direction some extensions of the matrix model description are performed by starting with the two dimensional conformal field theory.

Another type of the matrix model is also investigated so far [27, 28, 30, 52, 53]. This is apparently different from the Dotsenko-Fateev type matrix models, but both of them correctly reproduce the results of the four dimensional gauge theory, e.g. Seiberg-Witten curve. While these studies mainly focus on rederiving the gauge theory results, the present author reveals the new kind of Seiberg-Witten curve by studying the corresponding new matrix model [27, 28]. Such a matrix models is directly derived from the combinatorial representation of the partition function by considering its asymptotic behavior. This treatment is quite analogous to the matrix integral representation of the combinatorial object, for example, the longest increasing subsequences in random permutations [3], the non-equilibrium stochastic model, so-called TASEP [26], and so on (see also [46]). Their remarkable connection to the Tracy-Widom distribution [56] can be understood from the viewpoint of the random matrix theory through the Robinson-Schensted-Knuth (RSK) correspondence (see e.g. [51]).

In this article we review such a universal relation between combinatorics and the matrix model, and discuss its relation to the gauge theory. The gauge theory consequence can be naturally extacted from such a matrix model description. Actually the spectral curve of the matrix model can be interpreted as Seiberg-Witten curve for $\mathcal{N} = 2$ supersymmetric gauge theory. This identification suggests some aspects of the gauge theory are also described by the significant universality of the matrix model.

This article is organized as follows. In section 2 we introduce statistical models defined in a combinaorial manner. These models are based on the Plancherel measure on a combinatorial object, and its origin from the gauge theory perspective is also discussed. In section 3 it is shown that the matrix model is derived from the combinatorial model by considering its asymptotic limit. There are various matrix integral representations, corresponding to some deformations of the combinatorial model. In section 4 we investigate the large matrix size limit of the matrix model. It is pointed out that the algebraic curve is quite useful to study one-point function. Its relation to Seiberg-Witten theory is also discussed. Section 5 is devoted to conclusion.

2. Combinatorial partition function

In this section we introduce several kinds of combinatorial models. Their partition functions are defined as summation over partitions with a certain weight function, which is called *Plancherel measure*. It is also shown that such a combinatorial partition function is obtained by performing the path integral for supersymmetric gauge theories.

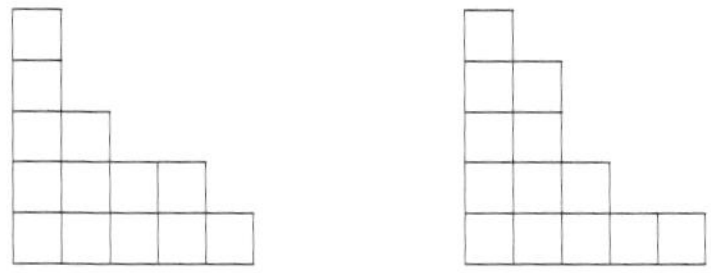

Figure 1. Graphical representation of a partition $\lambda = (5,4,3,1,1)$ and its transposed partition $\check{\lambda} = (5,3,2,2,1)$ by the associated Young diagrams. There are 5 non-zero entries in both of them, $\ell(\lambda) = \check{\lambda}_1 = 5$ and $\ell(\check{\lambda}) = \lambda_1 = 5$.

2.1. Random partition model

Let us first recall a partition of a positive integer n: it is a way of writing n as a sum of positive integers

$$\lambda = (\lambda_1, \lambda_2, \cdots, \lambda_{\ell(\lambda)}) \tag{1}$$

satisfying the following conditions,

$$n = \sum_{i=1}^{\ell(\lambda)} \lambda_i \equiv |\lambda|, \qquad \lambda_1 \geq \lambda_2 \geq \cdots \lambda_{\ell(\lambda)} > 0 \tag{2}$$

Here $\ell(\lambda)$ is the number of non-zero entries in λ. Now it is convenient to define $\lambda_i = 0$ for $i > \ell(\lambda)$. Fig. 2 shows *Young diagram,* which graphically describes a partition $\lambda = (5,4,2,1,1)$ with $\ell(\lambda) = 5$.

It is known that the partition is quite usefull for representation theory. We can obtain an irreducible representation of symmetric group $\mathfrak{S}_n$, which is in one-to-one correspondence with a partition λ with $|\lambda| = n$. For such a finite group, one can define a natural measure, which is called *Plancherel measure,*

$$\mu_n(\lambda) = \frac{(\dim \lambda)^2}{n!} \tag{3}$$

This measure is normalized as

$$\sum_{\lambda \text{ s.t. } |\lambda|=n} \mu_n(\lambda) = 1 \tag{4}$$

It is also interpreted as Fourier transform of Haar measure on the group. This measure has another useful representation, which is described in a combinatorial way,

$$\mu_n(\lambda) = n! \prod_{(i,j)\in\lambda} \frac{1}{h(i,j)^2} \tag{5}$$

This $h(i,j)$ is called *hook length,* which is defined with *arm length* and *leg length,*

$$\begin{aligned} h(i,j) &= a(i,j) + l(i,j) + 1, \\ a(i,j) &= \lambda_i - j, \\ l(i,j) &= \check{\lambda}_j - i \end{aligned} \tag{6}$$

Here $\check{\lambda}$ stands for the transposed partition. Thus the height of a partition λ can be explicitly written as $\ell(\lambda) = \check{\lambda}_1$.

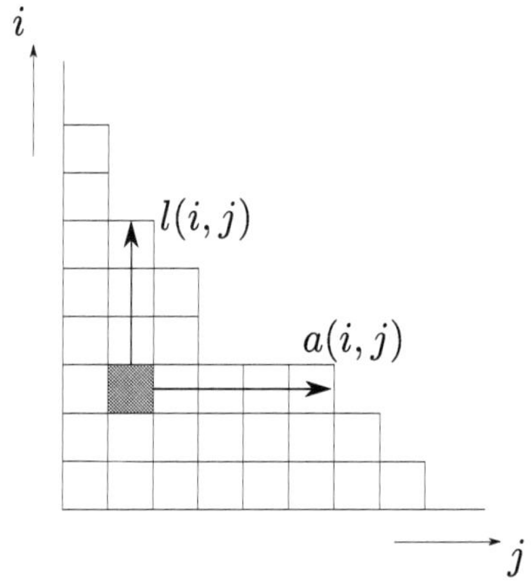

Figure 2. Combinatorics of Young diagram. Definitions of hook, arm and leg lengths are shown in (6). For the shaded box in this figure, $a(2,3) = 4$, $l(2,3) = 3$, and $h(2,3) = 8$.

With this combinatorial measure, we now introduce the following partition function,

$$Z_{\mathrm{U}(1)} = \sum_{\lambda} \left(\frac{\Lambda}{\hbar}\right)^{2|\lambda|} \prod_{(i,j)\in\lambda} \frac{1}{h(i,j)^2} \tag{7}$$

This model is often called *random partition model*. Here Λ is regarded as a parameter like a *chemical potential*, or a *fugacity*, and $\hbar$ stands for the size of boxes.

Note that a deformed model, which includes higher Casimir potentials, is also investigated in detail [19],

$$Z_{\text{higher}} = \sum_{\lambda} \prod_{(i,j)\in\lambda} \frac{1}{h(i,j)^2} \prod_{k=1} e^{-g_k C_k(\lambda)} \tag{8}$$

In this case the chemical potential term is absorbed by the linear potential term. There is an interesting interpretation of this deformation in terms of topological string, gauge theory and so on [18, 38].

In order to compute the U(1) partition function it is useful to rewrite it in a "canonical form" instead of the "grand canonical form" which is originally shown in (7),

$$Z_{\mathrm{U}(1)} = \sum_{n=0} \sum_{\lambda \text{ s.t. } |\lambda|=n} \left(\frac{\Lambda}{\hbar}\right)^{2n} \prod_{(i,j)\in\lambda} \frac{1}{h(i,j)^2} \tag{9}$$

Due to the normalization condition (4), this partition function can be computed as

$$Z_{\mathrm{U}(1)} = \exp\left(\frac{\Lambda}{\hbar}\right)^2 \tag{10}$$

Although this is explicitly solvable, its universal property and explicit connections to other models are not yet obvious. We will show, in section 3 and section 4, the matrix model description plays an important role in discussing such an interesting aspect of the combinatorial model.

Now let us remark one interesting observation, which is partially related to the following discussion. The combinatorial partition function (7) has another field theoretical representation using the free boson field [44]. We now consider the following coherent state,

$$|\psi\rangle = \exp\left(\frac{\Lambda}{\hbar}a_{-1}\right)|0\rangle \tag{11}$$

Here we introduce Heisenberg algebra, satisfying the commutation relation, $[a_n, a_m] = n\delta_{n+m,0}$, and the vacuum $|0\rangle$ annihilated by any positive modes, $a_n|0\rangle = 0$ for $n > 0$. Then it is easy to show the norm of this state gives rise to the partition function,

$$Z_{\mathrm{U}(1)} = \langle\psi|\psi\rangle \tag{12}$$

Similar kinds of observation is also performed for generalized combinatorial models introduced in section 2.2 [22, 44, 55].

Let us then introduce some generalizations of the $\mathrm{U}(1)$ model. First is what we call *β-deformed model* including an arbitrary parameter $\beta \in \mathbb{R}$,

$$Z^{(\beta)}_{\mathrm{U}(1)} = \sum_{\lambda}\left(\frac{\Lambda}{\hbar}\right)^{2|\lambda|}\prod_{(i,j)\in\lambda}\frac{1}{h_{\beta}(i,j)h^{\beta}(i,j)} \tag{13}$$

Here we involve the deformed hook lengths,

$$h_{\beta}(i,j) = a(i,j) + \beta l(i,j) + 1, \qquad h^{\beta}(i,j) = a(i,j) + \beta l(i,j) + \beta \tag{14}$$

This generalized model corresponds to Jack polynomial, which is a kind of symmetric polynomial obtained by introducing a free parameter to Schur polynomial [34]. This Jack polynomial is applied to several physical theories: quantum integrable model called Calogero-Sutherland model [10, 54], quantum Hall effect [4–6] and so on.

Second is a further generalized model involving two free parameters,

$$Z^{(q,t)}_{\mathrm{U}(1)} = \sum_{\lambda}\left(\frac{\Lambda}{\hbar}\right)^{2|\lambda|}\prod_{(i,j)\in\lambda}\frac{(1-q)(1-q^{-1})}{(1-q^{a(i,j)+1}t^{l(i,j)})(1-q^{-a(i,j)}t^{-l(i,j)-1})} \tag{15}$$

This is just a q-analog of the previous combinatorial model. One can see this is reduced to the β-deformed model (13) in the limit of $q \to 1$ with fixing $t = q^{\beta}$. This generalization is also related to the symmetric polynomial, which is called *Macdonald polynomial* [34]. This symmetric polynomial is used to study Ruijsenaars-Schneider model [45], and the stochastic process based on this function has been recently proposed [8].

Next is $\mathbb{Z}_r$-generalization of the model, which is defined as

$$Z_{\mathrm{orbifold,U}(1)} = \sum_{\lambda}\left(\frac{\Lambda}{\hbar}\right)^{2|\lambda|}\prod_{\Gamma\text{-inv}\subset\lambda}\frac{1}{h(i,j)^2} \tag{16}$$

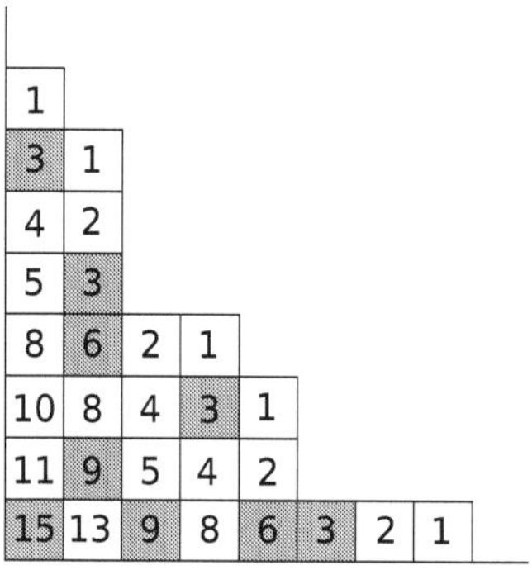

Figure 3. Γ-invariant sector for U(1) theory with $\lambda = (8, 5, 5, 4, 2, 2, 2, 1)$. Numbers in boxes stand for their hook lengths $h(i, j) = \lambda_i - j + \check{\lambda}_j - i + 1$. Shaded boxes are invariant under the action of $\Gamma = \mathbb{Z}_3$.

Here the product is taken only for the Γ-invariant sector as shown in Fig. 3,

$$h(i, j) = a(i, j) + l(i, j) + 1 \equiv 0 \pmod r \tag{17}$$

This restriction is considered in order to study the four dimensional supersymmetric gauge theory on orbifold $\mathbb{R}^4/\mathbb{Z}_r \cong \mathbb{C}^2/\mathbb{Z}_r$ [11, 20, 27], thus we call this *orbifold partition function*. This also corresponds to a certain symmetric polynomial [57] (see also [32]), which is related to the Calogero-Sutherland model involving spin degrees of freedom. We can further generalize this model (16) to the β- or the q-deformed $\mathbb{Z}_r$-orbifold model, and the generic toric orbifold model [28].

Let us comment on a relation between the orbifold partition function and the q-deformed model. Taking the limit of $q \to 1$, the latter is reduced to the U(1) model because the q-integer is just replaced by the usual integer in such a limit,

$$[x]_q \equiv \frac{1 - q^{-x}}{1 - q^{-1}} \longrightarrow x \tag{18}$$

This can be easily shown by l'Hopital's rule and so on. On the other hand, parametrizing $q \to \omega_r q$ with $\omega_r = \exp(2\pi i/r)$ being the primitive r-th root of unity, we have

$$\frac{1 - (\omega_r q)^{-x}}{1 - (\omega_r q)^{-1}} \stackrel{q \to 1}{\longrightarrow} \begin{cases} x & (x \equiv 0, \bmod r) \\ 1 & (x \not\equiv 0, \bmod r) \end{cases} \tag{19}$$

Therefore the orbifold partition function (16) is derived from the q-deformed one (15) by taking this root of unity limit. This prescription is useful to study its asymptotic behavior.

2.2. Gauge theory partition function

The path integral in quantum field theory involves some kinds of divergence, which are due to infinite degrees of freedom in the theory. On the other hand, we can exactly perform the path integral for several highly supersymmetric theories. We now show that the gauge theory partition function can be described in a combinatorial way, and yields some extended versions of the model we have introduced in section 2.1.

The main part of the gauge theory path integral is just evaluation of the moduli space volume for a topological excitation, for example, a vortex in two dimensional theory and an instanton in four dimensional theory. Here we concentrate on the four dimentional case. See [13, 21, 50] for the two dimensional vortex partition function. The most usuful method to deal with the instanton is ADHM construction [2]. According to this, the instanton moduli space for k-instanton in $\mathrm{SU}(n)$ gauge theory on $\mathbb{R}^4$, is written as a kind of hyper-Kähler quotient,

$$\mathcal{M}_{n,k} = \{(B_1, B_2, I, J) | \mu_{\mathbb{R}} = 0, \mu_{\mathbb{C}} = 0\} / \mathrm{U}(k) \tag{20}$$

$$B_{1,2} \in \mathrm{Hom}(\mathbb{C}^k, \mathbb{C}^k), \quad I \in \mathrm{Hom}(\mathbb{C}^n, \mathbb{C}^k), \quad J \in \mathrm{Hom}(\mathbb{C}^k, \mathbb{C}^n) \tag{21}$$

$$\mu_{\mathbb{R}} = [B_1, B_1^\dagger] + [B_2, B_2^\dagger] + II^\dagger - J^\dagger J, \tag{22}$$

$$\mu_{\mathbb{C}} = [B_1, B_2] + IJ \tag{23}$$

The $k \times k$ matrix condition $\mu_{\mathbb{R}} = \mu_{\mathbb{C}} = 0$, and parameters (B_1, B_2, I, J) satisfying this condition are called ADHM equation and ADHM data. Note that they are identified under the following $\mathrm{U}(k)$ transformation,

$$(B_1, B_2, I, J) \sim (gB_1g^{-1}, gB_2g^{-1}, gI, Jg^{-1}), \qquad g \in \mathrm{U}(k) \tag{24}$$

Thus all we have to do is to estimate the volume of this parameter space. However it is well known that there are some singularities in this moduli space, so that one has to regularize it in order to obtain a meaningful result. Its regularized volume had been derived by applying the localization formula to the moduli space integral [41], and it was then shown that the partition function correctly reproduces Seiberg-Witten theory [43].

We then consider the action of isometries on $\mathbb{C}^2 \cong \mathbb{R}^4$ for the ADHM data. If we assign $(z_1, z_2) \to (e^{i\epsilon_1} z_1, e^{i\epsilon_2} z_2)$ for the spatial coordinate of $\mathbb{C}^2$, and $\mathrm{U}(1)^{n-1}$ rotation coming from the gauge symmetry $\mathrm{SU}(n)$, ADHM data transform as

$$(B_1, B_2, I, J) \quad \longrightarrow \quad \left(T_1 B_1, T_2 B_2, I T_a^{-1}, T_1 T_2 T_a J\right) \tag{25}$$

where we define the torus actions as $T_a = \mathrm{diag}(e^{ia_1}, \cdots, e^{ia_n}) \in \mathrm{U}(1)^{n-1}$, $T_\alpha = e^{i\epsilon_\alpha} \in \mathrm{U}(1)^2$. Note that these toric actions are based on the maximal torus of the gauge theory symmetry, $\mathrm{U}(1)^2 \times \mathrm{U}(1)^{n-1} \subset \mathrm{SO}(4) \times \mathrm{SU}(n)$. We have to consider the fixed point of these isometries up to gauge transformation $g \in \mathrm{U}(k)$ to perform the localization formula.

The localization formula in the instanton moduli space is based on the vector field ξ^*, which is associated with $\xi \in \mathrm{U}(1)^2 \times \mathrm{U}(1)^{n-1}$. It generates the one-parameter flow $e^{t\xi}$ on the moduli space $\mathcal{M}$, corresponding to the isometries. The vector field is represented by the element of the maximal torus of the gauge theory symmetry under the Ω-background deformation. The gauge theory action is invariant under the deformed BRST transformation, whose generator satisfies $\xi^* = \{Q^*, Q^*\}/2$. Thus this generator can be interpreted as the equivariant derivative $d_\xi = d + i_{\xi^*}$ where i_{ξ^*} stands for the contraction with the vector field ξ^*. The localization formula is given by

$$\int_{\mathcal{M}} \alpha(\xi) = (-2\pi)^{n/2} \sum_{x_0} \frac{\alpha_0(\xi)(x_0)}{\det^{1/2} \mathcal{L}_{x_0}} \tag{26}$$

where $\alpha(\xi)$ is an equivariant form, which is related to the gauge theory action. $\alpha_0(\xi)$ is zero degree part and $\mathcal{L}_{x_0} : T_{x_0}\mathcal{M} \to T_{x_0}\mathcal{M}$ is the map generated by the vector field ξ^* at the fixed points x_0. These fixed points are defined as $\xi^*(x_0) = 0$ up to $\mathrm{U}(k)$ transformation of the instanton moduli space.

Let us then study the fixed point in the moduli space. The fixed point condition for them are obtained from the infinitesimal version of (24) and (25) as

$$(\phi_i - \phi_j + \epsilon_\alpha)B_{\alpha,ij} = 0, \qquad (\phi_i - a_l)I_{il} = 0, \qquad (-\phi_i + a_l + \epsilon)J_{li} = 0 \tag{27}$$

where the element of $\mathrm{U}(k)$ gauge transformation is diagonalized as $e^{i\phi} = \mathrm{diag}(e^{i\phi_1}, \cdots, e^{i\phi_k}) \in \mathrm{U}(k)$ with $\epsilon = \epsilon_1 + \epsilon_2$. We can show that an eigenvalue of ϕ turns out to be

$$a_l + (j-1)\epsilon_1 + (i-1)\epsilon_2 \tag{28}$$

and the corresponding eigenvector is given by

$$B_1^{j-1} B_2^{i-1} I_l \tag{29}$$

Since ϕ is a finite dimensional matrix, we can obtain k_l independent vectors from (29) with $k_1 + \cdots + k_n = k$. This means that the solution of this condition can be characterized by n-tuple Young diagrams, or partitions $\vec{\lambda} = (\lambda^{(1)}, \cdots, \lambda^{(n)})$ [42]. Thus the characters of the vector spaces are yielding

$$V = \sum_{l=1}^{n} \sum_{(i,j)\in\lambda^{(l)}} T_{a_l} T_1^{-j+1} T_2^{-i+1}, \qquad W = \sum_{l=1}^{n} T_{a_l} \tag{30}$$

and that of the tangent space at the fixed point under the isometries can be represented in terms of the n-tuple partition as

$$\begin{aligned} \chi_{\vec{\lambda}} &= -V^* V(1-T_1)(1-T_2) + W^* V + V^* W T_1 T_2 \\ &= \sum_{l,m}^{n} \sum_{(i,j)\in\lambda^{(l)}} \left(T_{a_{ml}} T_1^{-\check{\lambda}_j^{(l)}+i} T_2^{\lambda_i^{(m)}-j+1} + T_{a_{lm}} T_1^{\check{\lambda}_j^{(l)}-i+1} T_2^{-\lambda_i^{(m)}+j} \right) \end{aligned} \tag{31}$$

Here $\check{\lambda}$ is a conjugated partition. Therefore the instanton partition function is obtained by reading the weight function from the character [43, 44],

$$Z_{\mathrm{SU}(n)} = \sum_{\vec{\lambda}} \Lambda^{2n|\vec{\lambda}|} Z_{\vec{\lambda}} \tag{32}$$

$$Z_{\vec{\lambda}} = \prod_{l,m}^{n} \prod_{(i,j)\in\lambda^{(l)}} \frac{1}{a_{ml} + \epsilon_2(\lambda_i^{(m)} - j + 1) - \epsilon_1(\check{\lambda}_j^{(l)} - i)} \frac{1}{a_{lm} - \epsilon_2(\lambda_i^{(m)} - j) + \epsilon_1(\check{\lambda}_j^{(l)} - i + 1)} \tag{33}$$

This is regarded as a generalized model of (7) or (13). Furthermore by lifting it to the five dimensional theory on $\mathbb{R}^4 \times S^1$, one can obtain a generalized version of the q-deformed partition function (15). Actually it is easy to see these $\mathrm{SU}(n)$ models are reduced to the $\mathrm{U}(1)$ models in the case of $n = 1$. Note, if we take into account other matter contributions in addition to the vector multiplet, this partition function involves the associated combinatorial factors. We can extract various properties of the gauge theory from these partition functions, especially its asymptotic behavior.

3. Matrix model description

In this section we discuss the matrix model description of the combinatorial partition function. The matrix integral representation can be treated in a standard manner, which is developed in the random matrix theory [40].

3.1. Matrix integral

Let us consider the following $N \times N$ matrix integral,

$$Z_{\text{matrix}} = \int \mathcal{D}X \, e^{-\frac{1}{\hbar}\operatorname{Tr} V(X)} \tag{34}$$

Here X is an hermitian matrix, and $\mathcal{D}X$ is the associated matrix measure. This matrix can be diagonalized by a unitary transformation, $gXg^{-1} = \operatorname{diag}(x_1, \cdots, x_N)$ with $g \in \mathrm{U}(N)$, and the integrand is invariant under this transformation, $\operatorname{Tr} V(X) = \operatorname{Tr} V(gXg^{-1}) = \sum_{i=1}^N V(x_i)$. On the other hand, we have to take care of the matrix measure in (34): the non-trivial Jacobian is arising from the matrix diagonalization (see, e.g. [40]),

$$\mathcal{D}X = \mathcal{D}x \, \mathcal{D}U \, \Delta(x)^2 \tag{35}$$

The Jacobian part is called *Vandermonde determinant*, which is written as

$$\Delta(x) = \prod_{i<j}^{N} (x_i - x_j) \tag{36}$$

and $\mathcal{D}U$ is the Haar measure, which is invariant under unitary transformation, $\mathcal{D}(gU) = \mathcal{D}U$. The diagonal part is simply given by $\mathcal{D}x \equiv \prod_{i=1}^N dx_i$. Therefore, by integrating out the off-diagonal part, the matrix integral (34) is reduced to the integral over the matrix eigenvalues,

$$Z_{\text{matrix}} = \int \mathcal{D}x \, \Delta(x)^2 \, e^{-\frac{1}{\hbar}\sum_{i=1}^N V(x_i)} \tag{37}$$

This expression is up to a constant factor, associated with the volume of the unitary group, $\operatorname{vol}(\mathrm{U}(N))$, coming from the off-diagonal integral.

When we consider a real symmetric or a quaternionic self-dual matrix, it can be diagonalized by orthogonal/symplectic transformation. In these cases, the Jacobian part is slightly modified,

$$Z_{\text{matrix}} = \int \mathcal{D}x \, \Delta(x)^{2\beta} \, e^{-\frac{1}{\hbar}\sum_{i=1}^N V(x_i)} \tag{38}$$

The power of the Vandermonde determinant is given by $\beta = \frac{1}{2}, 1, 2$ for symmetric, hermitian and self-dual, respcecively.[1] They correspond to orthogonal, unitary, symplectic ensembles in random matrix theory, and the model with a generic $\beta \in \mathbb{R}$ is called *β-ensemble matrix model*.

[1] This notation is different from the standard one: $2\beta \to \beta = 1, 2, 4$ for symmetric, hermitian and self-dual matrices.

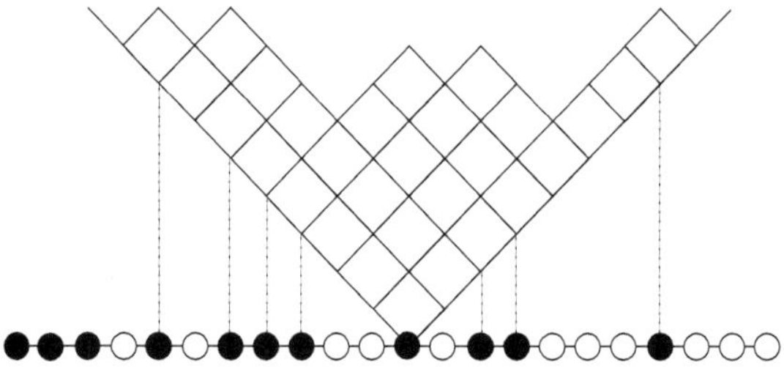

Figure 4. Shape of Young diagram can be represented by introducing one-dimensional exclusive particles. Positions of particles would be interpreted as eigenvalues of the matrix.

3.2. $\mathrm{U}(1)$ partition function

We would like to show an essential connection between the combinatorial partition function and the matrix model. By considering the thermodynamical limit of the partition function, it can be represented as a matrix integral discussed above.

Let us start with the most fundamental partition function (7). The main part of its partition function is the product all over the boxes in the partition λ. After some calculations, we can show this combinatorial factor is rewritten as

$$\prod_{(i,j)\in\lambda} \frac{1}{h(i,j)} = \prod_{i<j}^{N} (\lambda_i - \lambda_j + j - i) \prod_{i=1}^{N} \frac{1}{\Gamma(\lambda_i + N - i + 1)} \tag{39}$$

where N is an arbitrary integer satisfying $N > \ell(\lambda)$. This can be also represented in an infinite product form,

$$\prod_{(i,j)\in\lambda} \frac{1}{h(i,j)} = \prod_{i<j}^{\infty} \frac{\lambda_i - \lambda_j + j - i}{j - i} \tag{40}$$

These expressions correspond to an embedding of the finite dimensional symmetric group $\mathfrak{S}_N$ into the infinite dimensional one $\mathfrak{S}_\infty$.

By introducing a new set of variables $\xi_i = \lambda_i + N - i + 1$, we have another representation of the partition function,

$$Z_{\mathrm{U}(1)} = \sum_{\lambda} \left(\frac{\Lambda}{\hbar}\right)^{2\sum_{i=1}^{N} \xi_i - N(N+1)} \prod_{i<j}^{N} (\xi_i - \xi_j)^2 \prod_{i=1}^{N} \frac{1}{\Gamma(\xi_i)^2} \tag{41}$$

These new variables satisfy $\xi_i > \xi_2 > \cdots > \xi_{\ell(\lambda)}$ while the original ones satisfy $\lambda_1 \geq \lambda_2 \geq \cdots \geq \lambda_{\ell(\lambda)}$. This means $\{\xi_i\}$ and $\{\lambda_i\}$ are interpreted as fermionic and bosonic degrees of freedom. Fig. 4 shows the correspondence between the bosinic and fermionic variables. The bosonic excitation is regarded as density fluctuation of the fermionic particles around the Fermi energy. This is just the bosonization method, which is often used to study quantum one-dimensional systems (For example, see [24]). Especially we concentrate only on either of the Fermi points. Thus it yields the chiral conformal field theory.

We would like to show that the matrix integral form is obtained from the expression (41). First we rewrite the summation over partitions as

$$\sum_{\lambda} = \sum_{\lambda_1 \geq \cdots \geq \lambda_N} = \sum_{\xi_1 > \cdots > \xi_N} = \frac{1}{N!} \sum_{\xi_1, \cdots, \xi_N} \tag{42}$$

Then, introducing another variable defined as $x_i = \hbar \xi_i$, it can be regarded as a continuous variable in the large N limit,

$$N \longrightarrow \infty, \qquad \hbar \longrightarrow 0, \qquad \hbar N = \mathcal{O}(1) \tag{43}$$

This is called 't Hooft limit. The measure for this variable is given by

$$dx_i \approx \hbar \sim \frac{1}{N} \tag{44}$$

Therefore the partition function (41) is rewritten as the following matrix integral,

$$Z_{\mathrm{U}(1)} \approx \int \mathcal{D}x \, \Delta(x)^2 \, e^{-\frac{1}{\hbar} \sum_{i=1}^{N} V(x_i)} \tag{45}$$

Here the matrix potential is derived from the asymptotic behavior of the Γ-function,

$$\hbar \log \Gamma(x/\hbar) \longrightarrow x \log x - x, \qquad \hbar \longrightarrow 0 \tag{46}$$

Since this variable can take a negative value, the potential term should be simply extended to the region of $x < 0$. Thus, taking into account the fugacity parameter Λ, the matrix potential is given by

$$V(x) = 2 \left[x \log \left| \frac{x}{\Lambda} \right| - x \right] \tag{47}$$

This is the simplest version of the $\mathbf{CP}^1$ matrix model [18]. If we start with the partition function including the higher Casimir operators (8), the associated integral expression just yields the $\mathbf{CP}^1$ matrix model.

Let us comment on other possibilities to obtain the matrix model. It is shown that the matrix integral form can be derived without taking the large N limit [19]. Anyway one can see that it is reduced to the model we discussed above in the large N limit. There is another kind of the matrix model derived from the combinatorial partition function by *poissonizing* the probability measure. In this case, only the linear potential is arising in the matrix potential term. Such a matrix model is called Bessel-type matrix model, where its short range fluctuation is described by the Bessel kernel.

Next we shall derive the matrix model corresponding to the β-deformed $\mathrm{U}(1)$ model (13). The combinatorial part of the partition function is similarly given by

$$\begin{aligned} \prod_{(i,j) \in \lambda} \frac{1}{h_\beta(i,j) h^\beta(i,j)} &= \Gamma(\beta)^N \prod_{i<j}^{N} \frac{\Gamma(\lambda_i - \lambda_j + \beta(j-i) + \beta)}{\Gamma(\lambda_i - \lambda_j + \beta(j-i))} \frac{\Gamma(\lambda_i - \lambda_j + \beta(j-i) + 1)}{\Gamma(\lambda_i - \lambda_j + \beta(j-i) + 1 - \beta)} \\ &\times \prod_{i=1}^{N} \frac{1}{\Gamma(\lambda_i + \beta(N-i) + \beta)} \frac{1}{\Gamma(\lambda_i + \beta(N-i) + 1)} \end{aligned} \tag{48}$$

In this case we shall introduce the following variables, $\xi_i^{(\beta)} = \lambda_i + \beta(N-i) + 1$ or $\xi_i^{(\beta)} = \lambda_i + \beta(N-i) + \beta$, satisfying $\xi_i^{(\beta)} - \xi_{i+1}^{(\beta)} \geq \beta$. This means the parameter β characterizes how they are exclusive. They satisfy the generalized fractional exclusive statistics for $\beta \neq 1$ [25] (see also [32]). They are reduced to fermions and bosons for $\beta = 1$ and $\beta = 0$, respectively. Then, rescaling the variables, $x_i = \hbar \xi_i^{(\beta)}$, the combinatorial part (48) in the 't Hooft limit yields

$$\prod_{(i,j)\in\lambda} \frac{1}{h_\beta(i,j) h^\beta(i,j)} \longrightarrow \Delta(x)^{2\beta}\, e^{-\frac{1}{\hbar}\sum_{i=1}^N V(x_i)} \tag{49}$$

Here we use $\Gamma(\alpha+\beta)/\Gamma(\alpha) \sim \alpha^\beta$ with $\alpha \to \infty$. The matrix potential obtained here is the same as (47). Therefore the matrix model associated with the β-deformed partition function is given by

$$Z_{\mathrm{U}(1)}^{(\beta)} \approx \int \mathcal{D}x\, \Delta(x)^{2\beta}\, e^{-\frac{1}{\hbar}\sum_{i=1}^N V(x_i)} \tag{50}$$

This is just the β-ensemble matrix model shown in (38).

We can consider the matrix model description of the (q,t)-deformed partition function. In this case the combinatorial part of (15) is written as

$$\prod_{(i,j)\in\lambda} \frac{1-q}{1-q^{a(i,j)+1} t^{l(i,j)}} = (1-q)^{|\lambda|} \prod_{i<j}^N \frac{(q^{\lambda_i-\lambda_j+1} t^{j-i-1};q)_\infty}{(q^{\lambda_i-\lambda_j+1} t^{j-i};q)_\infty} \prod_{i=1}^N \frac{(q^{\lambda_i+1} t^{N-i};q)_\infty}{(q;q)_\infty} \tag{51}$$

$$\prod_{(i,j)\in\lambda} \frac{1-q^{-1}}{1-q^{-a(i,j)} t^{-l(i,j)-1}} = (1-q^{-1})^{|\lambda|} \prod_{i<j}^N \frac{(q^{-\lambda_i+\lambda_j+1} t^{-j+i-1};q)_\infty}{(q^{-\lambda_i+\lambda_j+1} t^{-j+i};q)_\infty} \prod_{i=1}^N \frac{(qt^{-1};q)_\infty}{(q^{-\lambda_i+1} t^{-N+i-1};q)_\infty} \tag{52}$$

Here $(x;q)_n = \prod_{m=0}^{n-1}(1-xq^m)$ is the q-Pochhammer symbol. When we parametrize $q = e^{-\hbar R}$ and $t = q^\beta$, a set of the variables $\{\xi_i^{(\beta)}\}$ plays an important role in considering the large N limit as well as the β-deformed model. Thus, rescaling these as $x_i = \hbar \xi_i^{(\beta)}$ and taking the 't Hooft limit, we obtain the integral expression of the q-deformed partition function,

$$Z_{\mathrm{U}(1)}^{(q,t)} \approx \int \mathcal{D}x\, (\Delta_R(x))^{2\beta}\, e^{-\frac{1}{\hbar}\sum_{i=1}^N V_R(x_i)} \tag{53}$$

The matrix measure and potential are given by

$$\Delta_R(x) = \prod_{i<j}^N \frac{2}{R} \sinh \frac{R}{2}(x_i - x_j) \tag{54}$$

$$V_R(x) = -\frac{1}{R}\left[\mathrm{Li}_2\left(e^{Rx}\right) - \mathrm{Li}_2\left(e^{-Rx}\right)\right] \tag{55}$$

We will discuss how to obtain these expressions below. We can see they are reduced to the standard ones in the limit of $R \to 0$,

$$\Delta_R(x) \longrightarrow \Delta(x), \qquad V_R(x) \longrightarrow V(x) \tag{56}$$

Note that this hyperbolic-type matrix measure is also investigated in the Chern-Simons matrix model [35], which is extensively involved with the recent progress on the three dimensional supersymmetric gauge theory via the localization method [36].

Let us comment on useful formulas to derive the integral expression (53). The measure part is relevant to the asymptotic form of the following function,

$$\frac{(x;q)_\infty}{(tx;q)_\infty} \longrightarrow \left.\frac{(x;q)_\infty}{(tx;q)_\infty}\right|_{q\to 1} = (1-x)^\beta, \qquad x \longrightarrow \infty \tag{57}$$

This essentially corresponds to the $q \to 1$ limit of the q-Vandermonde determinant[2],

$$\Delta_{q,t}^2(x) = \prod_{i\neq j}^{N} \frac{(x_i/x_j;q)_\infty}{(tx_i/x_j;q)_\infty} \tag{58}$$

Then, to investigate the matrix potential term, we now introduce the quantum dilogarithm function,

$$g(x;q) = \prod_{n=1}^{\infty} \left(1 - \frac{1}{x} q^n\right) \tag{59}$$

Its asymptotic expansion is given by (see, e.g. [19])

$$\log g(x; q = e^{-\hbar R}) = -\frac{1}{\hbar R} \sum_{m=0}^{\infty} \mathrm{Li}_{2-m}\left(x^{-1}\right) \frac{B_m}{m!} (\hbar R)^m \tag{60}$$

where B_m is the m-th Bernouilli number, and $\mathrm{Li}_m(x) = \sum_{k=1}^{\infty} x^k/k^m$ is the polylogarithm function. The potential term is coming from the leading term of this expression.

3.3. SU(n) partition function

Generalizing the result shown in section 3.2, we deal with the combinatorial partition function for SU(n) gauge theory (32). Its matrix model description is evolved in [30].

The combinatorial factor of the SU(n) partition function (33) can be represented as

$$Z_{\vec{\lambda}} = \frac{1}{\epsilon_2^{2n|\vec{\lambda}|}} \prod_{(l,i)\neq(m,j)} \frac{\Gamma(\lambda_i^{(l)} - \lambda_j^{(m)} + \beta(j-i) + b_{lm} + \beta)}{\Gamma(\lambda_i^{(l)} - \lambda_j^{(m)} + \beta(j-i) + b_{lm})} \frac{\Gamma(\beta(j-i) + b_{lk})}{\Gamma(\beta(j-i) + b_{lk} + \beta)} \tag{61}$$

where we define parameters as $\beta = -\epsilon_1/\epsilon_2$, $b_{lm} = a_{lm}/\epsilon_2$. This is an infinite product expression of the partition function. Anyway in this case one can see it is useful to introduce n kinds of fermionic variables, corresponding to the n-tupe partition,

$$\xi_i^{(l)} = \lambda_i^{(l)} + \beta(N-i) + 1 + b_l \tag{62}$$

[2] This expression is up to logarithmic term, which can be regarded as the zero mode contribution of the free boson field. See [28, 29] for details.

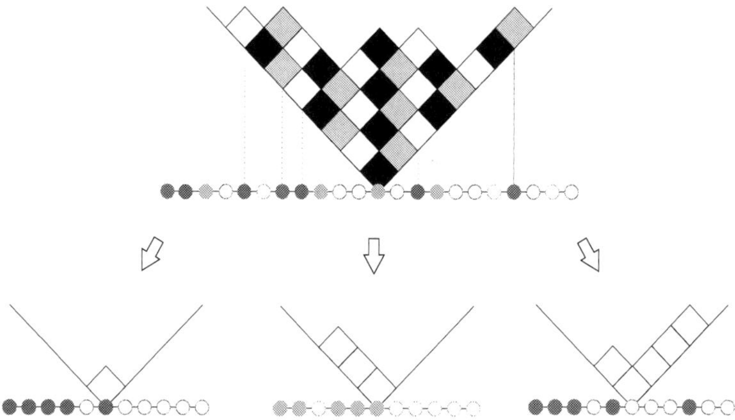

Figure 5. The decomposition of the partition for $\mathbb{Z}_{r=3}$. First suppose the standard correspondence between the one-dimensional particles and the original partition, and then rearrange them with respect to mod r.

Then, assuming $b_{lm} \gg 1$, let us introduce a set of variables,

$$(\zeta_1, \zeta_2, \cdots, \zeta_{nN}) = (\xi_1^{(n)}, \cdots, \xi_N^{(n)}, \xi_1^{(n-1)}, \cdots\cdots, \xi_N^{(2)}, \xi_1^{(1)}, \cdots, \xi_N^{(1)}) \tag{63}$$

satisfying $\zeta_1 > \zeta_2 > \cdots > \zeta_{nN}$. The combinatorial factor (61) is rewritten with these variables as

$$Z_{\vec{\lambda}} = \frac{1}{\epsilon_2^{2n|\vec{\lambda}|}} \prod_{i<j}^{nN} \frac{\Gamma(\zeta_i - \zeta_j + \beta)}{\Gamma(\zeta_i - \zeta_j)} \prod_{i=1}^{nN} \prod_{l=1}^{n} \frac{\Gamma(-\zeta_i + b_l + 1)}{\Gamma(\zeta_i - b_l - 1 + \beta)} \tag{64}$$

From this expression we can obtain the matrix model description for $\mathrm{SU}(n)$ gauge theory partition function, by rescaling $x_i = \hbar\zeta_i$ with reparametrizing $\hbar = \epsilon_2$,

$$Z_{\mathrm{SU}(n)} \approx \int \mathcal{D}x\, \Delta(x)^{2\beta} e^{-\frac{1}{\hbar}\sum_{i=1}^{nN} V_{\mathrm{SU}(n)}(x_i)} \tag{65}$$

In this case the matrix potential is given by

$$V_{\mathrm{SU}(n)}(x) = 2\sum_{l=1}^{n} \left[(x - a_l) \log \left| \frac{x - a_l}{\Lambda} \right| - (x - a_l) \right] \tag{66}$$

Note that this matrix model is regarded as the $\mathrm{U}(1)$ matrix model with external fields a_l. We will discuss how to extract the gauge theory consequences from this matrix model in section 4.

3.4. Orbifold partition function

The matrix model description for the random partition model is also possible for the orbifold theory. We would like to derive another kind of the matrix model from the combinatorial orbifold partition function (16). We now concentrate on the $\mathrm{U}(1)$ orbifold partition function for simplicity. See [27, 28] for details of the $\mathrm{SU}(n)$ theory.

To obtain the matrix integral representation of the combinatorial partition function, we have to find the associated one-dimensional particle description of the combinatorial factor. In this case, although the combinatorial weight itself is the same as the standard U(1) model, there is restriction on its product region. Thus it is useful to introduce another basis obtained by dividing the partition as follows,

$$\left\{ r\left(\lambda_i^{(u)} + N^{(u)} - i\right) + u \middle| i = 1, \cdots, N^{(u)}, u = 0, \cdots, r-1 \right\} = \{\lambda_i + N - i | i = 1, \cdots, N\} \tag{67}$$

Fig.5 shows the meaning of this procedure graphically. We now assume $N^{(u)} = N$ for all u. With these one-dimensional particles, we now utilize the relation between the orbifold partition function and the q-deformed model as discussed in section 2.1. Its calculation is quite straightforward, but a little bit complicated. See [27, 28] for details.

After some computations, we finally obtain the matrix model for the β-deformed orbifold partition function,

$$Z^{(\beta)}_{\text{orbifold,U(1)}} \approx \int \mathcal{D}\vec{x} \left(\Delta^{(\beta)}_{\text{orb}}(x)\right)^2 e^{-\frac{1}{\hbar}\sum_{u=0}^{r-1}\sum_{i=1}^{N} V(x_i^{(u)})} \tag{68}$$

In this case, we have a multi-matrix integral representation, since we introduce r kinds of partitions from the original partition. The matrix measure and the matrix potential are given as follows,

$$\mathcal{D}\vec{x} = \prod_{u=0}^{r-1}\prod_{i=1}^{N} dx_i^{(u)} \tag{69}$$

$$\left(\Delta^{(\beta)}_{\text{orb}}(x)\right)^2 = \prod_{u=0}^{r-1}\prod_{i<j}^{N}(x_i^{(u)} - x_j^{(u)})^{2(\beta-1)/r+2} \prod_{u<v}^{r-1}\prod_{i,j}^{N}(x_i^{(u)} - x_j^{(v)})^{2(\beta-1)/r} \tag{70}$$

$$V(x) = \frac{2}{r}\left[x \log\left|\frac{x}{\Lambda}\right| - x\right] \tag{71}$$

The matrix measure consists of two parts, interaction between eigenvalues from the same matrix and that between eigenvalues from different matrices. Note that in the case of $\beta = 1$, because the interaction part in the matrix measure beteen different matrices is vanishing, this multi-matrix model is simply reduced to the one-matrix model.

4. Large N analysis

One of the most important aspects of the matrix model is universality arising in the large N limit. The universality class described by the matrix model covers huge kinds of the statistical models, in particular its characteristic fluctuation rather than the eigenvalue density function. In the large N limit, which is regarded as a justification to apply a kind of the mean field approximation, anslysis of the matrix model is extremely reduced to the saddle point equation and a simple fluctuation around it.

4.1. Saddle point equation and spectral curve

Let us first define the prepotential, which is also interpreted as the effective action for the eigenvalues, from the matrix integral representation

(37),

$$-\frac{1}{\hbar^2}\mathcal{F}(\{x_i\}) = -\frac{1}{\hbar}\sum_{i=1}^{N} V(x_i) + 2\sum_{i<j}^{N}\log(x_i - x_j) \tag{72}$$

This is essentially the genus zero part of the prepotential. In the large N limit, in particular 't Hooft limit (43) with $N\hbar \equiv t$, we shall investigate the saddle point equation for the matrix integral. We can obtain the condition for criticality by differentiating the prepotential,

$$V'(x_i) = 2\hbar \sum_{j(\neq i)}^{N} \frac{1}{x_j - x_i}, \qquad \text{for all } i \tag{73}$$

This is also given by the extremal condition of the effective potential defined as

$$V_{\text{eff}}(x_i) = V(x_i) - 2\hbar \sum_{j(\neq i)}^{N} \log(x_i - x_j) \tag{74}$$

This potential involves a logarithmic Coulomb repulsion between eigenvalues. If the 't Hooft coupling is small, the potential term dominates the Coulomb interaction and eigenvalues concentrate on extrema of the potential $V'(x) = 0$. On the other hand, as the coupling gets bigger, the eigenvalue distribution is extended.

To deal with such a situation, we now define the density of eigenvalues,

$$\rho(x) = \frac{1}{N}\sum_{i=1}^{N}\delta(x - x_i) \tag{75}$$

where x_i is the solution of the criticality condition (73). In the large N limit, it is natural to think this eigenvalue distribution is smeared, and becomes a continuous function. Furthermore, we assume the eigenvalues are distributed around the critical points of the potential $V(x)$ as linear segments. Thus we generically denote the l-th segment for $\rho(x)$ as $\mathcal{C}_l$, and the total number of eigenvalues N splits into n integers for these segments,

$$N = \sum_{l=1}^{n} N_l \tag{76}$$

where N_l is the number of eigenvalues in the interval $\mathcal{C}_l$. The density of eigenvalues $\rho(x)$ takes non-zero value only on the segment $\mathcal{C}_l$, and is normalized as

$$\int_{\mathcal{C}_l} dx\, \rho(x) = \frac{N_l}{N} \equiv \nu_l \tag{77}$$

where we call it *filling fraction*. According to these fractions, we can introduce the partial 't Hooft parameters, $t_l = N_l\hbar$. Note there are n 't Hooft couplings and filling fractions, but only $n-1$ fractions are independent since they have to satisfy $\sum_{l=1}^{n}\nu_l = 1$ while all the 't Hooft couplings are independent.

We then introduce the resolvent for this model as an auxiliary function, a kind of Green function. By taking the large N limit, it can be given by the integral representation,

$$\omega(x) = t \int dy \, \frac{\rho(y)}{x-y} \tag{78}$$

This means that the density of states is regarded as the Hilbert transformation of this resolvent function. Indeed the density of states is associated with the discontinuities of the resolvent,

$$\rho(x) = -\frac{1}{2\pi i t} \left(\omega(x+i\epsilon) - \omega(x-i\epsilon) \right) \tag{79}$$

Thus all we have to do is to determine the resolvent instead of the density of states with satisfying the asymptotic behavior,

$$\omega(x) \longrightarrow \frac{1}{x}, \qquad x \longrightarrow \infty \tag{80}$$

Writing down the prepotential with the density of states,

$$\mathcal{F}(\{x_i\}) = t \int dx \, \rho(x) V(x) - t^2 \mathrm{P} \int dx dy \, \rho(x)\rho(y) \log(x-y) \tag{81}$$

the criticality condition is given by

$$\frac{1}{2t} V'(x) = \mathrm{P} \int dy \, \frac{\rho(y)}{x-y} \tag{82}$$

Here P stands for the principal value. Thus this saddle point equation can be also written in the following convenient form to discuss its analytic property,

$$V'(x) = \omega(x+i\epsilon) + \omega(x-i\epsilon) \tag{83}$$

On the other hand, we have another convenient form to treat the saddle point equation, which is called *loop equation*, given by

$$y^2(x) - V'(x)^2 + R(x) = 0 \tag{84}$$

where we denote

$$y(x) = V'(x) - 2\omega(x) = -2\omega_{\text{sing}}(x) \tag{85}$$

$$R(x) = \frac{4t}{N} \sum_{i=1}^{N} \frac{V'(x) - V'(x_i)}{x - x_i} \tag{86}$$

It is obtained from the saddle point equation by multiplying $1/(x-x_i)$ and taking their summation and the large N limit. This representation (84) is more appropriate to reveal its geometric meaning. Indeed this algebraic curve is interpreted as the hyperelliptic curve which is given by resolving the singular form,

$$y^2(x) - V'(x)^2 = 0 \tag{87}$$

The genus of the Riemann surface is directly related to the number of cuts of the corresponding resolvent. The filling fraction, or the partial 't Hooft coupling, is simply given by the contour integral on the hyperelliptic curve

$$t_l = \frac{1}{2\pi i} \oint_{C_l} dx\, \omega_{\text{sing}}(x) = -\frac{1}{4\pi i} \oint_{C_l} dx\, y(x) \tag{88}$$

4.2. Relation to Seiberg-Witten theory

We now discuss the relation between Seiberg-Witten curve and the matrix model. In the first place, the matrix model captures the asymptotic behavior of the combinatorial representation of the partition function. The energy functional, which is derived from the asymptotics of the partition function [44], in terms of the profile function

$$\mathcal{E}_\Lambda(f) = \frac{1}{4} \mathrm{P} \int_{y<x} dxdy\, f''(x) f''(y) (x-y)^2 \left(\log\left(\frac{x-y}{\Lambda}\right) - \frac{3}{2} \right) \tag{89}$$

can be rewritten as

$$\mathrm{E}_\Lambda(\varrho) = -\mathrm{P} \int_{x \neq y} dxdy\, \frac{\varrho(x)\varrho(y)}{(x-y)^2} - 2 \int dx\, \varrho(x) \log \prod_{l=1}^{N} \left(\frac{x-a_l}{\Lambda} \right) \tag{90}$$

up to the perturbative contribution

$$\frac{1}{2} \sum_{l,m} (a_l - a_m)^2 \log\left(\frac{a_l - a_m}{\Lambda} \right) \tag{91}$$

by identifying

$$f(x) - \sum_{l=1}^{n} |x - a_l| = \varrho(x) \tag{92}$$

Then integrating (90) by parts, we have

$$\mathrm{E}_\Lambda(\varrho) = -\mathrm{P} \int_{x \neq y} dxdy\, \varrho'(x)\varrho'(y) \log(x-y) + 2 \int dx\, \varrho'(x) \sum_{l=1}^{n} \left[(x-a_l) \log\left(\frac{x-a_l}{\Lambda} \right) - (x - a_l) \right] \tag{93}$$

This is just the matrix model discussed in section 3.3 if we identify $\varrho'(x) = \rho(x)$. Therefore analysis of this matrix model is equivalent to that of [?]. But in this section we reconsider the result of the gauge theory from the viewpoint of the matrix model.

We can introduce a regular function on the complex plane, except at the infinity,

$$P_n(x) = \Lambda^n \left(e^{y/2} + e^{-y/2} \right) \equiv \Lambda^n \left(w + \frac{1}{w} \right) \tag{94}$$

It is because the saddle point equation (83) yields the following equation,

$$e^{y(x+i\epsilon)/2} + e^{-y(x+i\epsilon)/2} = e^{y(x-i\epsilon)/2} + e^{-y(x-i\epsilon)/2} \tag{95}$$

This entire function turns out to be a monic polynomial $P_n(x) = x^n + \cdots$, because it is an analytic function with the following asymptotic behavior,

$$\Lambda^n e^{y/2} = \Lambda^n e^{-\omega(x)} \prod_{l=1}^{n} \left(\frac{x - a_l}{\Lambda} \right) \longrightarrow x^n, \qquad x \longrightarrow \infty \tag{96}$$

Here w should be the smaller root with the boundary condition as

$$w \longrightarrow \frac{\Lambda^n}{x^n}, \qquad x \longrightarrow \infty \tag{97}$$

thus we now identify

$$w = e^{-y/2} \tag{98}$$

Therefore from the hyperelliptic curve (94) we can relate Seiberg-Witten curve to the spectral curve of the matrix model,

$$\begin{aligned} dS &= \frac{1}{2\pi i} x \frac{dw}{w} \\ &= -\frac{1}{2\pi i} \log w \, dx \\ &= \frac{1}{4\pi i} y(x) dz \end{aligned} \tag{99}$$

Note that it is shown in [37, 38] we have to take the vanishing fraction limit to obtain the Coulomb moduli from the matrix model contour integral. This is the essential difference between the profile function method and the matrix model description.

4.3. Eigenvalue distribution

We now demonstrate that the eigenvalue distribution function is indeed derived from the spectral curve of the matrix model. The spectral curve (94) in the case of $n = 1$ with setting $\Lambda = 1$ and $P_{n=1}(x) = x$ is written as

$$x = w + \frac{1}{w} \tag{100}$$

From this relation the singular part of the resolvent can be extracted as

$$\omega_{\text{sing}}(x) = \operatorname{arccosh}\left(\frac{x}{2}\right) \tag{101}$$

This has a branch cut only on $x \in [-2, 2]$, namely a one-cut solution. Thus the eigenvalue distribution function is witten as follows at least on $x \in [-2, 2]$,

$$\rho(x) = \frac{1}{\pi} \arccos\left(\frac{x}{2}\right) \tag{102}$$

Note that this function has a non-zero value at the left boundary of the cut, $\rho(-2) = 1$, while at the right boundary we have $\rho(2) = 0$. Equivalently we now choose the cut of arccos function in this way. This seems a little bit strange because the eigenvalue density has to vanish except for on the cut. On the other hand, recalling the meaning of the eigenvalues, i.e. positions of one-dimensional particles, as shown in Fig. 4, this situation is quite reasonable. The region below the Fermi level is filled of the particles, and thus the density has to be a non-zero constant in such a region. This is just a property of the Fermi distribution function. ($1/N$ correction could be interpreted as a finite temperature effect.) Therefore the total eigenvalue

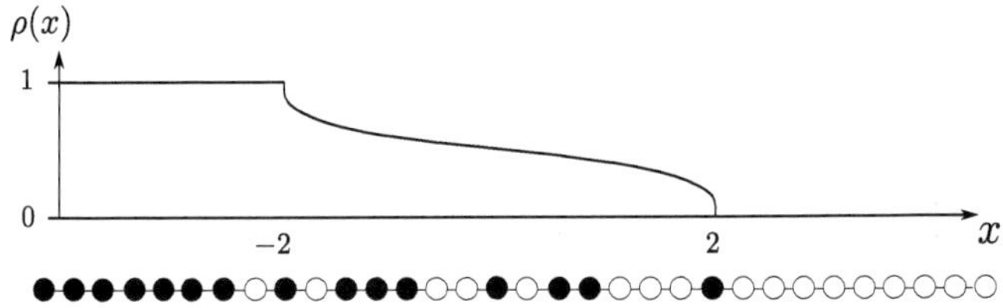

Figure 6. The eigenvalue distribution function for the U(1) model.

distribution function is given by

$$\rho(x) = \begin{cases} 1 & x < -2 \\ \frac{1}{\pi} \arccos\left(\frac{x}{2}\right) & |x| < 2 \\ 0 & x > 2 \end{cases} \tag{103}$$

Remark the eigenvalue density (103) is quite similar to the Wigner's semi-circle distribution function, especially its behavior around the edge,

$$\rho_{\text{circ}}(x) = \frac{1}{\pi}\sqrt{1 - \left(\frac{x}{2}\right)^2} \longrightarrow \frac{1}{\pi}\sqrt{2 - x}, \qquad x \longrightarrow 2 \tag{104}$$

The fluctuation at the spectral edge of the random matrix obeys Tracy-Widom distribution [56], thus it is natural that the edge fluctuation of the combinatorial model is also described by Tracy-Widom distribution. This remarkable fact was actually shown by [9]. Evolving such a similarity to the gaussian random matrix theory, the kernel of this model is also given by the following sine kernel,

$$K(x,y) = \frac{\sin \rho_0 \pi (x-y)}{\pi (x-y)} \tag{105}$$

where ρ_0 is the averaged density of eigenvalues. This means the U(1) combinatorial model belongs to the GUE random matrix universal class [40]. Then all the correlation functions can be written as a determinant of this kernel,

$$\rho(x_1, \cdots, x_k) = \det\left[K(x_i, x_j)\right]_{1 \le i,j, \le k} \tag{106}$$

Let us then remark a relation to the profile function of the Young diagram. It was shown that the shape of the Young diagram goes to the following form in the thermodynamical limit [33, 58, 59],

$$\Omega(x) = \begin{cases} \frac{2}{\pi}\left(x \arcsin\frac{x}{2} + \sqrt{4 - x^2}\right) & |x| < 2 \\ |x| & |x| > 2 \end{cases} \tag{107}$$

Rather than this profile function itself, the derivative of this function is more relevant to our study,

$$\Omega'(x) = \begin{cases} -1 & x < -2 \\ \frac{2}{\pi} \arcsin\left(\frac{x}{2}\right) & |x| < 2 \\ 1 & x > 2 \end{cases} \tag{108}$$

One can see the eigenvalue density (103) is directly related to this derivative function (108) as

$$\rho(x) = \frac{1 - \Omega'(x)}{2} \tag{109}$$

This relation is easily obtained from the correspondence between the Young diagram and the one-dimensional particle as shown in Fig. 4.

5. Conclusion

In this article we have investigated the combinatorial statistical model through its matrix model description. Starting from the $\mathrm{U}(1)$ model, which is motivated by representation theory, we have dealt with its β-deformation and q-deformation. We have shown that its non-Abelian generalization, including external field parameters, is obtained as the four dimensional supersymmetric gauge theory partition function. We have also referred to the orbifold partition function, and its relation to the q-deformed model through the root of unity limit.

We have then shown the matrix integral representation is derived from such a combinatorial partition function by considering its asymptotic behavior in the large N limit. Due to variety of the combinatorial model, we can obtain the β-ensemble matrix model, the hyperbolic matrix model, and those with external fields. Furthermore from the orbifold partition function the multi-matrix model is derived.

Based on the matrix model description, we have study the asymptotic behavior of the combinatorial models in the large N limit. In this limit we can extract various important properties of the matrix model by analysing the saddle point equation. Introducing the resolvent as an auxiliary function, we have obtained the algebraic curve for the matrix model, which is called the spectral curve. We have shown it can be interpreted as Seiberg-Witten curve, and then the eigenvalue distribution function is also obtained from this algebraic curve.

Let us comment on some possibilities of generalization and perspective. As discussed in this article we can obtain various interesting results from Macdonald polynomial by taking the corresponding limit. It is interesting to research its matrix model consequence from the exotic limit of Macdonald polynomial. For example, the $q \to 0$ limit of Macdonald polynomial, which is called Hall-Littlewood polynomial, is not investigated with respect to its connection with the matrix model. We also would like to study properties of the BC-type polynomial [31], which is associated with the corresponding root system. Recalling the meaning of the q-deformation in terms of the gauge theory, namely lifting up to the five dimensional theory $\mathbb{R}^4 \times S^1$ by taking into account all the Kaluza-Klein modes, it seems interesting to study the six dimensional theory on $\mathbb{R}^4 \times T^2$. In this case it is natural to obtain the elliptic generalization of the matrix model. It can not be interpreted as matrix integral representation any longer, however the large N analysis could be anyway performed in the standard manner. We would like to expect further develpopment beyond this work.

Author details

Taro Kimura
Mathematical Physics Laboratory, RIKEN Nishina Center, Japan

6. References

[1] Alday, L. F., Gaiotto, D. & Tachikawa, Y. [2010]. Liouville Correlation Functions from Four-dimensional Gauge Theories, *Lett.Math.Phys.* 91: 167–197.

[2] Atiyah, M. F., Hitchin, N. J., Drinfeld, V. G. & Manin, Y. I. [1978]. Construction of instantons, *Phys. Lett.* A65: 185–187.

[3] Baik, J., Deift, P. & Johansson, K. [1999]. On the Distribution of the Length of the Longest Increasing Subsequence of Random Permutations, *J. Amer. Math. Soc.* 12: 1119–1178.

[4] Bernevig, B. A. & Haldane, F. D. M. [2008a]. Generalized clustering conditions of Jack polynomials at negative Jack parameter α, *Phys. Rev.* B77: 184502.

[5] Bernevig, B. A. & Haldane, F. D. M. [2008b]. Model Fractional Quantum Hall States and Jack Polynomials, *Phys. Rev. Lett.* 100: 246802.

[6] Bernevig, B. A. & Haldane, F. D. M. [2008c]. Properties of Non-Abelian Fractional Quantum Hall States at Filling $\nu = k/r$, *Phys. Rev. Lett.* 101: 246806.

[7] Bonelli, G., Maruyoshi, K., Tanzini, A. & Yagi, F. [2011]. Generalized matrix models and AGT correspondence at all genera, *JHEP* 07: 055.

[8] Borodin, A. & Corwin, I. [2011]. Macdonald processes, arXiv:1111.4408 [math.PR].

[9] Borodin, A., Okounkov, A. & Olshanski, G. [2000]. On asymptotics of the plancherel measures for symmetric groups, *J. Amer. Math. Soc.* 13: 481–515.

[10] Calogero, F. [1969]. Ground state of one-dimensional N body system, *J. Math. Phys.* 10: 2197.

[11] Dijkgraaf, R. & Sułkowski, P. [2008]. Instantons on ALE spaces and orbifold partitions, *JHEP* 03: 013.

[12] Dijkgraaf, R. & Vafa, C. [2009]. Toda Theories, Matrix Models, Topological Strings, and $\mathcal{N} = 2$ Gauge Systems, arXiv:0909.2453 [hep-th].

[13] Dimofte, T., Gukov, S. & Hollands, L. [2010]. Vortex Counting and Lagrangian 3-manifolds, *Lett. Math. Phys.* 98: 225–287.

[14] Dotsenko, V. S. & Fateev, V. A. [1984]. Conformal algebra and multipoint correlation functions in 2D statistical models, *Nucl. Phys.* B240: 312–348.

[15] Dotsenko, V. S. & Fateev, V. A. [1985]. Four-point correlation functions and the operator algebra in 2D conformal invariant theories with central charge $c \leq 1$, *Nucl. Phys.* B251: 691–734.

[16] Eguchi, T. & Maruyoshi, K. [2010a]. Penner Type Matrix Model and Seiberg-Witten Theory, *JHEP* 02: 022.

[17] Eguchi, T. & Maruyoshi, K. [2010b]. Seiberg-Witten theory, matrix model and AGT relation, *JHEP* 07: 081.

[18] Eguchi, T. & Yang, S.-K. [1994]. The Topological CP^1 model and the large N matrix integral, *Mod. Phys. Lett.* A9: 2893–2902.

[19] Eynard, B. [2008]. All orders asymptotic expansion of large partitions, *J. Stat. Mech.* 07: P07023.

[20] Fucito, F., Morales, J. F. & Poghossian, R. [2004]. Multi instanton calculus on ALE spaces, *Nucl. Phys.* B703: 518–536.

[21] Fujimori, T., Kimura, T., Nitta, M. & Ohashi, K. [2012]. Vortex counting from field theory, arXiv:1204.1968 [hep-th].

[22] Gaiotto, D. [2009a]. Asymptotically free $\mathcal{N} = 2$ theories and irregular conformal blocks, arXiv:0908.0307 [hep-th].

[23] Gaiotto, D. [2009b]. $\mathcal{N} = 2$ dualities, arXiv:0904.2715 [hep-th].

[24] Giamarchi, T. [2003]. *Quantum Physics in One Dimension*, Oxford University Press.

[25] Haldane, F. D. M. [1991]. "Fractional statistics" in arbitrary dimensions: A generalization of the Pauli principle, *Phys. Rev. Lett.* 67: 937–940.
[26] Johansson, K. [2000]. Shape Fluctuations and Random Matrices, *Commun. Math. Phys.* 209: 437–476.
[27] Kimura, T. [2011]. Matrix model from $\mathcal{N} = 2$ orbifold partition function, *JHEP* 09: 015.
[28] Kimura, T. [2012a]. β-ensembles for toric orbifold partition function, *Prog. Theor. Phys.* 127: 271–285.
[29] Kimura, T. [2012b]. Spinless basis for spin-singlet FQH states, arXiv:1201.1903 [cond-mat.mes-hall].
[30] Klemm, A. & Sułkowski, P. [2009]. Seiberg-Witten theory and matrix models, *Nucl. Phys.* B819: 400–430.
[31] Koornwinder, T. [1992]. Askey-Wilson polynomials for root system of type *BC*, *Contemp. Math.* 138: 189–204.
[32] Kuramoto, Y. & Kato, Y. [2009]. *Dynamics of One-Dimensional Quantum Systems: Inverse-Square Interaction Models*, Cambridge University Press.
[33] Logan, B. & Shepp, L. [1977]. A variational problem for random Young tableaux, *Adv. Math.* 26: 206 – 222.
[34] Macdonald, I. G. [1997]. *Symmetric Functions and Hall Polynomials*, 2nd edn, Oxford University Press.
[35] Mariño, M. [2004]. Chern-Simons theory, matrix integrals, and perturbative three-manifold invariants, *Commun. Math. Phys.* 253: 25–49.
[36] Mariño, M. [2011]. Lectures on localization and matrix models in supersymmetric Chern-Simons-matter theories, *J. Phys.* A44: 463001.
[37] Marshakov, A. [2011]. Gauge Theories as Matrix Models, *Theor. Math. Phys.* 169: 1704–1723.
[38] Marshakov, A. & Nekrasov, N. A. [2007]. Extended Seiberg-Witten theory and integrable hierarchy, *JHEP* 01: 104.
[39] Maruyoshi, K. & Yagi, F. [2011]. Seiberg-Witten curve via generalized matrix model, *JHEP* 01: 042.
[40] Mehta, M. L. [2004]. *Random Matrices*, 3rd edn, Academic Press.
[41] Moore, G. W., Nekrasov, N. & Shatashvili, S. [2000]. Integrating over Higgs branches, *Commun. Math. Phys.* 209: 97–121.
[42] Nakajima, H. [1999]. *Lectures on Hilbert Schemes of Points on Surfaces*, American Mathematical Society.
[43] Nekrasov, N. A. [2004]. Seiberg-Witten Prepotential From Instanton Counting, *Adv. Theor. Math. Phys.* 7: 831–864.
[44] Nekrasov, N. A. & Okounkov, A. [2006]. Seiberg-Witten Theory and Random Partitions, *in* P. Etingof, V. Retakh & I. M. Singer (eds), *The Unity of Mathematics*, Vol. 244 of *Progress in Mathematics*, Birkhäuser Boston, pp. 525–596.
[45] Ruijsenaars, S. & Schneider, H. [1986]. A new class of integrable systems and its relation to solitons, *Annals of Physics* 170: 370 – 405.
[46] Sasamoto, T. [2007]. Fluctuations of the one-dimensional asymmetric exclusion process using random matrix techniques, *J. Stat. Mech.* 07: P07007.
[47] Schiappa, R. & Wyllard, N. [2009]. An A_r threesome: Matrix models, 2d CFTs and 4d $\mathcal{N} = 2$ gauge theories, arXiv:0911.5337 [hep-th].
[48] Seiberg, N. & Witten, E. [1994a]. Monopole condensation, and confinement in $\mathcal{N} = 2$ supersymmetric Yang-Mills theory, *Nucl. Phys.* B426: 19–52.

[49] Seiberg, N. & Witten, E. [1994b]. Monopoles, duality and chiral symmetry breaking in $\mathcal{N} = 2$ supersymmetric QCD, *Nucl. Phys.* B431: 484–550.

[50] Shadchin, S. [2007]. On F-term contribution to effective action, *JHEP* 08: 052.

[51] Stanley, R. P. [2001]. *Enumerative Combinatorics: Volume 2*, Cambridge Univ. Press.

[52] Sułkowski, P. [2009]. Matrix models for 2^* theories, *Phys. Rev.* D80: 086006.

[53] Sułkowski, P. [2010]. Matrix models for β-ensembles from Nekrasov partition functions, *JHEP* 04: 063.

[54] Sutherland, B. [1971]. Quantum many body problem in one-dimension: Ground state, *J. Math. Phys.* 12: 246.

[55] Taki, M. [2011]. On AGT Conjecture for Pure Super Yang-Mills and W-algebra, *JHEP* 05: 038.

[56] Tracy, C. & Widom, H. [1994]. Level-spacing distributions and the Airy kernel, *Commun. Math. Phys.* 159: 151–174.

[57] Uglov, D. [1998]. Yangian Gelfand-Zetlin bases, $\mathfrak{gl}_N$-Jack polynomials and computation of dynamical correlation functions in the spin Calogero-Sutherland model, *Commun. Math. Phys.* 193: 663–696.

[58] Vershik, A. & Kerov, S. [1977]. Asymptotics of the Plahcherel measure of the symmetric group and the limit form of Young tablezux, *Soviet Math. Dokl.* 18: 527–531.

[59] Vershik, A. & Kerov, S. [1985]. Asymptotic of the largest and the typical dimensions of irreducible representations of a symmetric group, *Func. Anal. Appl.* 19: 21–31.

[60] Witten, E. [1997]. Solutions of four-dimensional field theories via M-theory, *Nucl. Phys.* B500: 3–42.

Chapter 5

Nonnegative Inverse Eigenvalue Problem

Ricardo L. Soto

Additional information is available at the end of the chapter

http://dx.doi.org/10.5772/48279

1. Introduction

Nonnegative matrices have long been a sorce of interesting and challenging mathematical problems. They are real matrices with all their entries being nonnegative and arise in a number of important application areas: communications systems, biological systems, economics, ecology, computer sciences, machine learning, and many other engineering systems. Inverse eigenvalue problems constitute an important subclass of inverse problems that arise in the context of mathematical modeling and parameter identification. A simple application of such problems is the construction of Leontief models in economics [1]-[3].

The *nonnegative inverse eigenvalue problem* (*NIEP*) is the problem of characterizing those lists $\Lambda = \{\lambda_1, \lambda_2, ..., \lambda_n\}$ of complex numbers which can be the spectra of $n \times n$ entrywise nonnegative matrices. If there exists a nonnegative matrix A with spectrum Λ we say that Λ is realized by A and that A is the realizing matrix. A set K of conditions is said to be a *realizability criterion* if any list $\Lambda = \{\lambda_1, \lambda_2, ..., \lambda_n\}$, real or complex, satisfying conditions K is realizable. The *NIEP is an open problem.* A full solution is unlikely in the near future. The problem has only been solved for $n = 3$ by Loewy and London ([4], 1978) and for $n = 4$ by Meehan ([5], 1998) and Torre-Mayo et al.([6], 2007). The case $n = 5$ has been solved for matrices of trace zero in ([7], 1999). Other results, mostly in terms of sufficient conditions for the problem to have a solution (in the case of a complex list Λ), have been obtained, in chronological order, in [8]-[13].

Two main subproblems of the *NIEP* are of great interest: the *real nonnegative inverse eigenvalue problem* (*RNIEP*), in which Λ is a list of real numbers, and the *symmetric nonnegative inverse eigenvalue problem* (*SNIEP*), in which the realizing matrix must be symmetric. Both problems, *RNIEP* and *SNIEP* are equivalent for $n \leq 4$ (see [14]), but they are different otherwise (see [15]). Moreover, both problems remains unsolved for $n \geq 5$. The *NIEP* is also of interest for nonnegative matrices with a particular structure, like stochastic and doubly stochastic, circulant, persymmetric, centrosymmetric, Hermitian, Toeplitz, etc.

The first sufficient conditions for the existence of a nonnegative matrix with a given real spectrum (*RNIEP*) were obtained by Suleimanova ([16], 1949) and Perfect ([17, 18], 1953 and

1955). Other sufficient conditions have also been obtained, in chronological order in [19]-[26], (see also [27, 28], and references therein for a comprehensive survey).

The first sufficient conditions for the *SNIEP* were obtained by Fiedler ([29], 1974). Other results for symmetric realizability have been obtained in [8, 30] and [31]-[33]. Recently, new sufficient conditions for the *SNIEP* have been given in [34]-[37].

1.1. Necessary conditions

Let A be a nonnegative matrix with spectrum $\Lambda = \{\lambda_1, \lambda_2, ..., \lambda_n\}$. Then, from the Perron Frobenius theory we have the following basic necessary conditions

$$\begin{array}{l} (1)\ \overline{\Lambda} = \{\overline{\lambda_1}, \ldots, \overline{\lambda_n}\} = \Lambda \\ (2)\ \max_j\{\left|\lambda_j\right|\} \in \Lambda \\ (3)\ s_m(\Lambda) = \sum_{j=1}^{n} \lambda_j^m \geq 0,\ m = 1, 2, \ldots, \end{array} \tag{1}$$

where $\overline{\Lambda} = \Lambda$ means that Λ is closed under complex comjugation.

Moreover, we have

$$\begin{array}{l} (4)\ (s_k(\Lambda))^m \leq n^{m-1} s_{km}(\Lambda),\ \ k, m = 1, 2, \ldots \\ (5)\ (s_2(\Lambda))^2 \leq (n-1) s_4(\Lambda),\ \ n \text{ odd}, tr(A) = 0. \end{array} \tag{2}$$

Necessary condition (4) is due to Loewy and London [4]. Necessary condition (5), which is a refinement of (4), is due to Laffey and Meehan [38]. The list $\Lambda = \{5, 4, -3, -3, -3\}$ for instance, satisfies all above necessary conditions, except condition (5). Therefore Λ is not a realizable list. In [39] it was obtained a new necessary condition, which is independent of the previous ones. This result is based on the Newton's inequalities associated to the normalized coefficients of the characteristic polynomial of an M-matrix or an inverse M-matrix.

The chapter is organized as follows: In section 2 we introduce two important matrix results, due to Brauer and Rado, which have allowed to obtain many of the most general sufficient conditions for the *RNIEP*, the *SNIEP* and the complex case. In section 3 we consider the real case and we introduce, without proof (we indicate where the the proofs can be found), two sufficient conditions with illustrative examples. We consider, in section 4, the symmetric case. Here we introduce a symmetric version of the Rado result, Theorem 2, and we set, without proof (see the appropriate references), three sufficient conditions, which are, as far as we know, the most general sufficient conditions for the *SNIEP*. In section 5, we discuss the complex (non real) case. Here we present several results with illustrative examples. Section 6 is devoted to discuss some Fiedler results and Guo results, which are very related with the problem and have been employed with success to derive sufficient conditions. Finally, in section 7, we introduce some open questions.

2. Brauer and Rado Theorems

A real matrix $A = (a_{ij})_{i=1}^{n}$ is said to have *constant row sums* if all its rows sum up to the same constant, say, α, that is, $\sum\limits_{j=1}^{n} a_{ij} = \alpha, \quad i = 1, \ldots, n$. The set of all real matrices with constant row sums equal to α is denoted by $\mathcal{CS}_\alpha$. It is clear that any matrix in $\mathcal{CS}_\alpha$ has eigenvector

$\mathbf{e} = (1, 1, \ldots, 1)^T$ corresponding to the eigenvalue α. Denote by $\mathbf{e}_k$ the $n-$dimensional vector with one in the $k - th$ position and zeros elsewhere.

It is well known that the problem of finding a nonnegative matrix with spectrum $\Lambda = \{\lambda_1, \ldots, \lambda_n\}$ is equivalent to the problem of finding a nonnegative matrix in $\mathcal{CS}_{\lambda_1}$ with spectrum Λ (see [40]). This will allow us to exploit the advantages of two important theorems, Brauer Theorem and Rado Theorem, which will be introduced in this section.

The spectra of circulant nonnegative matrices have been characterized in [9], while in [10], a simple complex generalization of Suleimanova result has been proved, and efficient and general sufficient conditions for the realizability of partitioned spectra, with the partition allowing some of its pieces to be nonrealizable, provided there are other pieces, which are realizable and, in certain way, compensate the nonrealizability of the former, have been obtained. This is the procedure which we call *negativity compensation*. This strategy, based in the use of the following two perturbation results, together with the properties of real matrices with constant row sums, has proved to be successful.

Theorem 1. ***Brauer*** *[41] Let A be an $n \times n$ arbitrary matrix with eigenvalues $\lambda_1, \ldots, \lambda_n$. Let $\mathbf{v} = (v_1, ..., v_n)^T$ an eigenvector of A associated with the eigenvalue λ_k and let $\mathbf{q} = (q_1, ..., q_n)^T$ be any n-dimensional vector. Then the matrix $A + \mathbf{v}\mathbf{q}^T$ has eigenvalues $\lambda_1, \ldots, \lambda_{k-1}, \lambda_k + v^T q, \lambda_{k+1}, \ldots, \lambda_n$.*

Proof. Let U be an $n \times n$ nonsingular matrix such that

$$U^{-1}AU = \begin{bmatrix} \lambda_1 & * & \cdots & * \\ & \lambda_2 & \ddots & \vdots \\ & & \ddots & * \\ & & & \lambda_n \end{bmatrix}$$

is an upper triangular matrix, where we choose the first column of U as $\mathbf{v}$ (U there exists from a well known result of Schur). Then,

$$U^{-1}(A + \mathbf{v}\mathbf{q}^T)U = U^{-1}AU + \begin{bmatrix} q_1 \; q_2 \cdots q_n \\ \\ \\ \end{bmatrix} U = \begin{bmatrix} \lambda_1 + \mathbf{q}^T\mathbf{v} & * & \cdots & * \\ & \lambda_2 & \ddots & \vdots \\ & & \ddots & * \\ & & & \lambda_n \end{bmatrix}.$$

and the result follows. ■ This proof is due to Reams [42].

Theorem 2. ***Rado*** *[18] Let A be an $n \times n$ arbitrary matrix with eigenvalues $\lambda_1, \ldots, \lambda_n$ and let $\Omega = diag\{\lambda_1, \ldots, \lambda_r\}$ for some $r \leq n$. Let X be an $n \times r$ matrix with rank r such that its columns $x_1, x_2, \ldots, x_r$ satisfy $Ax_i = \lambda_i x_i$, $i = 1, \ldots, r$. Let C be an $r \times n$ arbitrary matrix. Then the matrix $A + XC$ has eigenvalues $\mu_1, \ldots, \mu_r, \lambda_{r+1}, \ldots, \lambda_n$, where $\mu_1, \ldots, \mu_r$ are eigenvalues of the matrix $\Omega + CX$.*

Proof. Let $S = [X \mid Y]$ a nonsingular matrix with $S^{-1} = [\begin{smallmatrix} U \\ V \end{smallmatrix}]$. Then $UX = I_r$, $VY = I_{n-r}$ and $VX = 0$, $UY = 0$. Let $C = [C_1 \mid C_2]$, $X = [\begin{smallmatrix} X_1 \\ X_2 \end{smallmatrix}]$, $Y = [\begin{smallmatrix} Y_1 \\ Y_2 \end{smallmatrix}]$. Then, since $AX = X\Omega$,

$$S^{-1}AS = \begin{bmatrix} U \\ V \end{bmatrix} [X\Omega \mid AY] = \begin{bmatrix} \Omega & UAY \\ 0 & VAY \end{bmatrix}$$

and

$$S^{-1}XCS = \begin{bmatrix} I_r \\ 0 \end{bmatrix} [C_1 \mid C_2]\, S = \begin{bmatrix} C_1 & C_2 \\ 0 & 0 \end{bmatrix} \begin{bmatrix} X_1 & Y_1 \\ X_2 & Y_2 \end{bmatrix} = \begin{bmatrix} CX & CY \\ 0 & 0 \end{bmatrix}.$$

Thus,

$$S^{-1}(A+XC)S = S^{-1}AS + S^{-1}XCS = \begin{bmatrix} \Omega + CX & UAY + CY \\ 0 & VAY \end{bmatrix},$$

and we have $\sigma(A+XC) = \sigma(\Omega + CX) + \sigma(A) - \sigma(\Omega)$. ∎

3. Real nonnegative inverse eigenvalue problem.

Regarding the *RNIEP*, by applying Brauer Theorem and Rado Theorem, efficient and general sufficient conditions have been obtained in [18, 22, 24, 36].

Theorem 3. *[24] Let $\Lambda = \{\lambda_1, \lambda_2, ..., \lambda_n\}$ be a given list of real numbers. Suppose that:*
i) There exists a partition $\Lambda = \Lambda_1 \cup ... \cup \Lambda_t$, where

$$\Lambda_k = \{\lambda_{k1}, \lambda_{k2}, \dots \lambda_{kp_k}\}, \ \lambda_{11} = \lambda_1, \ \lambda_{k1} \geq \cdots \geq \lambda_{kp_k}, \ \lambda_{k1} \geq 0,$$

$k = 1, \dots, t$, such that for each sublist Λ_k we associate a corresponding list

$$\Gamma_k = \{\omega_k, \lambda_{k2}, ..., \lambda_{kp_k}\}, \ 0 \leq \omega_k \leq \lambda_1,$$

which is realizable by a nonnegative matrix $A_k \in CS_{\omega_k}$ of order p_k.
ii) There exists a nonnegative matrix $B \in CS_{\lambda_1}$ with eigenvalues $\lambda_1, \lambda_{21}, ..., \lambda_{t1}$ (the first elements of the lists Λ_k) and diagonal entries $\omega_1, \omega_2, \dots, \omega_t$ (the first elements of the lists Γ_k).
Then Λ is realizable by a nonnegative matrix $A \in CS_{\lambda_1}$.

Perfect [18] gave conditions under which $\lambda_1, \lambda_2, ..., \lambda_t$ and $\omega_1, \omega_2, \dots, \omega_t$ are the eigenvalues and the diagonal entries, respectively, of a $t \times t$ nonnegative matrix $B \in CS_{\lambda_1}$. For $t = 2$ it is necessary and sufficient that $\lambda_1 + \lambda_2 = \omega_1 + \omega_2$, with $0 \leq \omega_i \leq \lambda_1$. For $t = 3$ Perfect gave the following result:

Theorem 4. *[18] The real numbers $\lambda_1, \lambda_2, \lambda_3$ and $\omega_1, \omega_2, \omega_3$ are the eigenvalues and the diagonal entries, respectively, of a 3×3 nonnegative matrix $B \in CS_{\lambda_1}$, if and only if:*

$$\begin{array}{ll} i) & 0 \leq \omega_i \leq \lambda_1, \ i = 1,2,3 \\ ii) & \lambda_1 + \lambda_2 + \lambda_3 = \omega_1 + \omega_2 + \omega_3 \\ iii) & \lambda_1\lambda_2 + \lambda_1\lambda_3 + \lambda_2\lambda_3 \leq \omega_1\omega_2 + \omega_1\omega_3 + \omega_2\omega_3 \\ iv) & max_k \omega_k \geq \lambda_2 \end{array} \tag{3}$$

Then, an appropriate 3×3 nonnegative matrix B is

$$B = \begin{bmatrix} \omega_1 & 0 & \lambda_1 - \omega_1 \\ \lambda_1 - \omega_2 - p & \omega_2 & p \\ 0 & \lambda_1 - \omega_3 & \omega_3 \end{bmatrix}, \tag{4}$$

where

$$p = \frac{1}{\lambda_1 - \omega_3}(\omega_1\omega_2 + \omega_1\omega_3 + \omega_2\omega_3 - \lambda_1\lambda_2 + \lambda_1\lambda_3 + \lambda_2\lambda_3).$$

For $t \geq 4$, we only have a sufficient condition:

$$\begin{array}{ll} i) & 0 \leq \omega_k \leq \lambda_1, \; k = 1,2,\ldots t, \\ ii) & \omega_1 + \omega_2 \cdots + \omega_t = \lambda_1 + \lambda_2 \cdots + \lambda_t, \\ iii) & \omega_k \geq \lambda_k, \; \omega_1 \geq \lambda_k, \; k = 2,3,\ldots,t, \end{array} \tag{5}$$

with the following matrix $B \in CS_{\lambda_1}$ having eigenvalues and diagonal entries $\lambda_1, \lambda_2, \ldots, \lambda_t$ and $\omega_1, \omega_2, \ldots, \omega_t$, respectively:

$$B = \begin{bmatrix} \omega_1 & \omega_2 - \lambda_2 & \cdots & \omega_r - \lambda_t \\ \omega_1 - \lambda_2 & \omega_2 & \cdots & \omega_r - \lambda_t \\ \vdots & \vdots & \ddots & \vdots \\ \omega_1 - \lambda_t & \omega_2 - \lambda_2 & \cdots & \omega_t \end{bmatrix}. \tag{6}$$

Example 1. *Let us consider the list $\Lambda = \{6, 1, 1, -4, -4\}$ with the partition*

$$\Lambda_1 = \{6, -4\}, \; \Lambda_2 = \{1, -4\}, \; \Lambda_3 = \{1\}$$

and the realizable associated lists

$$\Gamma_1 = \{4, -4\}, \; \Gamma_2 = \{4, -4\}, \; \Gamma_3 = \{0\}.$$

From (4) we compute the 3×3 nonnegative matrix

$$B = \begin{bmatrix} 4 & 0 & 2 \\ \frac{3}{2} & 4 & \frac{1}{2} \\ 0 & 6 & 0 \end{bmatrix}$$

with eigenvalues $6, 1, 1$, and diagonal entries $4, 4, 0$. Then

$$A = \begin{bmatrix} 0&4&0&0&0 \\ 4&0&0&0&0 \\ 0&0&0&4&0 \\ 0&0&4&0&0 \\ 0&0&0&0&0 \end{bmatrix} + \begin{bmatrix} 1&0&0 \\ 1&0&0 \\ 0&1&0 \\ 0&1&0 \\ 0&0&1 \end{bmatrix} \begin{bmatrix} 0&0&0&0&2 \\ \frac{3}{2}&0&0&0&\frac{1}{2} \\ 0&0&6&0&0 \end{bmatrix}$$

$$= \begin{bmatrix} 0&4&0&0&2 \\ 4&0&0&0&2 \\ \frac{3}{2}&0&0&4&\frac{1}{2} \\ \frac{3}{2}&0&4&0&\frac{1}{2} \\ 0&0&6&0&0 \end{bmatrix}$$

is nonnegative with spectrum Λ.

A map of sufficient conditions for the *RNIEP* it was constructed in [28], There, the sufficient conditions were compared to establish inclusion or independence relations between them. It is also shown in [28] that the criterion given by Theorem 3 contains all realizability criteria for lists of real numbers studied therein. In [36], from a new special partition, Theorem 3 is extended. Now, the first element λ_{k1} of the sublist Λ_k need not to be nonnegative and the realizable auxiliar list $\Gamma_k = \{\omega_k, \lambda_{k1}, ..., \lambda_{kp_k}\}$ contains one more element. Moreover, the number of lists of the partition depend on the number of elements of the first list Λ_1, and some lists Λ_k can be empty.

Theorem 5. *[36] Let* $\Lambda = \{\lambda_1, \lambda_2, \ldots, \lambda_n\}$ *be a list of real numbers and let the partition* $\Lambda = \Lambda_1 \cup \cdots \cup \Lambda_{p_1+1}$ *be such that*

$$\Lambda_k = \{\lambda_{k1}, \lambda_{k2}, \ldots \lambda_{kp_k}\}, \ \lambda_{11} = \lambda_1, \ \lambda_{k1} \geq \lambda_{k2} \geq \cdots \geq \lambda_{kp_k},$$

$k = 1, \ldots, p_1 + 1$, *where* p_1 *is the number of elements of the list* Λ_1 *and some of the lists* Λ_k *can be empty. Let* $\omega_2, \ldots, \omega_{p_1+1}$ *be real numbers satisfying* $0 \leq \omega_k \leq \lambda_1, k = 2, \ldots, p_1 + 1$. *Suppose that the following conditions hold:*
i) For each $k = 2, \ldots, p_1 + 1$, *there exists a nonnegative matrix* $A_k \in CS_{\omega_k}$ *with spectrum* $\Gamma_k = \{\omega_k, \lambda_{k1}, ..., \lambda_{kp_k}\}$,
ii) There exists a $p_1 \times p_1$ *nonnegative matrix* $B \in CS_{\lambda_1}$, *with spectrum* Λ_1 *and with diagonal entries* $\omega_2, \ldots, \omega_{p_1+1}$.
Then Λ *is realizable by a nonnegative matrix* $A \in CS_{\lambda_1}$.

Example 2. *With this extension, the authors show for instance, that the list*

$$\{5, 4, 0, -3, -3, -3\}$$

is realizable, which can not be done from the criterion given by Theorem 3. In fact, let the partition

$$\begin{aligned} &\Lambda_1 = \{5, 4, 0, -3\}, \ \Lambda_2 = \{-3\}, \ \Lambda_3 = \{-3\} \ \text{with} \\ &\Gamma_2 = \{3, -3\}, \ \Gamma_3 = \{3, -3\}, \ \Gamma_4 = \Gamma_5 = \{0\}. \end{aligned}$$

The nonnegative matrix

$$B = \begin{bmatrix} 3 & 0 & 2 & 0 \\ 0 & 3 & 0 & 2 \\ 3 & 0 & 0 & 2 \\ 0 & 3 & 2 & 0 \end{bmatrix}$$

has spectrum Λ_1 *and diagonal entries* $3, 3, 0, 0$. *It is clear that*

$$A_2 = A_3 = \begin{bmatrix} 0 & 3 \\ 3 & 0 \end{bmatrix} \ \text{realizes} \ \Gamma_2 = \Gamma_3.$$

Then

$$A = \begin{bmatrix} A_2 & & & \\ & A_3 & & \\ & & 0 & \\ & & & 0 \end{bmatrix} + \begin{bmatrix} 1 & 0 & 0 & 0 \\ 1 & 0 & 0 & 0 \\ 0 & 1 & 0 & 0 \\ 0 & 1 & 0 & 0 \\ 0 & 0 & 1 & 0 \\ 0 & 0 & 0 & 1 \end{bmatrix} \begin{bmatrix} 0 & 0 & 0 & 0 & 2 & 0 \\ 0 & 0 & 0 & 0 & 0 & 2 \\ 3 & 0 & 0 & 0 & 0 & 2 \\ 0 & 0 & 3 & 0 & 2 & 0 \end{bmatrix} = \begin{bmatrix} 0 & 3 & 0 & 0 & 2 & 0 \\ 3 & 0 & 0 & 0 & 2 & 0 \\ 0 & 0 & 0 & 3 & 0 & 2 \\ 0 & 0 & 3 & 0 & 0 & 2 \\ 3 & 0 & 0 & 0 & 0 & 2 \\ 0 & 0 & 3 & 0 & 2 & 0 \end{bmatrix}$$

has the desired spectrum $\{5, 4, 0, -3, -3, -3\}$.

4. Symmetric nonnegative inverse eigenvalue problem

Several realizability criteria which were first obtained for the *RNIEP* have later been shown to be symmetric realizability criteria as well. For example, Kellogg criterion [19] was showed by Fiedler [29] to imply symmetric realizability. It was proved by Radwan [8] that Borobia's criterion [21] is also a symmetric realizability criterion, and it was proved in [33] that Soto's criterion for the *RNIEP* is also a criterion for the *SNIEP*. In this section we shall consider the most general and efficient symmetric realizability criteria for the *SNIEP* (as far as we know they are). We start by introducing a symmetric version of the Rado Theorem:.

Theorem 6. *[34] Let A be an $n \times n$ symmetric matrix with spectrum $\Lambda = \{\lambda_1, \lambda_2, \ldots, \lambda_n\}$ and for some $r \leq n$, let $\{\mathbf{x}_1, \mathbf{x}_2, \ldots, \mathbf{x}_r\}$ be an orthonormal set of eigenvectors of A spanning the invariant subspace associated with $\lambda_1, \lambda_2, \ldots, \lambda_r$. Let X be the $n \times r$ matrix with $i-th$ column $\mathbf{x}_i$, let $\Omega = diag\{\lambda_1, \ldots, \lambda_r\}$, and let C be any $r \times r$ symmetric matrix. Then the symmetric matrix $A + XCX^T$ has eigenvalues $\mu_1, \mu_2, \ldots, \mu_r, \lambda_{r+1}, \ldots, \lambda_n$, where $\mu_1, \mu_2, \ldots, \mu_r$ are the eigenvalues of the matrix $\Omega + C$.*

Proof. Since the columns of X are an orthonormal set, we may complete X to an orthogonal matrix $W = [X\ Y]$, that is, $X^TX = I_r$, $Y^TY = I_{n-r}$, $X^TY = 0$, $Y^TX = 0$. Then

$$W^{-1}AW = \begin{bmatrix} X^T \\ Y^T \end{bmatrix} A\,[X\ Y] = \begin{bmatrix} \Omega & X^TAY \\ 0 & Y^TAY \end{bmatrix}$$

$$W^{-1}(XCX^T)W = \begin{bmatrix} I_r \\ 0 \end{bmatrix} C\,[I_r\ 0] = \begin{bmatrix} C & 0 \\ 0 & 0 \end{bmatrix}.$$

Therefore,

$$W^{-1}(A + XCX^T)W = \begin{bmatrix} \Omega + C & X^TAY \\ 0 & Y^TAY \end{bmatrix}$$

and $A + XCX^T$ is symmetric with eigenvalues $\mu_1, \ldots, \mu_r, \lambda_{r+1}, \ldots, \lambda_n$. ■

By using Theorem 6, the following sufficient condition was proved in [34]:

Theorem 7. *[34] Let $\Lambda = \{\lambda_1, \lambda_2, \ldots, \lambda_n\}$ be a list of real numbers with $\lambda_1 \geq \lambda_2 \geq \cdots \geq \lambda_n$ and, for some $t \leq n$, let $\omega_1, \ldots, \omega_t$ be real numbers satisfying $0 \leq \omega_k \leq \lambda_1$, $k = 1, \ldots, t$. Suppose there exists:*
i) a partition $\Lambda = \Lambda_1 \cup \cdots \cup \Lambda_t$ with

$$\Lambda_k = \{\lambda_{k1}, \lambda_{k2}, \ldots \lambda_{kp_k}\}, \ \lambda_{11} = \lambda_1, \ \ \lambda_{k1} \geq 0, \ \lambda_{k1} \geq \lambda_{k2} \geq \cdots \geq \lambda_{kp_k},$$

such that for each $k = 1, \ldots, t$, the list $\Gamma_k = \{\omega_k, \lambda_{k2}, ..., \lambda_{kp_k}\}$ is realizable by a symmetric nonnegative matrix A_k of order p_k, and
ii) a $t \times t$ symmetric nonnegative matrix B with eigenvalues $\lambda_{11}, \lambda_{21}, ..., \lambda_{t1}\}$ and with diagonal entries $\omega_1, \omega_2, \ldots, \omega_t$.
Then Λ is realizable by a symmetric nonnegative matrix.

Proof. Since A_k is a $p_k \times p_k$ symmetric nonnegative matrix realizing Γ_k, then $A = diag\{A_1, A_2, \ldots, A_t\}$ is symmetric nonnegative with spectrum $\Gamma_1 \cup \Gamma_2 \cup \cdots \cup \Gamma_t$. Let $\{\mathbf{x}_1, \ldots, \mathbf{x}_t\}$ be an orthonormal set of eigenvectors of A associated with $\omega_1, \ldots, \omega_t$, respectively. Then the $n \times t$ matrix X with $i-th$ column $\mathbf{x}_i$ satisfies $AX = X\Omega$ for $\Omega = dig\{\omega_1, \ldots, \omega_t\}$. Moreover, X is entrywise nonnegative, since each $\mathbf{x}_i$ contains the Perron eigenvector of A_i and zeros. Now, if we set $C = B - \Omega$, the matrix C is symmetric nonnegative and $\Omega + C$ has eigenvalues $\lambda_1, \ldots, \lambda_t$. Therefore, by Theorem 6 the symmetric matrix $A + XCX^T$ has spectrum Λ. Besides, it is nonnegative since all the entries of A, X, and C are nonnegative. ■

Theorem 7 not only ensures the existence of a realizing matrix, but it also allows to construct the realizing matrix. Of course, the key is to know under which conditions does there exists a $t \times t$ symmetrix nonnegative matrix B with eigenvalues $\lambda_1, \ldots, \lambda_t$ and diagonal entries $\omega_1, \ldots, \omega_t$.

The following conditions for the existence of a real symmetric matrix, not necessarily nonnegative, with prescribed eigenvalues and diagonal entries are due to Horn [43]: *There exists a real symmetric matrix with eigenvalues* $\lambda_1 \geq \lambda_2 \geq \cdots \geq \lambda_t$ *and diagonal entries* $\omega_1 \geq \omega_2 \geq \cdots \geq \omega_t$ *if and only if*

$$\left.\begin{array}{c} \sum_{i=1}^{k} \lambda_i \geq \sum_{i=1}^{k} \omega_i, \ k = 1, \ldots, t-1 \\ \sum_{i=1}^{t} \lambda_i = \sum_{i=1}^{t} \omega_i \end{array}\right\} \tag{7}$$

For $t = 2$, the conditions (7) become

$$\lambda_1 \geq \omega_1, \quad \lambda_1 + \lambda_2 = \omega_1 + \omega_2,$$

and they are also sufficient for the existence of a 2×2 symmetric nonnegative matrix B with eigenvalues $\lambda_1 \geq \lambda_2$ and diagonal entries $\omega_1 \geq \omega_2 \geq 0$, namely,

$$B = \begin{bmatrix} \omega_1 & \sqrt{(\lambda_1 - \omega_1)(\lambda_1 - \omega_2)} \\ \sqrt{(\lambda_1 - \omega_1)(\lambda_1 - \omega_2)} & \omega_2 \end{bmatrix}.$$

For $t = 3$, we have the following conditions:

Lemma 1. *[29] The conditions*

$$\left.\begin{array}{c} \lambda_1 \geq \omega_1 \\ \lambda_1 + \lambda_2 \geq \omega_1 + \omega_2 \\ \lambda_1 + \lambda_2 + \lambda_3 = \omega_1 + \omega_2 + \omega_3 \\ \omega_1 \geq \lambda_2 \end{array}\right\} \tag{8}$$

are necessary and sufficient for the existence of a 3×3 *symmetric nonnegative matrix* B *with eigenvalues* $\lambda_1 \geq \lambda_2 \geq \lambda_3$ *and diagonal entries* $\omega_1 \geq \omega_2 \geq \omega_3 \geq 0$.

In [34], the following symmetric nonnegative matrix B, satisfying conditions (8), it was constructed:

$$B = \begin{bmatrix} \omega_1 & \sqrt{\frac{\mu-\omega_3}{2\mu-\omega_2-\omega_3}}s & \sqrt{\frac{\mu-\omega_2}{2\mu-\omega_2-\omega_3}}s \\ \sqrt{\frac{\mu-\omega_3}{2\mu-\omega_2-\omega_3}}s & \omega_2 & \sqrt{(\mu-\omega_2)(\mu-\omega_3)} \\ \sqrt{\frac{\mu-\omega_2}{2\mu-\omega_2-\omega_3}}s & \sqrt{(\mu-\omega_2)(\mu-\omega_3)} & \omega_3 \end{bmatrix}, \tag{9}$$

where $\mu = \lambda_1 + \lambda_2 - \omega_1$; $s = \sqrt{(\lambda_1 - \mu)(\lambda_1 - \omega_1)}$.

For $t \geq 4$ we have only a sufficient condition:

Theorem 8. *Fiedler [29] If* $\lambda_1 \geq \cdots \geq \lambda_t$ *and* $\omega_1 \geq \cdots \geq \omega_t$ *satisfy*

$$\left.\begin{array}{l} i) \ \sum_{i=1}^{s} \lambda_i \geq \sum_{i=1}^{s} \omega_i, \ s = 1, \ldots, t-1 \\ ii) \ \sum_{i=1}^{t} \lambda_i = \sum_{i=1}^{t} \omega_i \\ iii) \ \omega_{k-1} \geq \lambda_k, \ k = 2, \ldots, t-1 \end{array}\right\}, \tag{10}$$

then there exists a $t \times t$ symmetric nonnegative matrix with eigenvalues $\lambda_1, \ldots, \lambda_t$ and diagonal entries $\omega_1, \ldots, \omega_t$.

Observe that

$$B = \begin{bmatrix} 5 & 2 & \frac{1}{2} & \frac{1}{2} \\ 2 & 5 & \frac{1}{2} & \frac{1}{2} \\ \frac{1}{2} & \frac{1}{2} & 5 & 2 \\ \frac{1}{2} & \frac{1}{2} & 2 & 5 \end{bmatrix}$$

has eigenvalues $8, 6, 3, 3$, but $\lambda_2 = 6 > 5 = \omega_1$.

Example 3. *Let us consider the list $\Lambda = \{7, 5, 1, -3, -4, -6\}$ with the partition*

$$\Lambda_1 = \{7, -6\}, \ \Lambda_2 = \{5, -4\}, \ \Lambda_3 = \{1, -3\} \ \textit{with}$$
$$\Gamma_1 = \{6, -6\}, \ \Gamma_2 = \{4, -4\}, \ \Gamma_3 = \{3, -3\}.$$

We look for a symmetric nonnegative matrix B with eigenvalues $7, 5, 1$ and diagonal entries $6, 4, 3$. Then conditions (8) are satisfied and from (9) we compute

$$B = \begin{bmatrix} 6 & \sqrt{\frac{3}{5}} & \sqrt{\frac{2}{5}} \\ \sqrt{\frac{3}{5}} & 4 & \sqrt{6} \\ \sqrt{\frac{2}{5}} & \sqrt{6} & 3 \end{bmatrix} \ \textit{and}\ C = B - \Omega,$$

where $\Omega = diag\{6, 4, 3\}$. The symmetric matrices

$$A_1 = \begin{bmatrix} 0 & 6 \\ 6 & 0 \end{bmatrix}, \ A_2 = \begin{bmatrix} 0 & 4 \\ 4 & 0 \end{bmatrix}, \ A_3 = \begin{bmatrix} 0 & 3 \\ 3 & 0 \end{bmatrix}$$

realize $\Gamma_1, \Gamma_2, \Gamma_3$. Then

$$A = \begin{bmatrix} A_1 & & \\ & A_2 & \\ & & A_3 \end{bmatrix} + XCX^T, \ \textit{where}\ X = \begin{bmatrix} \frac{\sqrt{2}}{2} & 0 & 0 \\ \frac{\sqrt{2}}{2} & 0 & 0 \\ 0 & \frac{\sqrt{2}}{2} & 0 \\ 0 & \frac{\sqrt{2}}{2} & 0 \\ 0 & 0 & \frac{\sqrt{2}}{2} \\ 0 & 0 & \frac{\sqrt{2}}{2} \end{bmatrix},$$

is symmetric nonnegative with spectrum Λ.

In the same way as Theorem 3 was extended to Theorem 5 (in the real case), Theorem 7 was also extended to the following result:

Theorem 9. *[36] Let $\Lambda = \{\lambda_1, \lambda_2, \ldots, \lambda_n\}$ be a list of real numbers and let the partition $\Lambda = \Lambda_1 \cup \cdots \cup \Lambda_{p_1+1}$ be such that*

$$\Lambda_k = \{\lambda_{k1}, \lambda_{k2}, \ldots \lambda_{kp_k}\}, \ \lambda_{11} = \lambda_1, \ \lambda_{k1} \geq \lambda_{k2} \geq \cdots \geq \lambda_{kp_k},$$

$k = 1, \ldots, p_1 + 1$, where Λ_1 is symmetrically realizable, p_1 is the number of elements of Λ_1 and some lists Λ_k can be empty. Let $\omega_2, \ldots, \omega_{p_1+1}$ be real numbers satisfying $0 \leq \omega_k \leq \lambda_1$, $k = 2, \ldots, p_1 + 1$.

Suppose that the following conditions hold:
i) For each $k = 2, \ldots, p_1 + 1$, *there exists a symmetric nonnegative matrix* A_k *with spectrum* $\Gamma_k = \{\omega_k, \lambda_{k1}, ..., \lambda_{kp_k}\}$,
ii) There exists a $p_1 \times p_1$ *symmetric nonnegative matrix B with spectrum* Λ_1 *and with diagonal entries* $\omega_2, \ldots, \omega_{p_1+1}$.
Then Λ *is symmetrically realizable.*

Example 4. *Now, from Theorem 9, we can see that there exists a symmetric nonnegative matrix with spectrum* $\Lambda = \{5, 4, 0, -3, -3, -3\}$, *which can not be seen from Theorem 7. Moreover, we can compute a realizing matrix. In fact, let the partition*

$$\Lambda_1 = \{5, 4, 0, -3\}, \ \Lambda_2 = \{-3\}, \ \Lambda_3 = \{-3\} \ \text{with}$$
$$\Gamma_2 = \{3, -3\}, \ \Gamma_3 = \{3, -3\}, \ \Gamma_4 = \Gamma_5 = \{0\}.$$

The symmetric nonnegative matrix

$$B = \begin{bmatrix} 3 & 0 & \sqrt{6} & 0 \\ 0 & 3 & 0 & \sqrt{6} \\ \sqrt{6} & 0 & 0 & 2 \\ 0 & \sqrt{6} & 2 & 0 \end{bmatrix}$$

has spectrum Λ_1 *and diagonal entries* $3, 3, 0, 0$. *Let* $\Omega = diag\{3, 3, 0, 0\}$ *and*

$$X = \begin{bmatrix} \frac{\sqrt{2}}{2} & 0 & 0 & 0 \\ \frac{\sqrt{2}}{2} & 0 & 0 & 0 \\ 0 & \frac{\sqrt{2}}{2} & 0 & 0 \\ 0 & \frac{\sqrt{2}}{2} & 0 & 0 \\ 0 & 0 & 1 & 0 \\ 0 & 0 & 0 & 1 \end{bmatrix}, \quad A_2 = A_3 = \begin{bmatrix} 0 & 3 \\ 3 & 0 \end{bmatrix}, \quad C = B - \Omega.$$

Then, from Theorem 6 we obtain

$$A = \begin{bmatrix} A_2 & & & \\ & A_3 & & \\ & & 0 & \\ & & & 0 \end{bmatrix} + XCX^T,$$

which is symmetric nonnegative with spectrum Λ.

The following result, although is not written in the fashion of a sufficient condition, is indeed a very general and efficient sufficient condition for the *SNIEP*.

Theorem 10. *[35] Let A be an* $n \times n$ *irreducible symmetric nonnegative matrix with spectrum* $\Lambda = \{\lambda_1, \lambda_2, \ldots, \lambda_n\}$, *Perron eigenvalue* λ_1 *and a diagonal element c. Let B be an* $m \times m$ *symmetric nonnegative matrix with spectrum* $\Gamma = \{\mu_1, \mu_2, \ldots, \mu_m\}$ *and Perron eigenvalue* μ_1.
i) If $\mu_1 \leq c$, *then there exists a symmetric nonnegative matrix C, of order* $(n + m - 1)$, *with spectrum* $\{\lambda_1, \ldots, \lambda_n, \mu_2, \ldots, \mu_m\}$.
ii) If $\mu_1 \geq c$, *then there exists a symmetric nonnegative matrix C, of order* $(n + m - 1)$, *with spectrum* $\{\lambda_1 + \mu_1 - c, \lambda_2, \ldots, \lambda_n, \mu_2, \ldots, \mu_m\}$.

Example 5. *The following example, given in [35], shows that* $\{7,5,0,-4,-4,-4\}$ *with the partition*

$$\Lambda = \{7,5,0,-4\}, \ \Gamma = \{4,-4\},$$

satisfies conditions of Theorem 10, where

$$A = \begin{bmatrix} 4 & 0 & b & 0 \\ 0 & 4 & 0 & d \\ b & 0 & 0 & \sqrt{6} \\ 0 & d & \sqrt{6} & 0 \end{bmatrix} \quad \text{with } b^2 + d^2 = 23, \ \ bd = 4\sqrt{6},$$

is symmetric nonnegative with spectrum Λ. *Then there exists a symmetric nonnegative matrix C with spectrum* $\{7,5,0,-4,-4\}$ *and a diagonal element* 4. *By applying again Theorem 10 to the lists* $\{7,5,0,-4,-4\}$ *and* $\{4,-4\}$, *we obtain the desired symmetric nonnegative matrix.*

It is not hard to show that both results, Theorem 9 and Theorem 10, are equivalent (see [44]). Thus, the list in the Example 4 is also realizable from Theorem 10, while the list in the example 5 is also realizable from Theorem 9.

5. List of complex numbers

In this section we consider lists of complex nonreal numbers. We start with a complex generalization of a well known result of Suleimanova, usually considered as one of the important results in the *RNIEP (see [16]): The list* $\lambda_1 > 0 > \lambda_2 \geq \cdots \geq \lambda_n$ is the spectrum of a nonnegative matrix if and only if $\lambda_1 + \lambda_2 + \cdots + \lambda_n \geq 0$.

Theorem 11. *[10] Let* $\Lambda = \{\lambda_0, \lambda_1, \ldots, \lambda_n\}$ *be a list of complex numbers closed under complex conjugation, with*

$$\Lambda' = \{\lambda_1, \ldots, \lambda_n\} \subset \{z \in \mathbb{C} : \operatorname{Re} z \leq 0; \ |\operatorname{Re} z| \geq |\operatorname{Im} z|\}.$$

Then Λ *is realizable if and only if* $\sum_{i=0}^{n} \lambda_i \geq 0$.

Proof. Suppose that the elements of Λ' are ordered in such a way that $\lambda_{2p+1}, \ldots, \lambda_n$ are real and $\lambda_1, \ldots, \lambda_{2p}$ are complex nonreal, with

$$x_k = \operatorname{Re} \lambda_{2k-1} = \operatorname{Re} \lambda_{2k} \ \text{ and } \ y_k = \operatorname{Im} \lambda_{2k-1} = \operatorname{Im} \lambda_{2k}$$

for $k = 1, \ldots, p$. Consider the matrix

$$B = \begin{bmatrix} 0 & 0 & 0 & \cdot & & & & & \\ -x_1 + y_1 & x_1 & -y_1 & \cdot & & & & & \\ -x_1 - y_1 & y_1 & x_1 & \cdot & & & & & \\ \vdots & \vdots & \vdots & \ddots & & & & & \\ -x_p + y_p & 0 & 0 & \cdot & x_p & -y_p & & & \\ -x_p - y_p & 0 & 0 & \cdot & y_p & x_p & & & \\ -\lambda_{2p+1} & 0 & 0 & \cdot & & & \lambda_{2p+1} & & \\ \vdots & \vdots & \vdots & \cdot & & & & \ddots & \\ -\lambda_n & 0 & & \cdot & & & & & \lambda_n \end{bmatrix}.$$

It is clear that $B \in \mathcal{CS}_0$ with spectrum $\{0, \lambda_1, \ldots, \lambda_n\}$ and all the entries on its first column are nonnegative. Define $\mathbf{q} = (q_0, q_1, \ldots, q_n)^T$ with $q_0 = \lambda_0 + \sum_{i=1}^{n} \lambda_i$ and

$$q_k = -\operatorname{Re}\lambda_k \text{ for } k = 1, \ldots, 2p \text{ and } q_k = -\lambda_k \text{ for } k = 2p+1, \ldots, n.$$

Then, from the Brauer Theorem 1 $A = B + \mathbf{e}\mathbf{q}^T$ is nonnegative with spectrum Λ. ■

In the case when all numbers in the given list, except one (the Perron eigenvalue), have real parts smaller than or equal to zero, remarkably simple necessary and sufficient conditions were obtained in [11].

Theorem 12. *[11] Let $\lambda_2, \lambda_3, \ldots, \lambda_n$ be nonzero complex numbers with real parts less than or equal to zero and let λ_1 be a positive real number. Then the list $\Lambda = \{\lambda_1, \lambda_2, \ldots, \lambda_n\}$ is the nonzero spectrum of a nonnegative matrix if the following conditions are satisfied:*

$$\begin{aligned} &i) \ \ \overline{\Lambda} = \Lambda \\ &ii) \ \ s_1 = \sum_{i=1}^{n} \lambda_i \geq 0 \\ &iii) \ s_2 = \sum_{i=1}^{n} \lambda_i^2 \geq 0 \end{aligned} \tag{11}$$

The minimal number of zeros that need to be added to Λ to make it realizable is the smallest nonnegative integer N for which the following inequality is satisfied:

$$s_1^2 \leq (n+N)s_2.$$

Furthermore, the list $\{\lambda_1, \lambda_2, \ldots, \lambda_n, 0, \ldots, 0\}$ can be realized by $C + \alpha I$, where C is a nonnegative companion matrix with trace zero, α is a nonnegative scalar and I is the $n \times n$ identity matrix.

Corollary 1. *[11] Let $\lambda_2, \lambda_3, \ldots, \lambda_n$ be complex numbers with real parts less than or equal to zero and let λ_1 be a positive real number. Then the list $\Lambda = \{\lambda_1, \lambda_2, \ldots, \lambda_n\}$ is the spectrum of a nonnegative matrix if and only if the following conditions are satisfied:*

$$\begin{aligned} &i) \ \ \overline{\Lambda} = \Lambda \\ &ii) \ \ s_1 = \sum_{i=1}^{n} \lambda_i \geq 0 \\ &iii) \ s_2 = \sum_{i=1}^{n} \lambda_i^2 \geq 0 \\ &iv) \ \ s_1^2 \leq n s_2 \end{aligned} \tag{12}$$

Example 6. *The list $\Lambda = \{8, -1+3i, -1-3i, -2+5i, -2-5i\}$ satisfies conditions (12). Then Λ is the spectrum of the nonnegative companion matrix*

$$C = \begin{bmatrix} 0 & 1 & 0 & 0 & 0 \\ 0 & 0 & 1 & 0 & 0 \\ 0 & 0 & 0 & 1 & 0 \\ 0 & 0 & 0 & 0 & 1 \\ 2320 & 494 & 278 & 1 & 2 \end{bmatrix}.$$

Observe that Theorem 11 gives no information about the realizability of Λ.
The list $\{19, -1+11i, -1-11i, -3+8i, -3-8i\}$ *was given in [11]. It does not satisfy conditions (12):* $s_1 = 11$, $s_2 = 11$ *and* $s_1^2 \nleq ns_2$. *The inequality* $11^2 \leq (5+N)11$ *is satisfied for* $N \geq 6$. *Then we need to add 6 zeros to the list to make it realizable.*

Theorem 3 (in section 3), can also be extended to the complex case:

Theorem 13. *[13] Let* $\Lambda = \{\lambda_2, \lambda_3, \ldots, \lambda_n\}$ *be a list of complex numbers such that* $\overline{\Lambda} = \Lambda$, $\lambda_1 \geq \max_i |\lambda_i|$, $i = 2, \ldots, n$, *and* $\sum_{i=1}^{n} \lambda_i \geq 0$. *Suppose that:*
i) there exists a partition $\Lambda = \Lambda_1 \cup \cdots \cup \Lambda_t$ *with*

$$\Lambda_k = \{\lambda_{k1}, \lambda_{k2}, \ldots \lambda_{kp_k}\}, \ \lambda_{11} = \lambda_1,$$

$k = 1, \ldots, t$, *such that* $\Gamma_k = \{\omega_k, \lambda_{k2}, ..., \lambda_{kp_k}\}$ *is realizable by a nonnegative matrix* $A_k \in \mathcal{CS}_{\omega_k}$, *and*
ii) there exists a $t \times t$ *nonnegative matrix* $B \in \mathcal{CS}_{\lambda_1}$, *with eigenvalues* $\lambda_1, \lambda_{21}, \ldots, \lambda_{t1}$ *(the first elements of the lists* Λ_k*) and with diagonal entries* $\omega_1, \omega_2, \ldots, \omega_t$ *(the Perron eigenvalues of the lists* Γ_k*).*
Then Λ *is realizable.*

Example 7. *Let* $\Lambda = \{7, 1, -2, -2, -2+4i, -2-4i\}$. *Consider the partition*

$$\Lambda_1 = \{7, 1, -2, -2\}, \ \Lambda_2 = \{-2+4i\}, \ \Lambda_3 = \{-2-4i\} \ \textit{with}$$
$$\Gamma_1 = \{3, 1, -2, -2\}, \ \Gamma_2 = \{0\}, \ \Gamma_3 = \{0\}.$$

We look for a nonnegative matrix $B \in \mathcal{CS}_7$ *with eigenvalues* $7, -2+4i, -2-4i$ *and diagonal entries* $3, 0, 0$, *and a nonnegative matrix* A_1 *realizing* Γ_1. *They are*

$$B = \begin{bmatrix} 3 & 0 & 4 \\ \frac{41}{7} & 0 & \frac{8}{7} \\ 0 & 7 & 0 \end{bmatrix} \quad \textit{and} \quad A_1 = \begin{bmatrix} 0 & 2 & 0 & 1 \\ 2 & 0 & 0 & 1 \\ 0 & 1 & 0 & 2 \\ 0 & 1 & 2 & 0 \end{bmatrix}.$$

Then

$$A = \begin{bmatrix} A_1 & & \\ & 0 & \\ & & 0 \end{bmatrix} + \begin{bmatrix} 1 & 0 & 0 \\ 1 & 0 & 0 \\ 1 & 0 & 0 \\ 1 & 0 & 0 \\ 0 & 1 & 0 \\ 0 & 0 & 1 \end{bmatrix} \begin{bmatrix} 0 & 0 & 0 & 0 & 0 & 4 \\ \frac{41}{7} & 0 & 0 & 0 & 0 & \frac{8}{7} \\ 0 & 0 & 0 & 0 & 7 & 0 \end{bmatrix} = \begin{bmatrix} 0 & 2 & 0 & 1 & 0 & 4 \\ 2 & 0 & 0 & 1 & 0 & 4 \\ 0 & 1 & 0 & 2 & 0 & 4 \\ 0 & 1 & 2 & 0 & 0 & 4 \\ \frac{41}{7} & 0 & 0 & 0 & 0 & \frac{8}{7} \\ 0 & 0 & 0 & 0 & 7 & 0 \end{bmatrix}$$

has the spectrum Λ.

6. Fiedler and Guo results

One of the most important works about the *SNIEP* is due to Fiedler [29]. In [29] Fiedler showed, as it was said before, that Kellogg sufficient conditions for the *RNIEP* are also sufficient for the *SNIEP*. Three important and very useful results of Fiedler are:

Lemma 2. *[29] Let A be a symmetric $m \times m$ matrix with eigenvalues $\alpha_1, \ldots, \alpha_m$, $A\mathbf{u} = \alpha_1 \mathbf{u}$, $\|\mathbf{u}\| = 1$. Let B be a symmetric $n \times n$ matrix with eigenvalues $\beta_1, \ldots, \beta_n$, $B\mathbf{v} = \beta_1 \mathbf{v}$, $\|\mathbf{v}\| = 1$. Then for any ρ, the matrix*

$$C = \begin{bmatrix} A & \rho \mathbf{u}\mathbf{v}^T \\ \rho \mathbf{v}\mathbf{u}^T & B \end{bmatrix}$$

has eigenvalues $\alpha_2, \ldots, \alpha_m, \beta_2, \ldots, \beta_n, \gamma_1, \gamma_2$, where γ_1, γ_2 are eigenvalues of the matrix

$$\widetilde{C} = \begin{bmatrix} \alpha_1 & \rho \\ \rho & \beta_1 \end{bmatrix}.$$

Lemma 3. *[29] If $\{\alpha_1, \ldots, \alpha_m\}$ and $\{\beta_1, \ldots, \beta_n\}$ are lists symmetrically realizable and $\alpha_1 \geq \beta_1$, then for any $t \geq 0$, the list*

$$\{\alpha_1 + t, \beta_1 - t, \alpha_2, \ldots, \alpha_m, \beta_2, \ldots, \beta_n\}$$

is also symmetrically realizable.

Lemma 4. *[29] If $\Lambda = \{\lambda_1, \lambda_2, \ldots, \lambda_n\}$ is symmetrically realizable by a nonnegative matrix and if $t > 0$, then*

$$\Lambda_t = \{\lambda_1 + t, \lambda_2, \ldots, \lambda_n\}$$

is symmetrically realizable by a positive matrix.

Remark 1. *It is not hard to see that Lemma 2 can be obtained from Theorem 6. In fact, it is enough to consider*

$$\begin{aligned} C &= \begin{bmatrix} A & \\ & B \end{bmatrix} + \begin{bmatrix} \mathbf{u} & 0 \\ 0 & \mathbf{v} \end{bmatrix} \begin{bmatrix} 0 & \rho \\ \rho & 0 \end{bmatrix} \begin{bmatrix} \mathbf{u}^T & \mathbf{0}^T \\ \mathbf{0}^T & \mathbf{v}^T \end{bmatrix} \\ &= \begin{bmatrix} A & \rho \mathbf{u}\mathbf{v}^T \\ \rho \mathbf{v}\mathbf{u}^T & B \end{bmatrix}, \end{aligned}$$

which is symmetric with eigenvalues $\gamma_1, \gamma_2, \alpha_2, \ldots, \alpha_m, \beta_2, \ldots, \beta_n$, where γ_1, γ_2 are eigenvalues of

$$B = \begin{bmatrix} \alpha_1 & \rho \\ \rho & \beta_1 \end{bmatrix}.$$

Now we consider a relevant result due to Guo [45]:

Theorem 14. *[45] If the list of complex numbers $\Lambda = \{\lambda_1, \lambda_2, \ldots, \lambda_n\}$ is realizable, where λ_1 is the Perron eigenvalue and $\lambda_2 \in \mathbb{R}$, then for any $t \geq 0$ the list $\Lambda_t = \{\lambda_1 + t, \lambda_2 \pm t, \lambda_3, \ldots, \lambda_n\}$ is also realizable.*

Corollary 2. *[45] If the list of real numbers $\Lambda = \{\lambda_1, \lambda_2, \ldots, \lambda_n\}$ is realizable and $t_1 = \sum_{i=2}^{n} |t_i|$ with $t_i \in \mathbb{R}$, $i = 2, \ldots, n$, then the list $\Lambda_t = \{\lambda_1 + t_1, \lambda_2 + t_2, \ldots, \lambda_n + t_n\}$ is also realizable.*

Example 8. *Let $\Lambda = \{8, 6, 3, 3, -5, -5, -5, -5\}$ be a given list. Since the lists $\Lambda_1 = \Lambda_2 = \{7, 3, -5, -5\}$ are both realizable (see [22] to apply a simple criterion, which shows the realizability of $\Lambda_1 = \Lambda_2$), then*

$$\Lambda_1 \cup \Lambda_2 = \{7, 7, 3, 3, -5, -5, -5, -5\}$$

is also realizable. Now, from Theorem 14, with $t = 1$, Λ is realizable.

Guo also sets the following two questions:
Question 1: Do complex eigenvalues of nonnegative matrices have a property similar to Theorem 14?
Question 2: If the list $\Lambda = \{\lambda_1, \lambda_2, \ldots, \lambda_n\}$ is symmetrically realizable, and $t > 0$, is the list $\Lambda_t = \{\lambda_1 + t, \lambda_2 \pm t, \lambda_3, \ldots, \lambda_n\}$ symmetrically realizable?.

It was shown in [12] and also in [46] that Question 1 has an affirmative answer.

Theorem 15. *[12] Let $\Lambda = \{\lambda_1, a+bi, a-bi, \lambda_4, \ldots, \lambda_n\}$ be a realizable list of complex numbers. Then for all $t \geq 0$, the perturbed list*

$$\Lambda_t = \{\lambda_1 + 2t, a - t + bi, a - t - bi, \lambda_4, \ldots, \lambda_n\}$$

is also realizable.

Question 2, however, remains open. An affirmative answer to Question 2, in the case that the symmetric realizing matrix is a nonnegative circulant matrix or it is a nonnegative left circulant matrix, it was given in [47]. The use of circulant matrices has been shown to be very useful for the *NIEP* [9, 24]. In [24] it was given a necessary and sufficient condition for a list of 5 real numbers, which corresponds to a even-conjugate vector, to be the spectrum of 5×5 symmetric nonnegative circulant matrix:

Lemma 5. *[24] Let $\boldsymbol{\lambda} = (\lambda_1, \lambda_2, \lambda_3, \lambda_3, \lambda_2)^T$ be a vector of real numbers (even-conjugate) such that*

$$\begin{aligned} \lambda_1 &\geq \left|\lambda_j\right|, \; j = 2,3 \\ \lambda_1 &\geq \lambda_2 \geq \lambda_3 \\ \lambda_1 + 2\lambda_2 + 2\lambda_3 &\geq 0 \end{aligned} \tag{13}$$

A necessary and sufficient condition for $\{\lambda_1, \lambda_2, \lambda_3, \lambda_3, \lambda_2\}$ to be the spectrum of a symmetric nonnegative circulant matrix is

$$\lambda_1 + (\lambda_3 - \lambda_2)\frac{\sqrt{5}-1}{2} - \lambda_2 \geq 0. \tag{14}$$

Example 9. *From Lemma 5 we may know, for instance, that the list $\{6, 1, 1, -4, -4\}$ is the spectrum of a symmetric nonnegative circulant matrix.*

7. Some open questions

We finish this chapter by setting two open questions:

Question 1: *If the list of real numbers $\Lambda = \{\lambda_1, \lambda_2, \ldots, \lambda_n\}$ is symmetrically realizable, and $t > 0$, is the list $\Lambda_t = \{\lambda_1 + t, \lambda_2 \pm t, \lambda_3, \ldots, \lambda_n\}$* also symmetrically realizable?

Some progress has been done about this question. In [47], it was given an affirmative answer to Question 1, in the case that the realizing matrix is symmetric nonnegative circulant matrix or it is nonnegative left circulant matrix. In [48] it was shown that if $1 > \lambda_2 \geq \cdots \geq \lambda_n \geq 0$, then Theorem 14 holds for positive stochastic, positive doubly stochastic and positive symmetric matrices.

Question 2: *How adding one or more zeros to a list can lead to its symmetric realizability by different symmetric patterned matrices?*

The famous Boyle-Handelman Theorem [49] gives a nonconstructive proof of the fact that if $s_k = \lambda_1^k + \lambda_2^k + \cdots + \lambda_n^k > 0$, for $k = 1, 2, \ldots$, then there exists a nonnegative number N for which the list $\{\lambda_1, \ldots, \lambda_n, 0, \ldots, 0\}$, with N zeros added, is realizable. In [11] Laffey and Šmigoc completely solve the *NIEP* for lists of complex numbers $\Lambda = \{\lambda_1, \ldots, \lambda_n\}$, closed under conjugation, with $\lambda_2, \ldots, \lambda_n$ having real parts smaller than or equal to zero. They show the existence of $N \geq 0$ for which Λ with N zeros added is realizable and show how to compute the least such N. The situation for symmetrically realizable spectra is different and even less is known.

8. Conclusion

The *nonnegative* inverse eigenvalue problem is an open and difficult problem. A full solution is unlikely in the near future. A number of partial results are known in the literature about the problem, most of them in terms of sufficient conditions. Some matrix results, like Brauer Theorem (Theorem 1), Rado Theorem (Theorem 2), and its symmetric version (Theorem 6) have been shown to be very useful to derive good sufficient conditions. This way, however, seems to be quite narrow and may be other techniques should be explored and applied.

Author details

Ricardo L. Soto
Department of Mathematics, Universidad Católica del Norte, Casilla 1280, Antofagasta, Chile.

9. References

[1] A. Berman, R. J. Plemmons (1994) Nonnegative Matrices in the Mathematical Sciences. In: Classics in Applied Mathematics 9, Society for Industrial and Applied Mathematics (SIAM), Philadelphia, PA.

[2] M. T. Chu, G. H. Golub (2005) Inverse eigenvalue problems: theory, algorithms and applications, Oxford University Press, New York.

[3] H. Minc (1988) Nonnegative Matrices, John Wiley & Sons, New York.

[4] R. Loewy, D. London (1978) A note on an inverse problem for nonnegative matrices. In: Linear and Multilinear Algebra 6 83-90.

[5] M.E. Meehan (1998) Some results on matrix spectra, Ph. D. Thesis, National University of Ireland, Dublin.

[6] J. Torre-Mayo, M.R. Abril-Raymundo, E. Alarcia-Estévez, C. Marijuán, M. Pisonero (2007) The nonnegative inverse eigenvalue problem from the coefficients of the characteristic polynomial. EBL digraphs In: Linear Algebra Appl. 426 729-773.

[7] T. J. Laffey, E. Meehan (1999) A characterization of trace zero nonnegative 5x5 matrices. In: Linear Algebra Appl. 302-303 295-302.

[8] N. Radwan (1996) An inverse eigenvalue problem for symmetric and normal matrices. In: Linear Algebra Appl. 248 101-109.

[9] O. Rojo, R. L. Soto (2003) Existence and construction of nonnegative matrices with complex spectrum. In: Linear Algebra Appl. 368 53-69

[10] A. Borobia, J. Moro, R. L. Soto (2004) Negativity compensation in the nonnegative inverse eigenvalue problem. In: Linear Algebra Appl. 393 73-89.

[11] T. J. Laffey, H. Šmigoc (2006) Nonnegative realization of spectra having negative real parts. In: Linear Algebra Appl. 416 148-159.

[12] T. J. Laffey (2005) Perturbing non-real eigenvalues of nonnegative real matrices. In: Electronic Journal of Linear Algebra 12 73-76.

[13] R.L. Soto, M. Salas, C. Manzaneda (2010) Nonnegative realization of complex spectra. In: Electronic Journal of Linear Algebra 20 595-609.

[14] W. Guo (1996) An inverse eigenvalue problem for nonnegative matrices. In: Linear Algebra Appl. 249 67-78.

[15] C. R. Johnson, T. J. Laffey, R. Loewy (1996) The real and the symmetric nonnegative inverse eigenvalue problems are different. In: Proc. AMS 124 3647-3651.

[16] H. R. Suleimanova (1949) Stochastic matrices with real characteristic values. In: Dokl. Akad. Nauk SSSR 66 343-345.

[17] H. Perfect (1953) Methods of constructing certain stochastic matrices. In: Duke Math. J. 20 395-404.

[18] H. Perfect (1955) Methods of constructing certain stochastic matrices II. In: Duke Math. J. 22 305-311.

[19] R. Kellogg (1971) Matrices similar to a positive or essentially positive matrix. In: Linear Algebra Appl. 4 191-204.

[20] F. Salzmann (1972) A note on eigenvalues of nonnegative matrices. In: Linear Algebra Appl. 5.329-338.

[21] A. Borobia (1995) On the Nonnegative Eigenvalue Problem. In: Linear Algebra Appl. 223-224 131-140.

[22] R. L. Soto (2003) Existence and construction of nonnegative matrices with prescribed spectrum. In: Linear Algebra Appl. 369 169-184.

[23] H. Šmigoc (2005) Construction of nonnegative matrices and the inverse eigenvalue problem. In: Linear and Multilinear Algebra 53 88-96.

[24] R. L. Soto, O. Rojo (2006) Applications of a Brauer Theorem in the nonnegative inverse eigenvalue problem. In: Linear Algebra Appl. 416 (2-3) (2006) 844-856.

[25] A. Borobia, J. Moro, R. L. Soto (2008) A unified view on compensation criteria the real nonnegative inverse eigenvalue problem. In: Linear Algebra Appl. 428 2574-2584.

[26] R.L. Soto (2011) A family of realizability criteria for the real and symmetric nonnegative inverse eigenvalue problem. In: Numerical Linear Algebra with Appl. (2011). DOI: 10.1002/nla.835.

[27] P. Egleston, T. Lenker, S. Narayan (2004) The nonnegative inverse eigenvalue problem. In: Linear Algebra Appl. 379 475-490.

[28] C. Marijuán, M. Pisonero, R. L. Soto (2007) A map of sufficient conditions for the real nonnegative inverse eigenvalue problem. In: Linear Algebra Appl. 426 (2007) 690-705.

[29] M. Fiedler (1974) Eigenvalues of nonnegative symmetric matrices. In: Linear Algebra Appl. 9 119-142.

[30] G. Soules (1983) Constructing symmetric nonnegative matrices. In: Linear and Multilinear Algebra 13 241-251.

[31] R. Reams (2002) Constructions of trace zero symmetric stochastic matrices for the inverse eigenvalue problem. In: Electronic Journal of Linear Algebra 9 270-275.

[32] R. Loewy, J. J. McDonald (2004) The symmetric nonnegative inverse eigenvalue problem for 5×5 matrices. In: Linear Algebra Appl. 393 275-298.

[33] R. L. Soto (2006) Realizability criterion for the symmetric nonnegative inverse eigenvalue problem. In: Linear Algebra Appl. 416 (2-3) 783-794.

[34] R. L. Soto, O. Rojo, J. Moro, A. Borobia (2007) Symmetric nonnegative realization of spectra. In: Electronic Journal of Linear Algebra 16 1-18.
[35] T. J. Laffey, H. Šmigoc (2007) Construction of nonnegative symmetric matrices with given spectrum. In: Linear Algebra Appl. 421 97-109.
[36] R.L. Soto, O. Rojo, C.B. Manzaneda (2011) On nonnegative realization of partitioned spectra. In: Electronic Journal of Linear Algebra 22 557-572.
[37] O. Spector (2011) A characterization of trace zero symmetric nonnegative 5×5 matrices. In: Linear Algebra Appl. 434 1000-1017.
[38] T. J. Laffey, E. Meehan (1998) A refinament of an inequality of Johnson, Loewy and London on nonnegative matrices and some applications. In: Electronic Journal of Linear Algebra 3 119-128.
[39] O. Holtz (2004) M-matrices satisfy Newton's inequalities. In: Proceedings of the AMS 133 (3) (2004) 711-717.
[40] C. R. Johnson (1981) Row stochastic matrices similar to doubly stochastic matrices. In: Linear and Multilinear Algebra 10 113-130.
[41] A. Brauer (1952) Limits for the characteristic roots of a matrix. IV: Aplications to stochastic matrices. In: Duke Math. J., 19 75-91.
[42] R. Reams (1996) An inequality for nonnegative matrices and the inverse eigenvalue problem. In: Linear and Multilinear Algebra 41 367-375.
[43] R. A. Horn, C. R. Johnson (1991) Matrix Analysis, Cambridge University Press, Cambridge.
[44] R.L. Soto, A.I. Julio (2011) A note on the symmetric nonnegative inverse eigenvalue problem. In: International Mathematical Forum 6 N° 50, 2447-2460.
[45] W. Guo (1997) Eigenvalues of nonnegative matrices. In: Linear Algebra Appl. 266 261-270.
[46] S. Guo, W. Guo (2007) Perturbing non-real eigenvalues of nonnegative real matrices. In: Linear Algebra Appl. 426 199-203.
[47] O. Rojo, R. L. Soto (2009) Guo perturbations for symmetric nonnegative circulant matrices. In: Linear Algebra Appl. 431 594-607.
[48] J. Ccapa, R.L. Soto (2009) On spectra perturbation and elementary divisors of positive matrices. In: Electron. J. of Linear Algebra 18 462-481.
[49] M. Boyle and D. Handelman, The spectra of nonnegative matrices via symbolic dynamics, Ann.of Math. 133: 249-316 (1991).

Chapter 6

Identification of Linear, Discrete-Time Filters via Realization

Daniel N. Miller and Raymond A. de Callafon

Additional information is available at the end of the chapter

http://dx.doi.org/10.5772/48293

1. Introduction

The realization of a discrete-time, linear, time-invariant (LTI) filter from its impulse response provides insight into the role of linear algebra in the analysis of both dynamical systems and rational functions. For an LTI filter, a sequence of output data measured over some finite period of time may be expressed as the linear combination of the past input and the input measured over that same period. For a finite-dimensional LTI filter, the mapping from past input to future output is a finite-rank linear operator, and the effect of past input, that is, the memory of the system, may be represented as a finite-dimensional vector. This vector is the *state* of the system.

The central idea of realization theory is to first identify the mapping from past input to future output and to then factor it into two parts: a map from the input to the state and another from the state to the output. This factorization guarantees that the resulting system representation is both casual and finite-dimensional; thus it can be physically constructed, or *realized*.

System identification is the science of constructing dynamic models from experimentally measured data. Realization-based identification methods construct models by estimating the mapping from past input to future output based on this measured data. The non-deterministic nature of the estimation process causes this mapping to have an arbitrarily large rank, and so a rank-reduction step is required to factor the mapping into a suitable state-space model. Both these steps must be carefully considered to guarantee unbiased estimates of dynamic systems.

The foundations of realization theory are primarily due to Kalman and first appear in the landmark paper of [1], though the problem is not defined explicitly until [2], which also coins the term "realization" as being the a state-space model of a linear system constructed from an experimentally measured impulse response. It was [3] that introduced the structured-matrix approach now synonymous with the term "realization theory" by re-interpreting a theorem originally due to [4] in a state-space LTI system framework.

Although Kalman's original definition of "realization" implied an identification problem, it was not until [5] proposed rank-reduction by means of the singular-value decomposition that Ho's method became feasible for use with non-deterministic data sets. The combination of Ho's method and the singular-value decomposition was finally generalized to use with experimentally measured data by Kung in [6].

With the arrival of Kung's method came the birth of what is now known as the field of *subspace identification* methods. These methods use structured matrices of arbitrary input and output data to estimate a state-sequence from the system. The system is then identified from the propagation of the state over time. While many subspace methods exist, the most popular are the Multivariable Output-Error State Space (MOESP) family, due to [7], and the Numerical Algorithms for Subspace State-Space System Identification (N4SID) family, due to [8]. Related to subspace methods is the Eigensystem Realization Algorithm [9], which applies Kung's algorithm to impulse-response estimates, which are typically estimated through an Observer/Kalman Filter Identification (OKID) algorithm [10].

This chapter presents the central theory behind realization-based system identification in a chronological context, beginning with Kronecker's theorem, proceeding through the work of Kalman and Kung, and presenting a generalization of the procedure to arbitrary sets of data. This journey provides an interesting perspective on the original role of linear algebra in the analysis of rational functions and highlights the similarities of the different representations of LTI filters. Realization theory is a diverse field that connects many tools of linear algebra, including structured matrices, the QR-decomposition, the singular-value decomposition, and linear least-squares problems.

2. Transfer-function representations

We begin by reviewing some properties of discrete-time linear filters, focusing on the role of infinite series expansions in analyzing the properties of rational functions. The reconstruction of a transfer function from an infinite impulse response is equivalent to the reconstruction of a rational function from its Laurent series expansion. The reconstruction problem is introduced and solved by forming structured matrices of impulse-response coefficients.

2.1. Difference equations and transfer functions

Discrete-time linear filters are most frequently encountered in the form of difference equations that relate an input signal u_k to an output signal y_k. A simple example is an output y_k determined by a weighted sum of the inputs from u_k to u_{k-m},

$$y_k = b_m u_k + b_{m-1} u_{k-1} + \cdots + b_0 u_{k-m}. \tag{1}$$

More commonly, the output y_k also contains a weighted sum of previous outputs, such as a weighted sum of samples from y_{k-1} to y_{k-n},

$$y_k = b_m u_k + b_{m-1} u_{k-1} + \cdots + b_0 u_{k-m} - a_{n-1} y_{k-1} - a_{n-2} y_{k-2} - \cdots + a_0 y_{k-n}. \tag{2}$$

The impulse response of a filter is the output sequence $g_k = y_k$ generated from an input

$$u_k = \begin{cases} 1 & k = 0, \\ 0 & k \neq 0. \end{cases} \tag{3}$$

The parameters g_k are the impulse-response coefficients, and they completely describe the behavior of an LTI filter through the convolution

$$y_k = \sum_{j=0}^{\infty} g_j u_{k-j}. \tag{4}$$

Filters of type (1) are called finite-impulse response (FIR) filters because g_k is a finite-length sequence that settles to 0 once $k > m$. Filters of type (2) are called infinite impulse response (IIR) filters since generally the impulse response will never completely settle to 0.

A system is stable if a bounded u_k results in a bounded y_k. Because the output of LTI filters is a linear combination of the input and previous output, any input-output sequence can be formed from a linear superposition of other input-output sequences. Hence proving that the system has a bounded output for a single input sequence is necessary and sufficient to prove the stability of an LTI filter. The simplest input to consider is an impulse, and so a suitable definition of system stability is that the absolute sum of the impulse response is bounded,

$$\sum_{k=0}^{\infty} |g_k| < \infty. \tag{5}$$

Though the impulse response completely describes the behavior of an LTI filter, it does so with an infinite number of parameters. For this reason, discrete-time LTI filters are often written as transfer functions of a complex variable z. This enables analysis of filter stability and computation of the filter's frequency response in a finite number of calculations, and it simplifies convolution operations into basic polynomial algebra.

The transfer function is found by grouping output and input terms together and taking the Z-transform of both signals. Let $Y(z) = \sum_{k=-\infty}^{\infty} y_k z^{-k}$ be the Z-transform of y_k and $U(z)$ be the Z-transform of u_k. From the property

$$\mathcal{Z}\left[y_{k-1}\right] = Y(z)z^{-1}$$

the relationship between $Y(z)$ and $U(z)$ may be expressed in polynomials of z as

$$a(z)Y(z) = b(z)U(z).$$

The ratio of these two polynomials is the filter's transfer function

$$G(z) = \frac{b(z)}{a(z)} = \frac{b_m z^m + b_{m-1} z^{m-1} + \cdots + b_1 z + b_0}{z^n + a_{n-1} z^{n-1} + \cdots + a_1 z + a_0}. \tag{6}$$

When $n \geq m$, $G(z)$ is *proper*. If the transfer function is not proper, then the difference equations will have y_k dependent on future input samples such as u_{k+1}. Proper transfer functions are required for causality, and thus all physical systems have proper transfer function representations. When $n > m$, the system is *strictly proper*. Filters with strictly proper transfer functions have no feed-through terms; the output y_k does not depend on u_k, only the preceding input $u_{k-1}, u_{k-2}, \ldots$. In this chapter, we assume all systems are causal and all transfer functions proper.

If $a(z)$ and $b(z)$ have no common roots, then the rational function $G(z)$ is *coprime*, and the order n of $G(z)$ cannot be reduced. Fractional representations are not limited to single-input-single-output systems. For vector-valued input signals $u_k \in \mathbb{R}^{n_u}$ and output signals $y_k \in \mathbb{R}^{n_y}$, an LTI filter may be represented as an $n_y \times n_u$ matrix of rational functions $G_{ij}(z)$, and the system will have matrix-valued impulse-response coefficients. For simplicity, we will assume that transfer function representations are single-input-single-output, though all results presented here generalize to the multi-input-multi-output case.

2.2. Stability of transfer function representations

Because the effect of $b(z)$ is equivalent to a finite-impulse response filter, the only requirement for $b(z)$ to produce a stable system is that its coefficients be bounded, which we may safely assume is always the case. Thus the stability of a transfer function $G(z)$ is determined entirely by $a(z)$, or more precisely, the roots of $a(z)$. To see this, suppose $a(z)$ is factored into its roots, which are the poles p_i of $G(z)$,

$$G(z) = \frac{b(z)}{\prod_{i=1}^{n}(z - p_i)}. \tag{7}$$

To guarantee a bounded y_k, it is sufficient to study a single pole, which we will denote simply as p. Thus we wish to determine necessary and sufficient conditions for stability of the system

$$G'(z) = \frac{1}{z - p}. \tag{8}$$

Note that p may be complex. Assume that $|z| > |p|$. $G'(z)$ then has the Laurent-series expansion

$$G'(z) = z^{-1}\left(\frac{1}{1 - pz^{-1}}\right) = z^{-1}\sum_{k=0}^{\infty} p^k z^{-k} = \sum_{k=1}^{\infty} p^{k-1} z^{-k}. \tag{9}$$

From the time-shift property of the z-transform, it is immediately clear that the sequence

$$g'_k = \begin{cases} 0 & k = 1, \\ p^{k-1} & k > 1, \end{cases} \tag{10}$$

is the impulse response of $G'(z)$. If we require that (9) is absolutely summable and let $|z| = 1$, the result is the original stability requirement (5), which may be written in terms of p as

$$\sum_{k=1}^{\infty} \left| p^{k-1} \right| < \infty.$$

This is true if and only if $|p| < 1$, and thus $G'(z)$ is stable if and only if $|p| < 1$. Finally, from (7) we may deduce that a system is stable if and only if all the poles of $G(z)$ satisfy the property $|p_i| < 1$.

2.3. Construction of transfer functions from impulse responses

Transfer functions are a convenient way of representing complex system dynamics in a finite number of parameters, but the coefficients of $a(z)$ and $b(z)$ cannot be measured directly. The impulse response of a system can be found experimentally by either direct measurement or from other means such as taking the inverse Fourier transform of a measured frequency response [11]. It cannot, however, be represented in a finite number of parameters. Thus the conversion between transfer functions and impulse responses is an extremely useful tool.

For a single-pole system such as (8), the expansion (9) provides an obvious means of reconstructing a transfer function from a measured impulse response: given any 2 sequential impulse-response coefficients g_k and g_{k+1}, the pole of $G'(z)$ may be found from

$$p = g_k^{-1} g_{k+1}. \tag{11}$$

Notice that this is true for any k, and the impulse response can be said to have a *shift-invariant* property in this respect.

Less clear is the case when an impulse response is generated by a system with higher-order $a(z)$ and $b(z)$. In fact, there is no guarantee that an arbitrary impulse response is the result of a linear system of difference equations at all. For an LTI filter, however, the coefficients of the impulse response exhibit a linear dependence which may be used to not only verify the linearity of the system, but to construct a transfer function representation as well. The exact nature of this linear dependence may be found by forming a structured matrix of impulse response coefficients and examining its behavior when the indices of the coefficients are shifted forward by a single increment, similar to the single-pole case in (11). The result is stated in the following theorem, originally due to Kronecker [4] and adopted from the English translation of [12].

Theorem 1 (Kronecker's Theorem). *Suppose $G(z) : \mathbb{C} \to \mathbb{C}$ is an infinite series of descending powers of z, starting with z^{-1},*

$$G(z) = g_1 z^{-1} + g_2 z^{-2} + g_3 z^{-3} + \cdots = \sum_{k=1}^{\infty} g_k z^{-k}. \tag{12}$$

Assume G(z) is analytic (the series converges) for all $|z| > 1$. Let H be an infinitely large matrix of the form

$$H = \begin{bmatrix} g_1 & g_2 & g_3 & \cdots \\ g_2 & g_3 & g_4 & \cdots \\ g_3 & g_4 & g_5 & \cdots \\ \vdots & \vdots & \vdots & \end{bmatrix} \tag{13}$$

Then H has finite rank n if and only if $G(z)$ is a strictly proper, coprime, rational function of degree n with poles inside the unit circle. That is, $G(z)$ has an alternative representation

$$G(z) = \frac{b(z)}{a(z)} = \frac{b_m z^m + b_{m-1} z^{m-1} + \cdots + b_1 z + b_0}{z^n + a_{n-1} z^{n-1} + \cdots + a_1 z + a_0}, \tag{14}$$

in which $m < n$, all roots of $a(z)$ satisfy $|z| < 1$, $a(z)$ and $b(z)$ have no common roots, and we have assumed without loss of generality that $a(z)$ is monic.

To prove Theorem 1, we first prove that for $k > n$, g_k must be linearly dependent on the previous n terms of the series for H to have finite rank.

Theorem 2. *The infinitely large matrix H is of finite rank n if and only if there exists a finite sequence $\alpha_1, \alpha_2, \cdots, \alpha_n$ such that for $k \geq n$,*

$$g_{k+1} = \sum_{j=1}^{n} \alpha_j g_{k-j+1}, \tag{15}$$

and n is the smallest number with this property.

Proof. Let h_k be the row of H beginning with g_k. If H has rank n, then the first $n+1$ rows of H are linearly dependent. This implies that for some $1 \le p \le n$, h_{p+1} is a linear combination of $h_1, \ldots, h_p$, and thus there exists some sequence α_k such that

$$h_{p+1} = \sum_{j=1}^{p} \alpha_j h_{p-j+1}. \tag{16}$$

The structure and infinite size of H imply that such a relationship must hold for all following rows of H, so that for $q \ge 0$

$$h_{q+p+1} = \sum_{j=1}^{p} \alpha_j h_{q+p-j+1}.$$

Hence any row h_k, $k > p$, can be expressed as a linear combination of the previous p rows. Since H has at least n linearly independent rows, $p = n$, and since this applies element-wise, $\text{rank}(H) = n$ implies (15).

Alternatively, (15) implies a relationship of the form (16) exists, and hence $\text{rank}(H) = p$. Since n is the smallest possible p, this implies $\text{rank}(H) = n$. □

We now prove Theorem 1.

Proof. Suppose $G(z)$ is a coprime rational function of the form (14) with series expansion (12), which we know exists, since $G(z)$ is analytic for $|z| < 1$. Without loss of generality, let $m = n-1$, since we may always let $b_k = 0$ for some k. Hence

$$\frac{b_{n-1}z^{n-1} + b_{n-2}z^{n-2} + \cdots + b_1 z + b_0}{z^n + a_{n-1}z^{n-1} + \cdots + a_1 z + a_0} = g_1 z^{-1} + g_2 z^{-2} + g_3 z^{-3} + \cdots$$

Multiplying both sides by the denominator of the left,

$$\begin{aligned} b_{n-1}z^{n-1} &+ b_{n-2}z^{n-2} + \cdots + b_1 z + b_0 \\ &= g_1 z^{n-1} + (g_2 + g_1 a_{n-1}) z^{n-2} + (g_3 + g_2 a_{n-1} + g_1 a_{n-2}) z^{n-3} + \cdots, \end{aligned}$$

and equating powers of z, we find

$$\begin{aligned} b_{n-1} &= g_1 \\ b_{n-2} &= g_2 + g_1 a_{n-1} \\ b_{n-3} &= g_3 + g_2 a_{n-1} + g_1 a_{n-2} \\ &\vdots \\ b_1 &= g_{n-1} + g_{n-2} a_{n-1} + \cdots + g_1 a_2 \\ b_0 &= g_n + g_{n-1} a_{n-1} + \cdots + g_1 a_1 \\ 0 &= g_{k+1} + g_k a_{n-1} + \cdots + g_{k-n+1} a_0 \qquad k \ge n. \end{aligned} \tag{17}$$

From this, we have, for $k \geq n$,

$$g_{k+1} = \sum_{j=1}^{n} -a_j g_{k-j+1},$$

which not only shows that (15) holds, but also shows that $\alpha_j = -a_j$. Hence by Theorem 2, H must have finite rank.

Conversely, suppose H has finite rank. Then (15) holds, and we may construct $a(z)$ from α_k and $b(z)$ from (17) to create a rational function. This function must be coprime since its order n is the smallest possible. □

The construction in Theorem 1 is simple to extend to the case in which $G(z)$ is only proper and not strictly proper; the additional coefficient b_n is simply the feed-through term in the impulse response, that is, g_0.

A result of Theorem 2 is that given finite-dimensional, full-rank matrices

$$H_k = \begin{bmatrix} g_k & g_{k+1} & \cdots & g_{k+n-1} \\ g_{k+1} & g_{k+2} & \cdots & g_{k+n} \\ \vdots & \vdots & & \vdots \\ g_{k+n-1} & g_{k+n} & \cdots & g_{k+2n-2} \end{bmatrix} \tag{18}$$

and

$$H_{k+1} = \begin{bmatrix} g_{k+1} & g_{k+2} & \cdots & g_{k+n} \\ g_{k+2} & g_{k+3} & \cdots & g_{k+n+1} \\ \vdots & \vdots & & \vdots \\ g_{k+n} & g_{k+n+1} & \cdots & g_{k+2n-1} \end{bmatrix}, \tag{19}$$

the coefficients of $a(z)$ may be calculated as

$$\begin{bmatrix} 0 & 0 & \cdots & 0 & -a_0 \\ 1 & 0 & \cdots & 0 & -a_1 \\ 0 & 1 & \cdots & 0 & -a_2 \\ \vdots & \vdots & \ddots & \vdots & \vdots \\ 0 & 0 & \cdots & 1 & -a_{n-1} \end{bmatrix} = H_k^{-1} H_{k+1}. \tag{20}$$

Notice that (11) is in fact a special case of (20). Thus we need only know the first $2n+1$ impulse-response coefficients to reconstruct the transfer function $G(z)$: $2n$ to form the matrices H_k and H_{k+1} from (18) and (19), respectively, and possibly the initial coefficient g_0 in case of an nth-order $b(z)$.

Matrices with the structure of H are useful enough to have a special name. A *Hankel matrix* H is a matrix constructed from a sequence $\{h_k\}$ so that each element $H_{(j,k)} = h_{j+k}$. For the Hankel matrix in (13), $h_k = g_{k-1}$. H_k also has an interesting property implied by (20): its row space is invariant under shifting of the index k. Because its symmetric, this is also true for its column space. Thus this matrix is also often referred to as being *shift-invariant*.

While (20) provides a potential method of identifying a system from a measured impulse response, this is not a reliable method to use with measured impulse response coefficients that are corrupted by noise. The exact linear dependence of the coefficients will not be identical for all k, and the structure of (20) will not be preserved. Inverting H_k will also invert any noise on g_k, potentially amplifying high-frequency noise content. Finally, the system order n is required to be known beforehand, which is usually not the case if only an impulse response is available. Fortunately, these difficulties may all be overcome by reinterpreting the results Kronecker's theorem in a state-space framework. First, however, we more carefully examine the role of the Hankel matrix in the behavior of LTI filters.

2.4. Hankel and Toeplitz operators

The Hankel matrix of impulse response coefficients (13) is more than a tool for computing the transfer function representation of a system from its impulse response. It also defines the mapping of past input signals to future output signals. To define exactly what this means, we write the convolution of (4) around sample $k = 0$ in matrix form as

$$\begin{bmatrix} \vdots \\ y_{-3} \\ y_{-2} \\ y_{-1} \\ \hline y_0 \\ y_1 \\ y_2 \\ \vdots \end{bmatrix} = \left[\begin{array}{ccc|ccc} \ddots & & & & \cdots & 0 \\ \cdots & g_0 & & & & \vdots \\ \cdots & g_1 & g_0 & & & \\ \cdots & g_2 & g_1 & g_0 & & \\ \hline \cdots & g_3 & g_2 & g_1 & g_0 & \\ \cdots & g_4 & g_3 & g_2 & g_1 \; g_0 & \\ \cdots & g_5 & g_4 & g_3 & g_2 \; g_1 & g_0 \\ & \vdots & \vdots & \vdots & \vdots \;\; \vdots & \vdots \;\; \ddots \end{array}\right] \begin{bmatrix} \vdots \\ u_{-3} \\ u_{-2} \\ u_{-1} \\ \hline u_0 \\ u_1 \\ u_2 \\ \vdots \end{bmatrix},$$

where the vectors and matrix have been partitioned into sections for $k < 0$ and $k \geq 0$. The output for $k \geq 0$ may then be split into two parts:

$$\underbrace{\begin{bmatrix} y_0 \\ y_1 \\ y_2 \\ \vdots \end{bmatrix}}_{y_f} = \underbrace{\begin{bmatrix} g_1 & g_2 & g_3 & \cdots \\ g_2 & g_3 & g_4 & \cdots \\ g_3 & g_4 & g_5 & \cdots \\ \vdots & \vdots & \vdots & \end{bmatrix}}_{H} \underbrace{\begin{bmatrix} u_{-1} \\ u_{-2} \\ u_{-3} \\ \vdots \end{bmatrix}}_{u_p} + \underbrace{\begin{bmatrix} g_0 & & \cdots & 0 \\ g_1 & g_0 & & \vdots \\ g_2 & g_1 & g_0 & \\ \vdots & \vdots & \vdots & \ddots \end{bmatrix}}_{T} \underbrace{\begin{bmatrix} u_0 \\ u_1 \\ u_2 \\ \vdots \end{bmatrix}}_{u_f}, \quad (21)$$

where the subscripts p and f denote "past" and "future," respectively. The system Hankel matrix H has returned to describe the effects of the past input u_p on the future output y_f. Also present is the matrix T, which represents the convolution of future input u_f with the impulse response. Matrices such as T with constant diagonals are called *Toeplitz* matrices.

From (21), it can be seen that H defines the effects of past input on future output. One interpretation of this is that H represents the "memory" of the system. Because H is a linear mapping from u_p to y_f, the induced matrix 2-norm of H, $||H||_2$, can be considered a function norm, and in a sense, $||H||_2$ is a measure of the "gain" of the system. $||H||_2$ is often called the *Hankel-norm* of a system, and it plays an important role in model reduction and in the analysis of anti-causal systems. More information on this aspect of linear systems can be found in the literature of robust control, for instance, [13].

3. State-space representations

Although transfer functions define system behavior completely with a finite number of parameters and simplify frequency-response calculations, they are cumbersome to manipulate when the input or output is multi-dimensional or when initial conditions must be considered. The other common representation of LTI filters is the state-space form

$$\begin{aligned} x_{k+1} &= Ax_k + Bu_k \\ y_k &= Cx_k + Du_k, \end{aligned} \tag{22}$$

in which $x_k \in \mathbb{R}^n$ is the system state. The matrices $A \in \mathbb{R}^{n\times n}$, $B \in \mathbb{R}^{n\times n_u}$, $C \in \mathbb{R}^{n_y\times n}$, and $D \in \mathbb{R}^{n_y\times n_u}$ completely parameterize the system. Only D uniquely defines the input-output behavior; any nonsingular matrix T' may be used to change the state basis via the relationships

$$x' = T'x \qquad A' = T'AT'^{-1} \qquad B' = T'B \qquad C' = CT'^{-1}.$$

The Z-transform may also be applied to the state-space equations (22) to find

$$\begin{aligned} \mathcal{Z}[x_{k+1}] &= A\mathcal{Z}[x_k] + B\mathcal{Z}[u_k] & \Rightarrow \qquad X(z)z &= AX(z) + BU(z) \\ \mathcal{Z}[y_k] &= C\mathcal{Z}[x_k] + D\mathcal{Z}[u_k] & \Rightarrow \qquad Y(z) &= CX(z) + DU(z) \end{aligned}$$

$$\frac{Y(z)}{U(z)} = G(z) \qquad\qquad G(z) = C\left(zI - A\right)^{-1} B + D, \tag{23}$$

and thus, if (22) is the state-space representation of the single-variable system (6), then $a(z)$ is the characteristic polynomial of A, $\det(zI - A)$.

Besides clarifying the effect of initial conditions on the output, state-space representations are inherently causal, and (23) will always result in a proper system (strictly proper if $D = 0$). For this reason, state-space representations are often called *realizable* descriptions; while the forward-time-shift of z is an inherently non-causal operation, state-space systems may always be constructed in reality.

3.1. Stability, controllability, and observability of state-space representations

The system impulse response is simple to formulate in terms of the state-space parameters by calculation of the output to a unit impulse with $x_0 = 0$:

$$g_k = \begin{cases} D & k = 0, \\ CA^{k-1}B & k > 0 \end{cases}. \tag{24}$$

Notice the similarity of (10) and (24). In fact, from the eigenvalue decomposition of A,

$$A = V\Lambda V^{-1},$$

we find

$$\sum_{k=1}^{\infty} |g_k| = \sum_{k=1}^{\infty} \left|CA^{k-1}B\right| = \sum_{k=1}^{\infty} |CV| \left(|\Lambda^{k-1}|\right) \left|V^{-1}B\right|.$$

The term $|\Lambda^{k-1}|$ will only converge if the largest eigenvalue of A is within the unit circle, and thus the condition that all eigenvalues λ_i of A satisfy $|\lambda_i| < 1$ is a necessary and sufficient condition for stability.

For state-space representations, there is the possibility that a combination of A and B will result in a system for which x_k cannot be entirely controlled by the input u_k. Expressing x_k in a matrix-form similar to (21) as

$$x_k = \mathcal{C} \begin{bmatrix} u_{k-1} \\ u_{k-2} \\ u_{k-3} \\ \vdots \end{bmatrix}, \qquad \mathcal{C} = \begin{bmatrix} B & AB & A^2B & \cdots \end{bmatrix} \tag{25}$$

demonstrates that x_k is in subspace $\mathbb{R}^n$ if and only if $\mathcal{C}$ has rank n. $\mathcal{C}$ is the *controllability matrix* and the system is *controllable* if it has full row rank.

Similarly, the state x_k may not uniquely determine the output for some combinations of A and C. Expressing the evolution of the output as a function of the state in matrix-form as

$$\begin{bmatrix} y_k \\ y_{k+1} \\ y_{k+2} \\ \vdots \end{bmatrix} = \mathcal{O} x_k, \qquad \mathcal{O} = \begin{bmatrix} C \\ CA \\ CA^2 \\ \vdots \end{bmatrix}$$

demonstrates that there is no nontrivial null space in the mapping from x_k to y_k if and only if $\mathcal{O}$ has rank n. $\mathcal{O}$ is the *observability matrix* and the system is *observable* if it has full column rank.

Systems that are both controllable and observable are called *minimal*, and for minimal systems, the dimension n of the state variable cannot be reduced. In the next section we show that minimal state-space system representations convert to coprime transfer functions that are found through (23).

3.2. Construction of state-space representations from impulse responses

The fact that the denominator of $G(z)$ is the characteristic polynomial of A not only allows for the calculation of a transfer function from a state-space representation, but provides an alternative version of Kronecker's theorem for state-space systems, known as the Ho-Kalman Algorithm [3]. From the Caley-Hamilton theorem, if $a(z)$ is the characteristic polynomial of A, then $a(A) = 0$, and

$$\begin{aligned} CA^k a(A) B &= CA^k \left(A^n + a_{n-1}A^{n-1} + \cdots + a_1 A + a_0 \right) B \\ &= CA^{k+n}B + \sum_{j=0}^{n-1} a_j CA^{k+j} B, \end{aligned}$$

which implies

$$CA^{k+n}B = -\sum_{j=0}^{n-1} a_j CA^{k+j} B. \tag{26}$$

Indeed, substitution of (24) into (26) and rearrangement of the indices leads to (15). Additionally, substitution of (24) into the product of $\mathcal{O}$ and $\mathcal{C}$ shows that

$$\mathcal{O}\mathcal{C} = \begin{bmatrix} CB & CAB & CA^2B & \cdots \\ CAB & CA^2B & CA^3B & \cdots \\ CA^2B & CA^3B & CA^4B & \cdots \\ \vdots & \vdots & \vdots & \end{bmatrix} = \begin{bmatrix} g_1 & g_2 & g_3 & \cdots \\ g_2 & g_3 & g_4 & \cdots \\ g_3 & g_4 & g_5 & \cdots \\ \vdots & \vdots & \vdots & \end{bmatrix} = H,$$

which confirms our previous statement that H effectively represents the memory of the system. Because

$$\text{rank}(H) = \min\{\text{rank}(\mathcal{O}), \text{rank}(\mathcal{C})\},$$

we see that $\text{rank}(H) = n$ implies the state-space system (22) is minimal.

If the entries of H are shifted forward by one index to form

$$\overline{H} = \begin{bmatrix} g_2 & g_3 & g_4 & \cdots \\ g_3 & g_4 & g_5 & \cdots \\ g_4 & g_5 & g_6 & \cdots \\ \vdots & \vdots & \vdots & \end{bmatrix},$$

then once again substituting (24) reveals

$$\overline{H} = \mathcal{O}A\mathcal{C}. \tag{27}$$

Thus the row space and column space of H are invariant under a forward-shift of the indices, implying the same shift-invariant structure seen in (20).

The appearance of A in (27) hints at a method for constructing a state-space realization from an impulse response. Suppose the impulse response is known exactly, and let H_r be a finite slice of H with r block rows and L columns,

$$H_r = \begin{bmatrix} g_1 & g_2 & g_3 & \cdots & g_L \\ g_2 & g_3 & g_4 & \cdots & g_{L+1} \\ g_3 & g_4 & g_5 & \cdots & g_{L+2} \\ \vdots & \vdots & \vdots & & \vdots \\ g_{r-1} & g_r & g_{r+1} & \cdots & g_{r+L-1} \end{bmatrix}.$$

Then any appropriately dimensioned factorization

$$H_r = \mathcal{O}_r\mathcal{C}_L = \begin{bmatrix} C \\ CA \\ CA^2 \\ \vdots \\ CA^{r-1} \end{bmatrix} \begin{bmatrix} B & AB & A^2B & \cdots & A^{L-1}B \end{bmatrix} \tag{28}$$

may be used to find A for some arbitrary state basis as

$$A = (\mathcal{O}_r)^{\dagger}\,\overline{H}_r\,(\mathcal{C}_L)^{\dagger} \tag{29}$$

where $\overline{H}_r$ is H_r with the indices shifted forward once and $(\cdot)^\dagger$ is the Moore-Penrose pseudoinverse. C taken from the first block row of $\mathcal{O}_r$, B taken from the first block column of $\mathcal{C}_L$, and D taken from g_0 then provides a complete and minimal state-space realization from an impulse response. Because H_r has rank n and $\det(zI - A)$ has degree n, we know from Kronecker's theorem that $G(z)$ taken from (23) will be coprime.

However, as mentioned before, the impulse response of the system is rarely known exactly. In this case only an estimate $\hat{H}_r$ with a non-deterministic error term is available:

$$\hat{H}_r = H_r + E.$$

Because E is non-deterministic, $\hat{H}$ will always have full rank, regardless of the number of rows r. Thus n cannot be determined from examining the rank of H, and even if n is known beforehand, a factorization (28) for $r > n$ will not exist. Thus we must find a way of reducing the rank of $\hat{H}_r$ in order to find a state-space realization.

3.3. Rank-reduction of the Hankel matrix estimate

If $\hat{H}_r$ has full rank, or if n is unknown, its rank must be reduced prior to factorization. The obvious tool for reducing the rank of matrices is the *singular-value decomposition* (SVD). Assume for now that n is known. The SVD of $\hat{H}_r$ is

$$\hat{H}_r = U\Sigma V^T$$

where U and V^T are orthogonal matrices and Σ is a diagonal matrix containing the nonnegative *singular values* σ_i ordered from largest to smallest. The SVD for a matrix is unique and guaranteed to exist, and the number of nonzero singular values of a matrix is equal to its rank [14].

Because U and V^T are orthogonal, the SVD satisfies

$$\hat{H}_r = \left|\left|U\Sigma V^T\right|\right|_2 = ||\Sigma||_2 = \sigma_1 \tag{30}$$

where $||\cdot||_2$ is the induced matrix 2-norm, and

$$\hat{H}_r = \left|\left|U\Sigma V^T\right|\right|_F = ||\Sigma||_F = \left(\sum_i^l \sigma_i^2\right)^{1/2} \tag{31}$$

where $||\cdot||_F$ is the Frobenius norm. Equation (30) also shows that the Hankel norm of a system is the maximum singular value of H_r. From (30) and (31), we can directly see that if the SVD of H_r is partitioned into

$$\hat{H}_r = \begin{bmatrix} U_n & U_s \end{bmatrix} \begin{bmatrix} \Sigma_n & 0 \\ 0 & \Sigma_s \end{bmatrix} \begin{bmatrix} V_n^T \\ V_s^T \end{bmatrix},$$

where U_n is the first n columns of U, Σ_n is the upper-left $n \times n$ block of Σ, and V_n^T is the first n rows of V^T, the solution to the rank-reduction problem is [14]

$$Q = \underset{\text{rank}(Q)=n}{\arg\min} \, ||Q - \hat{H}_r||_2 = \underset{\text{rank}(Q)=n}{\arg\min} \, ||Q - \hat{H}_r||_F = U_n \Sigma_n V_n^T.$$

Additionally, the error resulting from the rank reduction is

$$e = ||Q - \hat{H}_r||_2 = \sigma_{n+1},$$

which suggests that if the rank of H_r is not known beforehand, it can be determined by examining the nonzero singular values in the deterministic case or by searching for a significant drop-off in singular values if only a noise-corrupted estimate is available.

3.4. Identifying the state-space realization

From a rank-reduced $\hat{H}_r$, any factorization

$$\hat{H}_r = \hat{\mathcal{O}}_r \hat{\mathcal{C}}_L$$

can be used to estimate $\mathcal{O}_r$ and $\mathcal{C}_L$. The error in the state-space realization, however, will depend on the chosen state basis. Generally we would like to have a state variable with a norm $||x_k||_2$ in between $||u_k||_2$ and $||y_k||_2$. As first proposed in [5], choosing the factorization

$$\hat{\mathcal{O}}_r = U_n \Sigma_n^{1/2} \quad \text{and} \quad \hat{\mathcal{C}}_L = \Sigma_n^{1/2} V_n^T \tag{32}$$

results in

$$||\hat{\mathcal{O}}_r||_2 = ||\hat{\mathcal{C}}_L||_2 = \sqrt{||\hat{H}_r||_2}, \tag{33}$$

and thus, from a functional perspective, the mappings from input to state and state to output will have equal magnitudes, and each entry of the state vector x_k will have similar magnitudes. State-space realizations that satisfy (33) are sometimes called *internally balanced* realizations [11]. (Alternative definitions of a "balanced" realization exist, however, and it is generally wise to verify the definition in each context.)

Choosing the factorization (32) also simplifies computation of the estimate $\hat{A}$, since

$$\begin{aligned}\hat{A} &= \left(\hat{\mathcal{O}}_r\right)^{\dagger} \hat{\overline{H}}_r \left(\hat{\mathcal{C}}_L\right)^{\dagger} \\ &= \Sigma_n^{-1/2} U_n^T \hat{\overline{H}}_r V_n \Sigma_n^{-1/2}.\end{aligned}$$

By estimating $\hat{B}$ as the first block column of $\hat{\mathcal{C}}_L$, $\hat{C}$ as the first block row of $\hat{\mathcal{O}}_L$, and $\hat{D}$ as g_0, a complete state-space realization $(\hat{A},\ \hat{B},\ \hat{C},\ \hat{D})$ is identified from this method.

3.5. Pitfalls of direct realization from an impulse response

Even though the rank-reduction process allows for realization from a noise-corrupted estimate of an impulse response, identification methods that generate a system estimate from a Hankel matrix constructed from an estimated impulse response have numerous difficulties when applied to noisy measurements. Measuring an impulse response directly is often infeasible; high-frequency damping may result in a measurement that has a very brief response before the signal-to-noise ratio becomes prohibitively small, and a unit pulse will often excite high-frequency nonlinearities that degrade the quality of the resulting estimate.

Taking the inverse Fourier transform of the frequency response guarantees that the estimates of the Markov parameters will converge as the dataset grows only so long as the input is broadband. Generally input signals decay in magnitude at higher frequencies, and calculation of the frequency response by inversion of the input will amplify high-frequency noise. We would prefer an identification method that is guaranteed to provide a system estimate that converges to the true system as the amount of data measured increases and that avoids inverting the input. Fortunately, the relationship between input and output data in (21) may be used to formulate just such an identification procedure.

4. Realization from input-output data

To avoid the difficulties in constructing a system realization from an estimated impulse response, we will form a realization-based identification procedure applicable to measured input-output data. To sufficiently account for non-deterministic effects in measured data, we add a noise term $v_k \in \mathbb{R}^{n_y}$ to the output to form the noise-perturbed state-space equations

$$\begin{aligned} x_{k+1} &= Ax_k + Bu_k \\ y_k &= Cx_k + Du_k + v_k. \end{aligned} \tag{34}$$

We assume that the noise signal v_k is generated by a stationary stochastic process, which may be either white or colored. This includes the case in which the state is disturbed by process noise, so that the noise process may have the same poles as the deterministic system. (See [15] for a thorough discussion of representations of noise in the identification context.)

4.1. Data-matrix equations

The goal is to construct a state-space realization using the relationships in (21), but doing so requires a complete characterization of the row space of H_r. To this end, we expand a finite-slice of the future output vector to form a block-Hankel matrix of output data with r block rows,

$$Y = \begin{bmatrix} y_0 & y_1 & y_2 & \cdots & y_L \\ y_1 & y_2 & y_3 & \cdots & y_{L+1} \\ y_2 & y_3 & y_4 & \cdots & y_{L+2} \\ \vdots & \vdots & \vdots & & \vdots \\ y_{r-1} & y_r & y_{r+1} & \cdots & y_{r+L-1} \end{bmatrix}.$$

This matrix is related to a block-Hankel matrix of future input data

$$U_f = \begin{bmatrix} u_0 & u_1 & u_2 & \cdots & u_L \\ u_1 & u_2 & u_3 & \cdots & u_{L+1} \\ u_2 & u_3 & u_4 & \cdots & u_{L+2} \\ \vdots & \vdots & \vdots & & \vdots \\ u_{r-1} & u_r & u_{r+1} & \cdots & u_{r+L-1} \end{bmatrix},$$

a block-Toeplitz matrix of past input data

$$U_p = \begin{bmatrix} u_{-1} & u_0 & u_1 & \cdots & u_{L-1} \\ u_{-2} & u_{-1} & u_0 & \cdots & u_{L-2} \\ u_{-3} & u_{-2} & u_{-1} & \cdots & u_{L-3} \\ \vdots & \vdots & \vdots & & \vdots \end{bmatrix},$$

a finite-dimensional block-Toeplitz matrix

$$T = \begin{bmatrix} g_0 & & & \cdots & 0 \\ g_1 & g_0 & & & \vdots \\ g_2 & g_1 & g_0 & & \\ \vdots & \vdots & \vdots & \ddots & \\ g_{r-1} & g_{r-2} & g_{r-3} & \cdots & g_0 \end{bmatrix},$$

the system Hankel matrix H, and a block-Hankel matrix V formed from noise data v_k with the same indices as Y by the equation

$$Y = HU_p + TU_f + V. \tag{35}$$

If the entries of Y_f are shifted forward by one index to form

$$\overline{Y} = \begin{bmatrix} y_1 & y_2 & y_3 & \cdots & y_{L+1} \\ y_2 & y_3 & y_4 & \cdots & y_{L+2} \\ y_3 & y_4 & y_5 & \cdots & y_{L+3} \\ \vdots & \vdots & \vdots & & \vdots \\ y_r & y_{r+1} & y_{r+2} & \cdots & y_{r+L} \end{bmatrix},$$

then $\overline{Y}_f$ is related to the shifted system Hankel matrix $\overline{H}$, the past input data U_p, T with a block column appended to the left, and U_f with a block row appended to the bottom,

$$\overline{T} = \left[\begin{array}{c|c} g_1 & \\ g_2 & \\ g_3 & T \\ \vdots & \\ g_r & \end{array}\right], \qquad \overline{U}_f = \left[\begin{array}{ccccc} & & U_f & & \\ \hline u_r & u_{r+1} & u_{r+2} & \cdots & u_{r+L} \end{array}\right],$$

and a block-Hankel matrix $\overline{V}$ of noise data v_k with the same indices as $\overline{Y}$ by the equation

$$\overline{Y} = \overline{H}U_p + \overline{T}\,\overline{U}_f + \overline{V}. \tag{36}$$

From (25), the state is equal to the column vectors of U_p multiplied by the entries of the controllability matrix $\mathcal{C}$, which we may represent as the block-matrix

$$X = \begin{bmatrix} x_0 & x_1 & x_2 & \cdots & x_L \end{bmatrix} = \mathcal{C}U_p,$$

which is an alternative means of representing the memory of the system at sample $0, 1, \ldots$. The two data matrix equations (35) and (36) may then be written as

$$Y = \mathcal{O}_r X + TU_f + V, \tag{37}$$

$$\overline{Y} = \mathcal{O}_r AX + \overline{T}\,\overline{U}_f + \overline{V}. \tag{38}$$

Equation (37) is basis for the field of system identification methods known as subspace methods. Subspace identification methods typically fall into one of two categories. First, because a shifted observability matrix

$$\overline{\mathcal{O}} = \begin{bmatrix} CA \\ CA^2 \\ CA^3 \\ \vdots \end{bmatrix},$$

satisfies

$$\text{im}(\mathcal{O}) = \text{im}(\overline{\mathcal{O}}),$$

where im(·) of denotes the row space (often called the "image"), the row-space of $\mathcal{O}$ is shift-invariant, and A may be identified from estimates $\mathcal{O}_r$ and $\overline{\mathcal{O}}_r$ as

$$\hat{A} = \hat{\mathcal{O}}_r^{\dagger} \hat{\overline{\mathcal{O}}}_r.$$

Alternatively, because a forward-propagated sequence of states

$$\overline{X} = AX$$

satisfies

$$\text{im}(X^T) = \text{im}(\overline{X}^T),$$

the column-space of X is shift-invariant, and A may be identified from estimates $\hat{X}$ and $\hat{\overline{X}}$ as

$$\hat{A} = \hat{\overline{X}} \hat{X}^{\dagger}.$$

In both instances, the system dynamics are estimated by propagating the indices forward by one step and examining a propagation of linear dynamics, not unlike (20) from Kronecker's theorem. Details of these methods may be found in [16] and [17]. In the next section we present a system identification method that constructs system estimates from the shift-invariant structure of Y itself.

4.2. Identification from shift-invariance of output measurements

Equations (37) and (38) still contain the effects of the future input in $\overline{U}_f$. To remove these effects from the output, we must first add some assumptions about $\overline{U}_f$. First, we assume that $\overline{U}_f$ has full row rank. This is true for any U_f with a smooth frequency response or if $\overline{U}_f$ is generated from some pseudo-random sequence. Next, we assume that the initial conditions in X do not somehow cancel out the effects of future input. A sufficient condition for this is to require

$$\text{rank}\left(\begin{bmatrix} X \\ \overline{U}_f \end{bmatrix}\right) = n + rn_u$$

to have full row rank. Although these assumptions might appear restrictive at first, since it is impossible to verify without knowledge of X, it is generally true with the exception of some pathological cases.

Next, we form the null-space projector matrix

$$\Pi = I_{L+1} - \overline{U}_f^T \left(\overline{U}_f \overline{U}_f^T\right)^{-1} \overline{U}_f, \tag{39}$$

which has the property

$$\overline{U}_f \Pi = 0.$$

We know the inverse of $(\overline{U}_f \overline{U}_f^T)$ exists, since we assume $\overline{U}_f$ has full row rank. Projector matrices such as (39) have many interesting properties. Their eigenvalues are all 0 or 1, and if they are symmetric, they separate the subspace of real vectors — in this case, vectors in $\mathbb{R}^{L+1}$ — into a subspace and its orthogonal complement. In fact, it is simple to verify that the null space of $\overline{U}_f$ contains the null space of U_f as a subspace, since

$$\overline{U}_f \Pi = \begin{bmatrix} U_f \\ \hline \cdots \end{bmatrix} \Pi = 0.$$

Thus multiplication of (37) and (38) on the right by Π results in

$$Y\Pi = \mathcal{O}_r X \Pi + V\Pi, \tag{40}$$

$$\overline{Y}\Pi = \mathcal{O}_r A X \Pi + \overline{V}\Pi. \tag{41}$$

It is also unnecessary to compute the projected products $Y\Pi$ and $\overline{Y}\Pi$ directly, since from the QR-decomposition

$$\begin{bmatrix} \overline{U}^T & Y^T \end{bmatrix} = \begin{bmatrix} Q_1 & Q_2 \end{bmatrix} \begin{bmatrix} R_{11} & R_{12} \\ 0 & R_{22} \end{bmatrix},$$

we have

$$Y = R_{12}^T Q_1^T + R_{22}^T Q_2^T \tag{42}$$

and $U = R_{11}^T Q_1^T$. Substitution into (39) reveals

$$\Pi = I - Q_1 Q_1^T. \tag{43}$$

Because the columns of Q_1 and Q_2 are orthogonal, multiplication of (42) on the right by (43) results in

$$Y\Pi = R_{22}^T Q_2^T.$$

A similar result holds for $\overline{Y}\Pi$. Taking the QR-decomposition of the data can alternatively be thought of as using the principle of superposition to construct new sequences of input-output data through a Gram-Schmidt-type orthogonalization process. A detailed discussion of this interpretation can be found in [18].

Thus we have successfully removed the effects of future input on the output while retaining the effects of the past, which is the foundation of the realization process. We still must account for non-deterministic effects in V and $\overline{V}$. To do so, we look for some matrix Z such that

$$V\Pi Z^T \to 0,$$

$$\overline{V}\Pi Z^T \to 0.$$

This requires the content of Z to be statistically independent of the process that generates v_k. The input u_k is just such a signal, so long as the filter output is not a function of the input — that is, the data was measured in open-loop operation,. If we begin measuring input before

$k = 0$ at some sample $k = -\zeta$ and construct Z as a block-Hankel matrix of past input data,

$$Z = \frac{1}{L} \begin{bmatrix} u_{-\zeta} & u_{-\zeta+1} & u_{-\zeta+2} & \cdots & u_{-\zeta+L} \\ u_{-\zeta+1} & u_{-\zeta+2} & u_{-\zeta+3} & \cdots & u_{-\zeta+L+1} \\ u_{-\zeta+2} & u_{-\zeta+3} & u_{-\zeta+4} & \cdots & u_{-\zeta+L+2} \\ \vdots & \vdots & \vdots & & \vdots \\ u_{-1} & u_0 & u_1 & \cdots & u_{r+L-2} \end{bmatrix},$$

then multiplication of (40) and (41) on the right by Z^T results in

$$Y\Pi Z^T \to \mathcal{O}_r X\Pi Z^T, \tag{44}$$

$$\overline{Y}\Pi Z^T \to \mathcal{O}_r A\Pi Z^T, \tag{45}$$

as $L \to \infty$. Note the term $\frac{1}{L}$ in Z is necessary to keep (44) and (45) bounded.

Finally we are able to perform our rank-reduction technique directly on measured data without needing to first estimate the impulse response. From the SVD

$$Y\Pi Z^T = U\Sigma V^T,$$

we may estimate the order n by looking for a sudden decrease in singular values. From the partitioning

$$Y\Pi Z^T = \begin{bmatrix} U_n & U_s \end{bmatrix} \begin{bmatrix} \Sigma_n & 0 \\ 0 & \Sigma_s \end{bmatrix} \begin{bmatrix} V_n^T \\ V_s^T \end{bmatrix},$$

we may estimate $\mathcal{O}_r$ and $X\Pi Z^T$ from the factorization

$$\hat{\mathcal{O}}_r = U_n \Sigma_n^{1/2} \qquad \text{and} \qquad \hat{X}\Pi Z^T = \Sigma_n^{1/2} V_n^T.$$

A may then be estimated as

$$\begin{aligned} \hat{A} &= \left(\hat{\mathcal{O}}_r\right)^{\dagger} \overline{Y}\Pi Z^T \left(\hat{X}\Pi Z^T\right)^{\dagger} = \Sigma_n^{-1/2} U_n^T \overline{Y}\Pi Z^T V_n \Sigma_n^{-1/2} \\ &\approx (\mathcal{O}_r)^{\dagger} \left(\overline{H} U_p \Pi\right) \left(C_L U_p \Pi\right)^{\dagger} \approx (\mathcal{O}_r)^{\dagger} \, \overline{H} \, (C_L)^{\dagger}. \end{aligned}$$

And so we have returned to our original relationship (29).

While C may be estimated from the top block row of $\hat{\mathcal{O}}_r$, our projection has lost the column space of H_r that we previously used to estimate B, and initial conditions in X prevent us from estimating D directly. Fortunately, if A and C are known, then the remaining terms B, D, and an initial condition x_0 are linear in the input output data, and may be estimated by solving a linear-least-squares problem.

4.3. Estimation of B, D, and $x0$

The input-to-state terms B and D may be estimated by examining the convolution with the state-space form of the impulse response. Expanding (24) with the input and including an initial condition x_0 results in

$$y_k = CA^k x_0 + \sum_{j=0}^{k-1} CA^{k-j-1} B u_j + D u_k + v_k. \tag{46}$$

Factoring out B and D on the right provides

$$y_k = CA^k x_0 + \left(\sum_{j=0}^{k-1} u_k^T \otimes CA^{k-j-1} \right) \text{vec}(B) + \left(u_k^T \otimes I_{n_y} \right) \text{vec}(D) + v_k,$$

in which $\text{vec}(\cdot)$ is the operation that stacks the columns of a matrix on one another, $\otimes$ is the (coincidentally named) Kronecker product, and we have made use of the identity

$$\text{vec}(AXB) = (B^T \otimes A)\text{vec}(X).$$

Grouping the unknown terms together results in

$$y_k = \begin{bmatrix} CA^k & \sum_{j=0}^{k-1} u_k^T \otimes CA^{k-j-1} & u_k^T \otimes I_{n_y} \end{bmatrix} \begin{bmatrix} x_0 \\ \text{vec}(B) \\ \text{vec}(D) \end{bmatrix} + v_k.$$

Thus by forming the regressor

$$\phi_k = \begin{bmatrix} \hat{C}\hat{A}^k & \sum_{j=0}^{k-1} u_k^T \otimes \hat{C}\hat{A}^{k-j-1} & u_k^T \otimes I_{n_y} \end{bmatrix}$$

from the estimates $\hat{A}$ and $\hat{C}$, estimates of B and D may be found from the least-squares solution of the linear system of N equations

$$\begin{bmatrix} y_0 \\ y_1 \\ y_2 \\ \vdots \\ y_N \end{bmatrix} = \begin{bmatrix} \phi_0 \\ \phi_1 \\ \phi_2 \\ \vdots \\ \phi_N \end{bmatrix} \begin{bmatrix} \hat{x}_0 \\ \text{vec}(\hat{B}) \\ \text{vec}(\hat{D}) \end{bmatrix}.$$

Note that N is arbitrary and does not need to be related in any way to the indices of the data matrix equations. This can be useful, since for large-dimensional systems, the regressor ϕ_k may become very computationally expensive to compute.

5. Conclusion

Beginning with the construction of a transfer function from an impulse response, we have constructed a method for identification of state-space realizations of linear filters from measured input-output data, introducing the fundamental concepts of realization theory of linear systems along the way. Computing a state-space realization from measured input-output data requires many tools of linear algebra: projections and the QR-decomposition, rank reduction and the singular-value decomposition, and linear least squares. The principles of realization theory provide insight into the different representations of linear systems, as well as the role of rational functions and series expansions in linear algebra.

Author details

Daniel N. Miller and Raymond A. de Callafon
University of California, San Diego, USA

6. References

[1] Rudolf E. Kalman. On the General Theory of Control Systems. In *Proceedings of the First International Congress of Automatic Control*, Moscow, 1960. IRE.

[2] Rudolf E. Kalman. Mathematical Description of Linear Dynamical Systems. *Journal of the Society for Industrial and Applied Mathematics, Series A: Control*, 1(2):152, July 1963.

[3] B. L. Ho and Rudolf E. Kalman. Effective construction of linear state-variable models from input/output functions. *Regelungstechnik*, 14:545–548, 1966.

[4] Leopold Kronecker. Zur Theorie der Elimination einer Variabeln aus zwei Algebraischen Gleichungen. *Monatsberichte der Königlich Preussischen Akademie der Wissenschaften zu Berlin*, pages 535–600, 1881.

[5] Paul H. Zeiger and Julia A. McEwen. Approximate Linear Realizations of Given Dimension via Ho's Algorithm. *IEEE Transactions on Automatic Control*, 19(2):153 – 153, 1971.

[6] Sun-Yuan Kung. A new identification and model reduction algorithm via singular value decomposition. In *Proceedings of the 12th Asilomar Conference on Circuits, Systems, and Computers*, pages 705–714. IEEE, 1978.

[7] Michel Verhaegen. Identification of the deterministic part of MIMO state space models given in innovations form from input-output data. *Automatica*, 30(1):61–74, January 1993.

[8] Peter Van Overschee and Bart De Moor. N4SID: Subspace algorithms for the identification of combined deterministic-stochastic systems. *Automatica*, 30(1):75–93, January 1994.

[9] Jer-Nan Juang and R. S. Pappa. An Eigensystem Realization Algorithm (ERA) for Modal Parameter Identification and Model Reduction. *JPL Proc. of the Workshop on Identification and Control of Flexible Space Structures*, 3:299–318, April 1985.

[10] Jer-Nan Juang, Minh Q. Phan, Lucas G. Horta, and Richard W. Longman. Identification of Observer/Kalman Filter Markov Parameters: Theory and Experiments. *Journal of Guidance, Control, and Dynamics*, 16(2):320–329, 1993.

[11] Chi-Tsong Chen. *Linear System Theory and Design*. Oxford University Press, New York, 1st edition, 1984.

[12] Felix R. Gantmacher. *The Theory of Matrices - Volume Two*. Chelsea Publishing Company, New York, 1960.

[13] Kemin Zhou, John C. Doyle, and Keith Glover. *Robust and Optimal Control*. Prentice Hall, August 1995.

[14] G.H. Golub and C.F. Van Loan. *Matrix Computations*. The Johns Hopkins University Press, Baltimore, Maryland, USA, third edition, 1996.

[15] Lennart Ljung. *System Identification: Theory for the User*. PTR Prentice Hall Information and System Sciences. Prentice Hall PTR, Upper Saddle River, NJ, 2nd edition, 1999.

[16] Michel Verhaegen and Vincent Verdult. *Filtering and System Identification: A Least Squares Approach*. Cambridge University Press, New York, 1 edition, May 2007.

[17] Peter Van Overschee and Bart De Moor. *Subspace Identification for Linear Systems: Theory, Implementation, Applications*. Kluwer Academic Publishers, London, 1996.

[18] Tohru Katayama. Role of LQ Decomposition in Subspace Identification Methods. In Alessandro Chiuso, Stefano Pinzoni, and Augusto Ferrante, editors, *Modeling, Estimation and Control*, pages 207–220. Springer Berlin / Heidelberg, 2007.

Chapter 7

Partition-Matrix Theory Applied to the Computation of Generalized-Inverses for MIMO Systems in Rayleigh Fading Channels

P. Cervantes, L. F. González, F. J. Ortiz and A. D. García

Additional information is available at the end of the chapter

http://dx.doi.org/10.5772/48198

1. Introduction

Partition-Matrix Theory and Generalized-Inverses are interesting topics explored in linear algebra and matrix computation. Partition-Matrix Theory is associated with the problem of properly partitioning a matrix into block matrices (i.e. an array of matrices), and is a matrix computation tool widely employed in several scientific-technological application areas. For instance, blockwise Toeplitz-based covariance matrices are used to model structural properties for space-time multivariate adaptive processing in radar applications [1], Jacobian response matrices are partitioned into several block-matrix instances in order to enhance medical images for Electrical-Impedance-Tomography [2], design of state-regulators and partial-observers for non-controllable/non-observable linear continuous systems contemplates matrix blocks for controllable/non-controllable and observable/non-observable eigenvalues [3]. The Generalized-Inverse is a common and natural problem found in a vast of applications. In control robotics, non-collocated partial linearization is applied to underactuated mechanical systems through inertia-decoupling regulators which employ a pseudoinverse as part of a modified input control law [4]. At sliding-mode control structures, a Right-Pseudoinverse is incorporated into a state-feedback control law in order to stabilize electromechanical non-linear systems [5]. Under the topic of system identification, definition of a Left-Pseudoinverse is present in auto-regressive moving-average models (ARMA) for matching dynamical properties of unknown systems [6]. An interesting approach arises whenever Partition-Matrix Theory and Generalized-Inverse are combined together yielding attractive solutions for solving the problem of block matrix inversion [7-10]. Nevertheless, several assumptions and restrictions regarding numerical stability and structural properties are considered for these alternatives. For example, an attractive pivot-free block matrix inversion algorithm is proposed in [7], which unfortunately exhibits an overhead in matrix multiplications that are required in order to guarantee full-rank properties for particular blocks within it. For circumventing the expense in rank deficiency, [8] offers block-matrix completion strategies in order to find the Generalized-Inverse of any non-singular block matrix (irrespective of the singularity of their constituting sub-blocks). However, the existence of intermediate matrix inverses and pseudoinverses throughout this algorithm still rely on full-rank assumptions, as well as introducing more hardness to the problem. The proposals exposed in [9-10] avoid completion strategies and contemplate all possible scenarios for avoiding any rank deficiency among each matrix sub-block, yet demanding full-rank assumptions for each scenario. In this chapter, an iterative-recursive algorithm for computing a Left-Pseudoinverse (LPI) of a MIMO channel matrix is developed by combining Partition-Matrix Theory and Generalized-Inverse concepts. For this approach, no matrix-operations' overhead nor any particular block matrix full-rank assumptions are needed because of structural attributes of the MIMO channel matrix, which models dynamical properties of a Rayleigh fading channel (RFC) within wireless MIMO communication systems.

The content of this work is outlined as follows. Section 2 provides a description of the MIMO communication link, pointing out its principal physical effects and the mathematical model considered for RFC-based environments. Section 3 defines formally the problem of computing the Left-Pseudoinverse as the Generalized-Inverse for the MIMO channel matrix applying Partition-Matrix Theory concepts. Section 4 presents linear algebra and matrix computation concepts and tools needed for tracking a solution for the aforementioned problem. Section 5 analyzes important properties of the MIMO channel matrix derived from a Rayleigh fading channel scenario. Section 6 explains the proposed novel algorithm. Section 7 presents a brief analysis of VLSI (Very Large Scale of Integration) aspects towards implementation of arithmetic operations presented in this algorithm. Section 8 concludes the chapter. Due to the vast literature about MIMO systems, and to the best of the authors' knowledge, this chapter provides a nice and strategic list of references in order to easily correlate essential concepts between matrix theory and MIMO systems. For instance, [11-16] describe and analyze information and system aspects about MIMO communication systems, as well as studying MIMO channel matrix behavior under RFC-based environments; [17-18] contain all useful linear algebra and matrix computation theoretical concepts around the mathematical background immersed in MIMO systems; [19-21] provide practical guidelines and examples for MIMO channel matrix realizations comprising RFC scenarios; [22] treats the formulation and development of the algorithm presented in this chapter; [23-27] detail a splendid survey on architectural aspects for implementing several arithmetic operations.

2. MIMO systems

In the context of wireless communication systems, MIMO (Multiple-Input Multiple-Output) is an extension of the classical SISO (Single-Input Single-Output) communication paradigm, where instead of having a communication link composed of a single transmitter-end and a receiver-end element (or antenna), wireless MIMO communication systems (or just MIMO systems) consist of an array of multiple elements at both the transmission and reception parts [11-16,19-21]. Generally speaking, the MIMO communication link contains n_T transmitter-end and n_R receiver-end antennas sending-and-receiving information through a wireless channel. Extensive studies on MIMO systems and commercial devices already employing them reveal that these communication systems offer promising results in terms of: a) spectral efficiency and channel capacity enhancements (many user-end applications supporting high-data rates at limited available bandwidth); b) improvements on Bit-Error-Rate (BER) performance; and c) practical feasability already seen in several wireless communication standards. The conceptualization of this paradigm is illustrated in figure 1, where Tx is the transmitter-end, Rx the receiver-end, and Chx the channel.

Figure 1. The MIMO system: conceptualization for the MIMO communication paradigm.

Notice that information sent from the trasnmission part (Tx label on figure 1) will suffer from several degradative and distorional effects inherent in the channel (Chx label on figure 1), forcing the reception part (Rx label on figure 1) to decode information properly. Information at Rx will suffer from degradations caused by time, frequency, and spatial characteristics of the MIMO communication link [11-12,14]. These issues are directly related to: i) the presence of physical obstacles obstructing the Line-of-Sight (LOS) between Tx and Rx (existance of non-LOS); ii) time delays between received and transmitted information signals due to Tx and Rx dynamical properties (time-selectivity of Chx); iii) frequency distortion and interference among signal carriers through Chx (frequency-selectivity of Chx); iv) correlation of information between receiver-end elements. Fading (or fading mutlipath) and noise are the most common destructive phenomena that significantly affect information at Rx [11-16]. Fading is a combination of time-frequency replicas of the trasnmitted information as a consequence of the MIMO system phenomena i)-iv) exposed before, whereas noise affects information at every receiver-end element under an additve or multiplicative way. As a consequence, degradation of signal information rests mainly upon magnitude attenuation and time-frequency shiftings. The simplest treatable MIMO communication link has a slow-flat quasi-static fading channel (proper of a non-LOS indoor environment). For this type of scenario, a well-known dynamical-stochastic model considers a Rayleigh fading channel (RFC) [13,15-16,19-21], which gives a quantitative clue of how information has been degradated by means of Chx. Moreover, this type of channels allows to: a) distiguish among each information block tranmitted from the n_T elements at every Chx realization (i.e. the time during which the channel's properties remain unvariant); and b) implement easily symbol decoding tasks related to channel equalization (CE) techniques. Likewise, noise is commonly assumed to have additive effects over Rx. Once again, all of these assumptions provide a treatable information-decoding problem (refered as MIMO demodulation [12]), and the mathematical model that suits the aforementioned MIMO communication link characteristics will be represented by

$$y = Hx + \eta \tag{1}$$

where: $x \in \mathbb{Z}_{[j]}^{n_T \times 1} \subset \mathbb{C}^{n_T \times 1}$ is a complex-valued n_T - dimensional transmitted vector with entries drawn from a Gaussian-integer finite-lattice constellation (digital modulators, such as: q-QAM, QPSK); $y \in \mathbb{C}^{n_R \times 1}$ is a complex-valued n_R - dimensional received vector; $\eta \in \mathbb{C}^{n_R \times 1}$ is a n_R - dimensional independent-identically-distributed (idd) complex-circularly-symmetric (ccs) Additive White Gaussian Noise (AWGN) vector; and $H \in \mathbb{C}^{n_R \times n_T}$ is the $(n_R \times n_T)$ - dimensional MIMO channel matrix whose entries model: a) the RFC-based environment behavior according to a Gaussian probabilistic density function with zero-mean and 0.5-variance statistics; and b) the time-invariant transfer function (which measures the degradation of the signal information) between the i-th receiver-end and the j-th trasnmitter-end antennas [11-16,19-21]. Figure 2 gives a representation of (1). As shown therein, the MIMO communication link model stated in (1) can be also expressed as

$$\begin{bmatrix} y_1 \\ \vdots \\ y_{n_R} \end{bmatrix} = \begin{bmatrix} h_{11} & \cdots & h_{1n_T} \\ \vdots & & \vdots \\ h_{n_R 1} & \cdots & h_{n_R n_T} \end{bmatrix} \begin{bmatrix} x_1 \\ \vdots \\ x_{n_T} \end{bmatrix} + \begin{bmatrix} \eta_1 \\ \vdots \\ \eta_{n_R} \end{bmatrix} \tag{2}$$

Notice from (1-2) that an important requisite for CE purposes within RFC scenarios is that H is provided somehow to the Rx. This MIMO system requirement is classically known as Channel State Information (CSI) [11-16]. In the sequel of this work, symbol-decoding efforts will consider the problem of finding x from y regarding CSI at the Rx part within a slow-flat quasi-static RFC-based environment as modeled in (1-2). In simpler words, Rx must find x from degradated information y through calculating an inversion over H. Moreover, $n_R \geq n_T$ is commonly assumed for MIMO demodulation tasks [13-14] because it guarantees linear independency between row-entries of matrix H in (2), yielding a nonhomogeneous overdetermined system of linear equations.

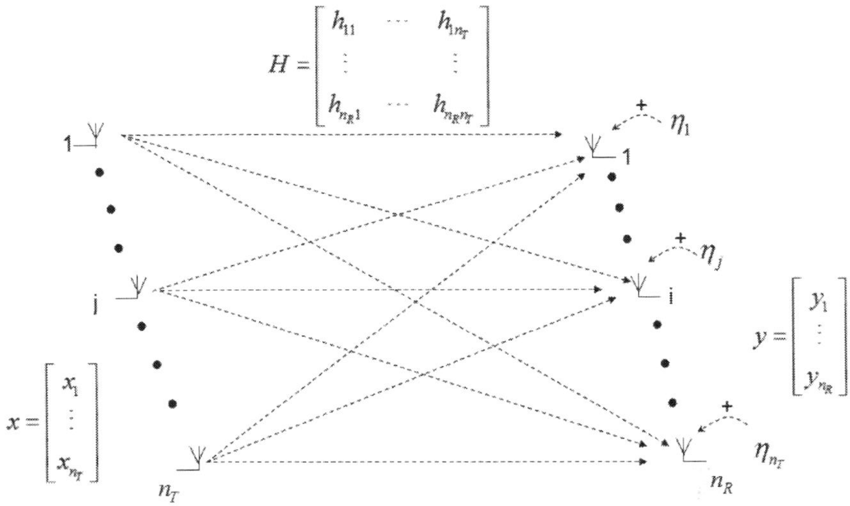

Figure 2. Representation for the MIMO communication link model according to $y = Hx + \eta$. Here, each dotted arrow represents an entry h_{ij} in H which determines channel degradation between the j-th transmitter and the i-th receiver elements. AWGN appears additively in each receiver-end antenna.

3. Problem definition

Recall for the moment the mathematical model provided in (1). Consider Φ^{r} and Φ^{i} to be the real and imaginary parts of a complex-valued matrix (vector) Φ, that is, $\Phi = \Phi^{r} + j\Phi^{i}$. Then, Equation (1) can be expanded as follows:

$$y^{r} + jy^{i} = \left(H^{r}x^{r} - H^{i}x^{i} + \eta^{r}\right) + j\left(H^{i}x^{r} + H^{r}x^{i} + \eta^{i}\right) \tag{3}$$

It can be noticed from Equation (3) that: $x^{r}, x^{i} \in \mathbb{Z}^{n_T \times 1}$; $y^{r}, y^{i} \in \mathbb{R}^{n_R \times 1}$; $\eta^{r}, \eta^{i} \in \mathbb{R}^{n_R \times 1}$; and $H^{r}, H^{i} \in \mathbb{R}^{n_R \times n_T}$. An alternative representation for the MIMO communication link model in (2) can be expressed as

$$\begin{bmatrix} y^{r} \\ y^{i} \end{bmatrix} = \begin{bmatrix} H^{r} & -H^{i} \\ H^{i} & H^{r} \end{bmatrix} \begin{bmatrix} x^{r} \\ x^{i} \end{bmatrix} + \begin{bmatrix} \eta^{r} \\ \eta^{i} \end{bmatrix} \tag{4}$$

where $\begin{bmatrix} y^{r} \\ y^{i} \end{bmatrix} \doteq \mathrm{Y} \in \mathbb{R}^{2n_R \times 1}$, $\begin{bmatrix} H^{r} & -H^{i} \\ H^{i} & H^{r} \end{bmatrix} \doteq \mathrm{h} \in \mathbb{R}^{2n_R \times 2n_T}$, $\begin{bmatrix} x^{r} \\ x^{i} \end{bmatrix} \doteq \mathrm{X} \in \mathbb{Z}^{2n_T \times 1}$, and $\begin{bmatrix} \eta^{r} \\ \eta^{i} \end{bmatrix} \doteq \mathrm{N} \in \mathbb{R}^{2n_R \times 1}$.

CSI is still needed for MIMO demodulation purposes involving (4). Moreover, if $N_r = 2n_R$ and $N_t = 2n_T$, then $N_r \geq N_t$. Obviously, while seeking for a solution of signal vector X from

(4), the reception part Rx will provide also the solution for signal vector x, and thus MIMO demodulation tasks will be fulfilled. This problem can be defined formally into the following manner:

Definition 1. Given parameters $N_r = 2n_R \in \mathbb{Z}^+$ and $N_t = 2n_T \in \mathbb{Z}^+$, and a block-matrix $\mathrm{h} \in \mathbb{R}^{N_r \times N_t}$, there exists an operator $\Gamma : \left(\mathbb{R}^{N_r \times 1} \times \mathbb{R}^{N_r \times N_t}\right) \mapsto \mathbb{R}^{N_t \times 1}$ which solves the matrix-block equation $\mathrm{Y=hX+N}$ so that $\Gamma\left[\mathrm{Y,h}\right] = \mathrm{X}$. ∎

From Definition 1, the following affirmations hold: i) CSI over h is a necessary condition as an input argument for the operator Γ; and ii) Γ can be naïvely defined as a Generalized-Inverse of the block-matrix h. In simpler terms, $\mathrm{X=h^{\dagger}Y}$ [1] is associated with $\Gamma\left[\mathrm{Y,h}\right]$ and $\mathrm{h}^{\dagger} \in \mathbb{R}^{N_t \times N_r}$ stands for the Generalized-Inverse of the block-matrix h, where $\mathrm{h}^{\dagger} = \left(\mathrm{h}^{\mathrm{T}}\mathrm{h}\right)^{-1}\mathrm{h}^{\mathrm{T}}$ [17-18]. Clearly, $\left[\circ\right]^{-1}$ and $\left[\circ\right]^{\mathrm{T}}$ represent the inverse and transpose matrix operations over real-valued matrices. As a concluding remark, computing the Generalized-Inverse $\mathrm{h}^{\dagger}$ can be separated into two operations: 1) a block-matrix inversion $\left(\mathrm{h}^{\mathrm{T}}\mathrm{h}\right)^{-1}$ [2]; 2) a typical matrix multiplication $\left(\mathrm{h}^{\mathrm{T}}\mathrm{h}\right)^{-1} \cdot \mathrm{h}^{\mathrm{T}}$. For these tasks, Partition-Matrix Theory will be employed in order to find a novel algorithm for computing a Generalized-Inverse related to (4).

4. Mathematical background

4.1. Partition-matrix theory

Partition-Matrix Theory embraces structures related to block matrices (or partition matrices: an array of matrices) [17-18]. Furthermore, a block-matrix L with $(n+q)\times(m+p)$ dimension can be constructed (or partitioned) consistently according to matrix sub-blocks A, B, C, and D of $n\times m$, $n\times p$, $q\times m$, and $q\times p$ dimensions, respectively, yielding

$$L = \begin{bmatrix} A & B \\ C & D \end{bmatrix} \tag{5}$$

An interesting operation to be performed for these structures given in (5) is the inversion, i.e. a blockwise inversion L^{-1}. For instance, let $L \in \mathbb{R}^{(n+m)\times(n+m)}$ be a full-rank real-valued block matrix (the subsequent treatment is also valid for complex-valued entities, i.e.

---

[1] In the context of MIMO systems, this matrix operation is commonly found in Babai estimators for symbol-decoding purposes at the Rx part [12,13]. For the reader's interest, refer to [11-16] for other MIMO demodulation techniques.

[2] Notice that $\mathrm{h}^{\mathrm{T}} \in \mathbb{R}^{N_t \times N_r}$ and $\left(\mathrm{h}^{\mathrm{T}}\mathrm{h}\right)^{-1} \in \mathbb{R}^{N_t \times N_t}$.

$L \in \mathbb{C}^{(n+m)\times(n+m)}$). An alternative partition can be performed with $A \in \mathbb{R}^{n\times n}$, $B \in \mathbb{R}^{n\times m}$, $C \in \mathbb{R}^{m\times n}$, and $D \in \mathbb{R}^{m\times m}$. Assume also A and D to be full-rank matrices. Then,

$$L^{-1} = \begin{bmatrix} \left(A - BD^{-1}C\right)^{-1} & -\left(A - BD^{-1}C\right)^{-1} BD^{-1} \\ -\left(D - CA^{-1}B\right)^{-1} CA^{-1} & \left(D - CA^{-1}B\right)^{-1} \end{bmatrix} \tag{6}$$

This strategy (to be proved in the next part) requires additonally and mandatorily full-rank over matrices $A - BD^{-1}C$ and $D - CA^{-1}B$. The simple case is defined for $L = \begin{bmatrix} a & b \\ c & d \end{bmatrix}$ (indistinctly for $\mathbb{R}^{2\times 2}$ or $\mathbb{C}^{2\times 2}$). Once again, assuming $\det(L) \neq 0$, $a \neq 0$, and $d \neq 0$ (related to full-rank restictions within block-matrix L):

$$L^{-1} = \begin{bmatrix} \left(a - bd^{-1}c\right)^{-1} & -\left(a - bd^{-1}c\right)^{-1} bd^{-1} \\ -\left(d - ca^{-1}b\right)^{-1} ca^{-1} & \left(d - ca^{-1}b\right)^{-1} \end{bmatrix} = \frac{1}{ad - bc}\begin{bmatrix} d & -b \\ -c & a \end{bmatrix},$$

where evidently $(ad - bc) \neq 0$, $\mathbb{RC}^{(n+m)\times(n+m)}$ $\left(a - bd^{-1}c\right) \neq 0$, and $\left(d - ca^{-1}b\right) \neq 0$.

4.2. Matrix Inversion Lemma

The Matrix Inversion Lemma is an indirect consequence of inverting non-singular block matrices [17-18], either real-valued or complex-valued, e.g., under certain restrictions [3]. Lemma 1 states this result.

Lemma 1. Let $\Psi \in \mathbb{RC}^{r\times r}$, $\Sigma \in \mathbb{RC}^{r\times s}$, $\Upsilon \in \mathbb{RC}^{s\times s}$, and $\Xi \in \mathbb{RC}^{s\times r}$ be real-valued or complex-valued matrices. Assume these matrices to be non-singular: Ψ, Υ, $(\Psi + \Sigma\Upsilon\Xi)$, and $\left(\Upsilon^{-1} + \Xi\Psi^{-1}\Sigma\right)$. Then,

$$\left(\Psi + \Sigma\Upsilon\Xi\right)^{-1} = \Psi^{-1} - \Psi^{-1}\Sigma\left(\Upsilon^{-1} + \Xi\Psi^{-1}\Sigma\right)^{-1}\Xi\Psi^{-1} \tag{7}$$

Proof. The validation of (7) must satisfy

i. $\left(\Psi + \Sigma\Upsilon\Xi\right)\cdot\left(\Psi^{-1} - \Psi^{-1}\Sigma\left(\Upsilon^{-1} + \Xi\Psi^{-1}\Sigma\right)^{-1}\Xi\Psi^{-1}\right) = I_r$, and

$\left(\Psi^{-1} - \Psi^{-1}\Sigma\left(\Upsilon^{-1} + \Xi\Psi^{-1}\Sigma\right)^{-1}\Xi\Psi^{-1}\right)\cdot\left(\Psi + \Sigma\Upsilon\Xi\right) = I_r$., where I_r represents the $r\times r$ identity matrix. Notice the existance of matrices Ψ^{-1}, Υ^{-1}, $\left(\Psi + \Sigma\Upsilon\Xi\right)^{-1}$ and $\left(\Upsilon^{-1} + \Xi\Psi^{-1}\Sigma\right)^{-1}$. Manipulating i) shows:

---

[3] Refer to [3,7-10,17,18] to review lemmata exposed for these issues and related results.

$$(\Psi+\Sigma\Upsilon\Xi)\cdot\left(\Psi^{-1}-\Psi^{-1}\Sigma\left(\Upsilon^{-1}+\Xi\Psi^{-1}\Sigma\right)^{-1}\Xi\Psi^{-1}\right)$$

$$=I_r-\Sigma\left(\Upsilon^{-1}+\Xi\Psi^{-1}\Sigma\right)^{-1}\Xi\Psi^{-1}+\Sigma\Upsilon\Xi\Psi^{-1}-\Sigma\Upsilon\Xi\Psi^{-1}\Sigma\left(\Upsilon^{-1}+\Xi\Psi^{-1}\Sigma\right)^{-1}\Xi\Psi^{-1}$$

$$=I_r+\Sigma\Upsilon\Xi\Psi^{-1}-\Sigma\Upsilon\left(\Upsilon^{-1}+\Xi\Psi^{-1}\Sigma\right)\left(\Upsilon^{-1}+\Xi\Psi^{-1}\Sigma\right)^{-1}\Xi\Psi^{-1}$$

$$=I_r+\Sigma\Upsilon\Xi\Psi^{-1}-\Sigma\Upsilon\Xi\Psi^{-1}=I_r.$$

Likewise for ii):

$$\left(\Psi^{-1}-\Psi^{-1}\Sigma\left(\Upsilon^{-1}+\Xi\Psi^{-1}\Sigma\right)^{-1}\Xi\Psi^{-1}\right)\cdot(\Psi+\Sigma\Upsilon\Xi)$$

$$=I_r+\Psi^{-1}\Sigma\Upsilon\Xi-\Psi^{-1}\Sigma\left(\Upsilon^{-1}+\Xi\Psi^{-1}\Sigma\right)^{-1}\Xi-\Psi^{-1}\Sigma\left(\Upsilon^{-1}+\Xi\Psi^{-1}\Sigma\right)^{-1}\Xi\Psi^{-1}\Sigma\Upsilon\Xi$$

$$=I_r+\Psi^{-1}\Sigma\Upsilon\Xi-\Psi^{-1}\Sigma\left(\Upsilon^{-1}+\Xi\Psi^{-1}\Sigma\right)^{-1}\left(\Upsilon^{-1}+\Xi\Psi^{-1}\Sigma\right)\Upsilon\Xi$$

$$=I_r+\Psi^{-1}\Sigma\Upsilon\Xi-\Psi^{-1}\Sigma\Upsilon\Xi=I_r.\blacksquare$$

Now it is pertinent to demonstrate (6) with the aid of Lemma 1. It must be verified that both LL^{-1} and $L^{-1}L$ must be equal to the $(n+m)\times(n+m)$ identity block matrix $I_{(n+m)}=\begin{bmatrix} I_n & 0_{n\times m} \\ 0_{m\times n} & I_m \end{bmatrix}$, with consistent-dimensional identity and zero sub-blocks: I_n, I_m; $0_{n\times m}$, $0_{m\times n}$, respectively. We start by calulating

$$LL^{-1}=\begin{bmatrix} A & B \\ C & D \end{bmatrix}\begin{bmatrix} \left(A-BD^{-1}C\right)^{-1} & -\left(A-BD^{-1}C\right)^{-1}BD^{-1} \\ -\left(D-CA^{-1}B\right)^{-1}CA^{-1} & \left(D-CA^{-1}B\right)^{-1} \end{bmatrix} \tag{8}$$

and

$$L^{-1}L=\begin{bmatrix} \left(A-BD^{-1}C\right)^{-1} & -\left(A-BD^{-1}C\right)^{-1}BD^{-1} \\ -\left(D-CA^{-1}B\right)^{-1}CA^{-1} & \left(D-CA^{-1}B\right)^{-1} \end{bmatrix}\begin{bmatrix} A & B \\ C & D \end{bmatrix} \tag{9}$$

by applying (7) in Lemma 1 to both matrices $\left(A-BD^{-1}C\right)^{-1}\in\mathbb{RC}^{n\times n}$ and $\left(D-CA^{-1}B\right)^{-1}\in\mathbb{RC}^{m\times m}$, which are present in (8) and (9), and recalling full-rank conditions not only over those matrices but also for A and D, yields the relations

$$\left(A - BD^{-1}C\right)^{-1} = A^{-1} + A^{-1}B\left(D - CA^{-1}B\right)^{-1}CA^{-1} \tag{10}$$

$$\left(D - CA^{-1}B\right)^{-1} = D^{-1} + D^{-1}C\left(A - BD^{-1}C\right)^{-1}BD^{-1} \tag{11}$$

Using (10-11) in (8-9), the following results arise:

a. for operations involved in sub-blocks of LL^{-1}:

$$A\left(A - BD^{-1}C\right)^{-1} - B\left(D - CA^{-1}B\right)^{-1}CA^{-1}$$

$$= A\left[A^{-1} + A^{-1}B\left(D - CA^{-1}B\right)^{-1}CA^{-1}\right] - B\left(D - CA^{-1}B\right)^{-1}CA^{-1}$$

$$= I_n + B\left(D - CA^{-1}B\right)^{-1}CA^{-1} - B\left(D - CA^{-1}B\right)^{-1}CA^{-1} = I_n;$$

$$-A\left(A - BD^{-1}C\right)^{-1}BD^{-1} + B\left(D - CA^{-1}B\right)^{-1}$$

$$= -A\left[A^{-1} + A^{-1}B\left(D - CA^{-1}B\right)^{-1}CA^{-1}\right]BD^{-1} + B\left(D - CA^{-1}B\right)^{-1}$$

$$= -BD^{-1} - B\left(D - CA^{-1}B\right)^{-1}CA^{-1}BD^{-1} + B\left(D - CA^{-1}B\right)^{-1}$$

$$= -BD^{-1} - B\left(D - CA^{-1}B\right)^{-1}\left(-CA^{-1}B + D\right)D^{-1} = 0_{n\times m};$$

$$C\left(A - BD^{-1}C\right)^{-1} - D\left(D - CA^{-1}B\right)^{-1}CA^{-1}$$

$$= C\left(A - BD^{-1}C\right)^{-1} - D\left[D^{-1} + D^{-1}C\left(A - BD^{-1}C\right)^{-1}BD^{-1}\right]CA^{-1}$$

$$= C\left(A - BD^{-1}C\right)^{-1} - CA^{-1} - C\left(A - BD^{-1}C\right)^{-1}BD^{-1}CA^{-1}$$

$$= C\left(A - BD^{-1}C\right)^{-1}\left[A - BD^{-1}C\right]A^{-1} - CA^{-1} = 0_{m\times n};$$

$$-C\left(A - BD^{-1}C\right)^{-1}BD^{-1} + D\left(D - CA^{-1}B\right)^{-1}$$

$$= -C\left(A - BD^{-1}C\right)^{-1}BD^{-1} + D\left[D^{-1} + D^{-1}C\left(A - BD^{-1}C\right)^{-1}BD^{-1}\right]$$

$$= -C\left(A - BD^{-1}C\right)^{-1}BD^{-1} + I_m + C\left(A - BD^{-1}C\right)^{-1}BD^{-1} = I_m;$$

thus, $LL^{-1} = I_{(n+m)}$.

b. for operations involved in sub-blocks of $L^{-1}L$:

$$\left(A - BD^{-1}C\right)^{-1}A - \left(A - BD^{-1}C\right)^{-1}BD^{-1}C$$

$$= \left(A - BD^{-1}C\right)^{-1}\left[A - BD^{-1}C\right] = I_n;$$

$$\left(A - BD^{-1}C\right)^{-1}B - \left(A - BD^{-1}C\right)^{-1}BD^{-1}D = 0_{n\times m};$$

$$-\left(D - CA^{-1}B\right)^{-1}CA^{-1}A + \left(D - CA^{-1}B\right)^{-1}C = 0_{m\times n};$$

$$-\left(D - CA^{-1}B\right)^{-1}CA^{-1}B + \left(D - CA^{-1}B\right)^{-1}D$$

$$= -\left(D - CA^{-1}B\right)^{-1}\left[-CA^{-1}B + D\right] = I_m;$$

thus, $L^{-1}L = I_{(n+m)}$.

4.3. Generalized-Inverse

The concept of Generalized-Inverse is an extension of a matrix inversion operations applied to non-singular rectangular matrices [17-18]. For notation purposes and without loss of generalization, $\rho(G)$ and $G^{\intercal}$ denote the rank of a rectangular matrix $G \in \mathrm{M}^{m\times n}$, and $G^{\intercal} = G^{\mathrm{H}}$ is the transpose-conjugate of G (when $\mathrm{M}=\mathbb{C} \to G \in \mathbb{C}^{m\times n}$) or $G^{\intercal} = G^{\mathrm{T}}$ is the transpose of G (when $\mathrm{M}=\mathbb{R} \to G \in \mathbb{R}^{m\times n}$), respectively.

Definition 2. Let $G \in \mathrm{M}^{m\times n}$ and $0 \le \rho(G) \le \min(m,n)$. Then, there exists a matrix $G^{\dagger} \in \mathrm{M}^{n\times m}$ (identified as the Generalized-Inverse), such that it satisfies several conditions for the following cases:

case i: if $m > n$ and $0 \le \rho(G) \le \min(m,n) \Rightarrow \rho(G) = n$, then there exists a unique matrix $G^{\dagger} \doteq G^{+} \in \mathrm{M}^{n\times m}$ (identified as Left-Pseudoinverse: LPI) such that $G^{+}G = I_n$, satisfying: a) $GG^{+}G = G$, and b) $G^{+}GG^{+} = G^{+}$. Therefore, the LPI matrix is proposed as $G^{+} = \left(G^{\intercal}G\right)^{-1}G^{\intercal}$.

case ii: if $m = n$ and $\det(G) \ne 0 \Leftrightarrow \rho(G) = n$, then there exists a unique matrix $G^{\dagger} \doteq G^{-1} \in \mathrm{M}^{n\times n}$ (identified as Inverse) such that $G^{-1}G = GG^{-1} = I_n$.

case iii: if $m < n$ and $0 \le \rho(G) \le \min(m,n) \Rightarrow \rho(G) = m$, then there exists a unique matrix $G^{\dagger} \doteq G^{-} \in \mathrm{M}^{n\times m}$ (identified as Right-Pseudoinverse: RPI) such that $GG^{-} = I_m$, satisfying: a) $GG^{-}G = G$, and b) $G^{-}GG^{-} = G^{-}$. Therefore, the RPI matrix is proposed as $G^{-} = G^{T}\left(GG^{T}\right)^{-1}$. ■

Given the mathematical structure for $G^{\dagger}$ provided in Definition 2, it can be easily validated that: 1) For a LPI matrix stipulated in case i, $GG^{\dagger}G = G$ and $G^{\dagger}GG^{\dagger} = G^{\dagger}$ with $G^{\dagger} = \left(G^{T}G\right)^{-1}G^{T}$; 2) For a RPI matrix stipulated in case iii, $GG^{\dagger}G = G$ and $G^{\dagger}GG^{\dagger} = G^{\dagger}$ with $G^{\dagger} = G^{T}\left(GG^{T}\right)^{-1}$; iii) For the Inverse in case ii, $G^{+} = \left(G^{T}G\right)^{-1}G^{T} = G^{T}\left(GG^{T}\right)^{-1} = G^{-}$. For a uniqueness test for all cases, assume the existance of matrices $G_1^{\dagger} \in \mathrm{M}^{n\times m}$ and $G_2^{\dagger} \in \mathrm{M}^{n\times m}$ such that $G_1^{\dagger}G = I_n$ and $G_2^{\dagger}G = I_n$ (for case i), and $GG_1^{\dagger} = I_m$ and $GG_2^{\dagger} = I_m$ (for case iii). Notice immediately, $\left(G_1^{\dagger} - G_2^{\dagger}\right)G = 0_n$ (for case i) and $G\left(G_1^{\dagger} - G_2^{\dagger}\right) = 0_m$ (for case iii), which obligates $G_1^{\dagger} = G_2^{\dagger}$ for both cases, because of full-rank properties over G. Clearly, case ii is a particular consequence of cases i and iii.

5. The MIMO channel matrix

The MIMO channel matrix is the mathematical representation for modeling the degradation phenomena presented in the RFC scenario presented in (2). The elements h_{ij} in $H \in \mathbb{C}^{n_R \times n_T}$ represent a time-invariant transfer function (possesing spectral information about magnitude and phase profiles) between a j-th transmitter and an i-th receiver antenna. Once again, dynamical properties of physical phenomena [4] such as path-loss, shadowing, multipath, Doppler spreading, coherence time, absorption, reflection, scattering, diffraction, basestation-user motion, antenna's physical properties-dimensions, information correlation, associated with a slow-flat quasi-static RFC scenario (proper of a non-LOS indoor wireless environments) are highlighted into a statistical model represented by matrix H. For $H^{\dagger}$ purposes, CSI is a necessary feature required at the reception part in (2), as well as the $n_R \geq n_T$ condition. Table 1 provides several $n_R > n_T$ MIMO channel matrix realizations for RFC-based environments [19-21]. On table 1: a) $\mathrm{MIMO}(n_R, n_T)$: refers to the MIMO communication link configuration, i.e. amount of receiver-end and transmitter-end elements; b) H_{m}: refers to a MIMO channel matrix realization; c) H_{m}^{+}: refers to the corresponding LPI, computed as $H_{\mathrm{m}}^{+} = \left(H_{\mathrm{m}}^{\mathrm{H}}H_{\mathrm{m}}\right)^{-1}H_{\mathrm{m}}^{\mathrm{H}}$; d) h: blockwise matrix version for H_{m}; e) h^{+}: refers to the corresponding LPI, computed as $\mathrm{h}^{+} = \left(\mathrm{h}^{\mathrm{T}}\mathrm{h}\right)^{-1}\mathrm{h}^{\mathrm{T}}$. As an additional point of analysis, full-rank properties over H and h (and thus the existance of matrices H^{+}, H^{-1}, h^{+}, and h^{-1}) are validated and corroborated through a MATLAB simulation-driven model regarding frequency-selective and time-invariant properties for several RFC-based scenarios at different MIMO configurations. Experimental data were generated upon 10^{6} MIMO channel matrix realizations. As illustrated in figure 3, a common pattern is found regarding the statistical evolution for full-rank properties of H and h with $n_R \geq n_T$ at several typical MIMO configurations, for instance, $\mathrm{MIMO}(2,2)$, $\mathrm{MIMO}(4,2)$, and $\mathrm{MIMO}(4,4)$. It is plotted therein REAL(H,h) against IMAG(H,h), where each axis label denote respectively the real and imaginary parts of: a) $\det(H)$ and $\det(\mathrm{h})$ when $n_R = n_T$, and b) $\det\left(H^{\mathrm{H}}H\right)$ and $\det\left(\mathrm{h}^{\mathrm{T}}\mathrm{h}\right)$ when . Blue crosses indicate the behavior of $\rho(H)$ related to $\det(H)$ and $\det\left(H^{\mathrm{H}}H\right)$

[4] We suggest the reader consulting references [11-16] for a detail and clear explanation on these narrowband and wideband physical phenomena presented in wireless MIMO communication systems.

(det(H) legend on top-left margin), while red crosses indicate the behavior of $\rho(\mathrm{h})$ related to $\det(\mathrm{h})$ and $\det\left(\mathrm{h}^{\mathrm{T}}\mathrm{h}\right)$ (det(h) legend on top-left margin). The black-circled zone intersected with black-dotted lines locates the $0 + j0$ value. As depicted on figures (4)-(5), a closer glance at this statistical behavior reveals a prevalence on full-rank properties of H and h, meaning that non of the determinants $\det(H)$, $\det(\mathrm{h})$, $\det\left(H^{\mathrm{H}}H\right)$ and $\det\left(\mathrm{h}^{\mathrm{T}}\mathrm{h}\right)$ is equal to zero (behavior enclosed by the light-blue region and delimited by blue/red-dotted lines).

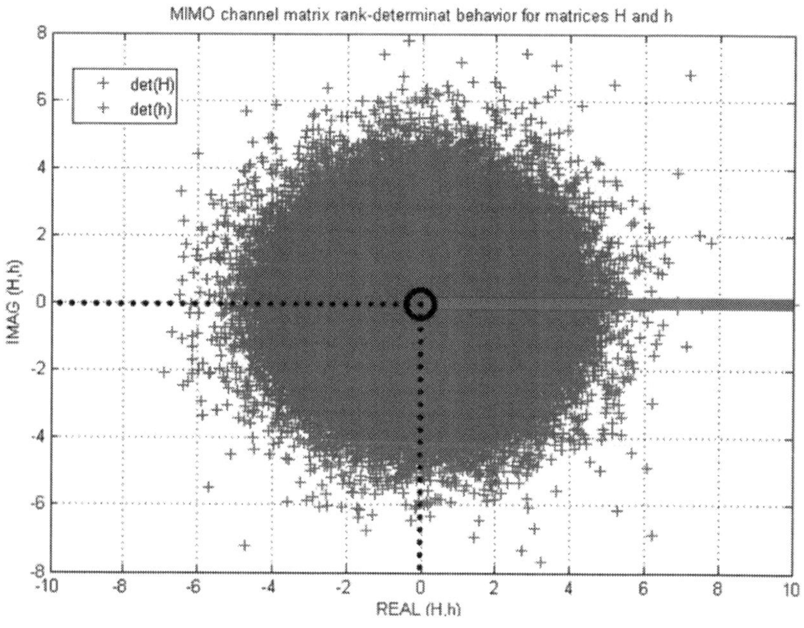

Figure 3. MIMO channel matrix rank-determinant behavior for several realizations for H and h. This statistical evolution is a common pattern found for several MIMO configurations involving slow-flat quasi-static RFC-based environments with $n_R \geq n_T$.

MIMO(n_R, n_T)	$H_{\mathbb{C}}$	$H^{+}_{\mathbb{C}}$
MIMO$(2,2)$	$\begin{bmatrix} 1.12 & 0.15 \\ 0.23 & 0.96 \end{bmatrix}$	$\begin{bmatrix} 0.922 & -0.144 \\ -0.221 & 1.076 \end{bmatrix}$
MIMO$(4,2)$	$\begin{bmatrix} 0.85 - j0.47 & -0.06 + j0.34 \\ -0.37 - j0.72 & 0.94 - j0.28 \\ -0.06 + j0.45 & 1.14 + j1.13 \\ -0.91 + j1.03 & 1.07 - j0.02 \end{bmatrix}$	$\begin{bmatrix} 0.256 + j0.115 & -0.06 + j0.26 & 0.088 - j0.15 & -0.206 - j0.258 \\ 0.051 - j0.082 & 0.217 + j0.116 & 0.234 - j0.275 & 0.15 - j0.018 \end{bmatrix}$

MIMO(n_R, n_T)	h	h^{+}
MIMO$(2,2)$	$\begin{bmatrix} 1.12 & 0.15 & 0 & 0 \\ 0.23 & 0.96 & 0 & 0 \\ 0 & 0 & 1.12 & 0.15 \\ 0 & 0 & 0.23 & 0.96 \end{bmatrix}$	$\begin{bmatrix} 0.922 & -0.144 & 0 & 0 \\ -0.221 & 1.076 & 0 & 0 \\ 0 & 0 & 0.922 & -0.144 \\ 0 & 0 & -0.221 & 1.076 \end{bmatrix}$
MIMO$(4,2)$	$\begin{bmatrix} 0.85 & -0.06 & 0.47 & -0.34 \\ -0.37 & 0.94 & 0.72 & 0.28 \\ -0.06 & 1.14 & -0.45 & -1.13 \\ -0.91 & 1.07 & -1.03 & 0.02 \\ -0.47 & 0.34 & 0.85 & -0.06 \\ -0.72 & -0.28 & -0.37 & 0.94 \\ 0.45 & 1.13 & -0.06 & 1.14 \\ 1.03 & -0.02 & -0.91 & 1.07 \end{bmatrix}$	$\begin{bmatrix} 0.256 & -0.06 & 0.088 & -0.206 & -0.115 & -0.26 & 0.15 & 0.258 \\ 0.051 & 0.217 & 0.234 & 0.15 & 0.082 & -0.116 & 0.275 & 0.018 \\ 0.115 & 0.26 & -0.15 & -0.258 & 0.256 & -0.06 & 0.088 & -0.206 \\ -0.082 & 0.116 & -0.275 & -0.018 & 0.051 & 0.217 & 0.234 & 0.15 \end{bmatrix}$

Table 1. MIMO channel matrix realizations for several MIMO communication link configurations at slow-flat quasi-static RFC scenarios.

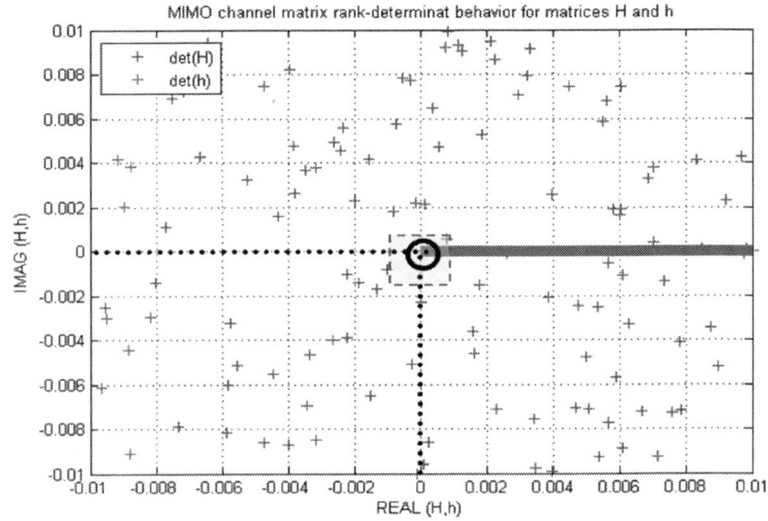

Figure 4. MIMO channel matrix rank-determinant behavior for several realizations for H. Full-rank properties for H and $H^{H}H$ preveal for RFC-based environments (light-blue region delimited by blue-dotted lines).

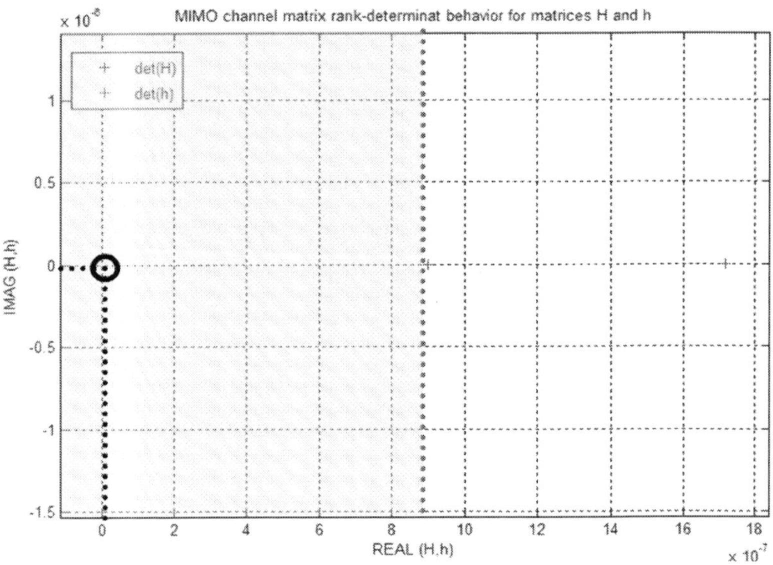

Figure 5. MIMO channel matrix rank-determinant behavior for several realizations for h. Full-rank properties for h and $\mathrm{h}^{T}\mathrm{h}$ preveal for RFC-based environments (light-blue region delimited by red-dotted line).

6. Proposed algorithm

The proposal for a novel algorithm for computing a LPI matrix $\mathrm{h}^{+} \in \mathbb{R}^{2n_T \times 2n_R}$ (with $n_R \geq n_T$) is based on the block-matrix structure of h as exhibited in (4). This idea is an extension of the approach presented in [22]. The existence for this Generalized-Inverse matrix is supported on the statistical properties of the slow-flat quasi-static RFC scenario which impact directly on the singularity of H at every MIMO channel matrix realization. Keeping in mind that other approaches attempting to solve the block-matrix inversion problem [7-10] requires several constraints and conditions, the subsequent proposal does not require any restriction at all mainly due to the aforementioned properties of H. From (4), it is suggested that $\begin{bmatrix} x^{\mathrm{r}} \\ x^{\mathrm{i}} \end{bmatrix}$ is somehow related to $\begin{bmatrix} \Re\{H^{+}\} & -\Im\{H^{+}\} \\ \Im\{H^{+}\} & \Re\{H^{+}\} \end{bmatrix} \cdot \mathrm{Y}$; hence, calculating h^{+} will lead to this solution. Let $A = H^{\mathrm{r}}$ and $B = H^{\mathrm{i}}$. It is kwon a priori that $\rho(A + jB) = n_T$. Then $\mathrm{h} = \begin{bmatrix} A & -B \\ B & A \end{bmatrix}$ with $\rho(\mathrm{h}) = 2n_T = N_t$. Define the matrix $\tilde{\Omega}$ as $\tilde{\Omega} \doteq \mathrm{h}^{\mathrm{T}}\mathrm{h} \in \mathbb{R}^{N_t \times N_t}$, where $\tilde{\Omega} = \begin{bmatrix} M & -L \\ L & M \end{bmatrix}$ with $M = A^{\mathrm{T}}A + B^{\mathrm{T}}B \in \mathbb{R}^{n_T \times n_T}$, $L = A^{\mathrm{T}}B - \left(A^{\mathrm{T}}B\right)^{\mathrm{T}} \in \mathbb{R}^{n_T \times n_T}$, and $\rho(\tilde{\Omega}) = N_t$ as a direct consequence from $2n_R \geq 2n_T \rightarrow N_r \geq N_t$. It can be seen that

$$\mathrm{h}^{+} = \tilde{\Omega}^{-1}\mathrm{h}^{\mathrm{T}} \in \mathbb{R}^{N_t \times N_r} \tag{12}$$

For simplicity, matrix operations involved in (12) require classic multiply-and-accumulate operations between row-entries of $\tilde{\Omega}^{-1} \in \mathbb{R}^{N_t \times N_t}$ and column-entries of $\mathrm{h}^{\mathrm{T}} \in \mathbb{R}^{N_t \times N_r}$. Notice immediately that the critical and essential task of computing h^{+} relies on finding the block matrix inverse $\tilde{\Omega}^{-1}$ [5]. The strategy to be followed in order to solve $\tilde{\Omega}^{-1}$ in (12) will consist of the following steps: 1) the proposition of partitioning $\tilde{\Omega}$ without any restriction on rank-defficiency over inner matrix sub-blocks; 2) the definition of iterative multiply-and-accumulate operations within sub-blocks comprised in $\tilde{\Omega}$; 3) the recursive definition for compacting the overall blockwise matrix inversion. Keep in mind that matrix Ω can be also viewed as $\tilde{\Omega} = \begin{bmatrix} \tilde{\omega}_{1,1} & \cdots & \tilde{\omega}_{1,N_t} \\ \vdots & \ddots & \vdots \\ \tilde{\omega}_{N_t,1} & \cdots & \tilde{\omega}_{N_t,N_t} \end{bmatrix}$. The symmetry presented in $\tilde{\Omega} = \begin{bmatrix} M & -L \\ L & M \end{bmatrix}$ will motivate the development for the pertinent LPI-based algorithm. From (12) and by the use of Lemma 1 it can be concluded that $\tilde{\Omega}^{-1} = \begin{bmatrix} Q & P \\ -P & Q \end{bmatrix}$, where $Q = \left(M + LM^{-1}L\right)^{-1} \in \mathbb{R}^{n_T \times n_T}$, $P = QX \in \mathbb{R}^{n_T \times n_T}$,

---

[5] Notice that $(A + jB)^{*} = (M + jL)^{-1}\left(A^{\mathrm{T}} - jB^{\mathrm{T}}\right)$. Moreover, $(M + jL)^{-1} = \zeta + j\xi \in \mathbb{C}^{n_T \times n_T}$, where $\zeta = \left(M + LM^{-1}L\right)^{-1} = \left(L + ML^{-1}M\right)^{-1} ML^{-1}$ and $\xi = -M^{-1}L\left(M + LM^{-1}L\right)^{-1} = -\left(L + ML^{-1}M\right)^{-1}$.

and $X = LM^{-1} \in \mathbb{R}^{n_T \times n_T}$. Interesting enough, full-rank is identified at each matrix sub-block in the main diagonal of $\tilde{\Omega}$ (besides $\rho(Q) = n_T$). This structural behavior serves as the leitmotiv for the construction of an algorithm for computing the blockwise inverse $\tilde{\Omega}^{-1}$. Basically speaking and concerning step 1) of this strategy, the matrix partition procedure obeys the assignments (13-16) defined as:

$$W_k = \begin{bmatrix} \tilde{\omega}_{N_t-(2k+1),N_t-(2k+1)} & \tilde{\omega}_{N_t-(2k+1),N_t-2k} \\ \tilde{\omega}_{N_t-2k,N_t-(2k+1)} & \tilde{\omega}_{N_t-2k,N_t-2k} \end{bmatrix} \in \mathbb{R}^{2\times 2} \tag{13}$$

$$X_k = \begin{bmatrix} \tilde{\omega}_{N_t-(2k+1),N_t-(2k-1)} & \cdots & \tilde{\omega}_{N_t-(2k+1),N_t} \\ \tilde{\omega}_{N_t-2k,N_t-(2k-1)} & \cdots & \tilde{\omega}_{N_t-2k,N_t} \end{bmatrix} \in \mathbb{R}^{2\times 2k} \tag{14}$$

$$Y_k = \begin{bmatrix} \tilde{\omega}_{N_t-(2k-1),N_t-(2k+1)} & \tilde{\omega}_{N_t-(2k-1),N_t-2k} \\ \vdots & \vdots \\ \tilde{\omega}_{N_t,N_t-(2k+1)} & \tilde{\omega}_{N_t,N_t-2k} \end{bmatrix} \in \mathbb{R}^{2k\times 2} \tag{15}$$

$$Z_0 = \begin{bmatrix} \tilde{\omega}_{N_t-1,N_t-1} & \tilde{\omega}_{N_T-1,N_t} \\ \tilde{\omega}_{N_t,N_t-1} & \tilde{\omega}_{N_t,N_t} \end{bmatrix} \in \mathbb{R}^{2\times 2} \tag{16}$$

The matrix partition over $\tilde{\Omega}$ obeys the index $k = 1:1:(N_t/2 \; -1)$. Because of the even-rectangular dimensions of $\tilde{\Omega}$, matirx $\tilde{\Omega}$ owns exactly an amount of $N_t/2 = n_T$ sub-block matrices of 2×2 dimension along its main diagonal. Interesting enough, due to RFC-based environment characteristics studied in (1) and (4), it is found that:

$$\rho(W_k) = \rho(Z_0) = 2 \tag{17}$$

After performing these structural characteristics for $\tilde{\Omega}$, and with the use of (13-16), step 2) of the strategy consists of the following iterative operations also indexed by $k = 1:1:(N_t/2 \; -1)$, in the sense of performing:

$$\phi_k = W_k - X_k Z_{k-1}^{-1} Y_k \tag{18}$$

$$\alpha_k = \phi_k^{-1} X_k Z_{k-1}^{-1} \tag{19}$$

$$\theta_k = Z_{k-1}^{-1} + Z_{k-1}^{-1} Y_k \alpha_k \tag{20}$$

Here: $Z_{k-1}^{-1} \in \mathbb{R}^{2k\times 2k}$, $\phi_k \in \mathbb{R}^{2\times 2}$, $\alpha_k \in \mathbb{R}^{2\times 2k}$, and $\theta_k \in \mathbb{R}^{2k\times 2k}$. Steps stated in (18-20) help to construct intermediate sub-blocks as

$$\underbrace{\tilde{\Omega}_k}_{2(k+1)\times 2(k+1)} = \begin{bmatrix} \underbrace{W_k}_{2\times 2} & \underbrace{X_k}_{2\times 2k} \\ \underbrace{Y_k}_{2k\times 2} & \underbrace{Z_{k-1}}_{2k\times 2k} \end{bmatrix} \rightarrow \underbrace{\tilde{\Omega}_k^{-1}}_{2(k+1)\times 2(k+1)} = \begin{bmatrix} \underbrace{\phi_k^{-1}}_{2\times 2} & \underbrace{-\alpha_k}_{2\times 2k} \\ \underbrace{-\theta_k Y_k W_k^{-1}}_{2k\times 2} & \underbrace{\theta_k}_{2k\times 2k} \end{bmatrix} \tag{21}$$

The dimensions of each real-valued sub-block in (21) are indicated consistently [6]. For step 3) of the strategy, a recursion step $Z_k^{-1}(Z_{k-1}^{-1})$ is provided in terms of the assignment $Z_k^{-1} = \tilde{\Omega}_k^{-1} \in \mathbb{R}^{2(k+1)\times 2(k+1)}$. Clearly, only inversions of W_k, Z_0, and ϕ_k (which are 2×2 matrices, yielding correspondingly W_k^{-1}, Z_0^{-1}, and ϕ_k^{-1}) are required to be performed throughout this iterative-recursive process, unlike the operation linked to Z_{k-1}^{-1}, which comes from a previous updating step associated with the recursion belonging to Z_k^{-1}. Although $\rho(\tilde{\Omega}) = N_t$ assures the existance of $\tilde{\Omega}^{-1}$, full-rank requirements outlined in (17) and non-zero determinants for (18) are strongly needed for this iterative-recursive algorithm to work accordingly. Also, full-rank is expected for every recursive outcome related to $Z_k^{-1}(Z_{k-1}^{-1})$. Again, thank to the characteristics of the slow-flat quasi-static RFC-based environment in which these operations are involved among every MIMO channel matrix realization, conditions in (17) and full-rank of (18) are always satisfied. These issues are corroborated with the aid of the same MATLAB-based simulation framework used to validate full-rank properties over H and h. The statistical evolution for the determinants for W_k, Z_0, and ϕ_k, and the behavior of singularity within the $Z_k^{-1}(Z_{k-1}^{-1})$ recursion are respectively illustrated in figures (6)-(8). $\mathrm{MIMO}(2,2)$, $\mathrm{MIMO}(4,2)$, and $\mathrm{MIMO}(4,4)$ were the MIMO communication link configurations considered for these tests. These simulation-driven outcomes provide supportive evidence for the proper functionality of the proposed iterative-recursive algorithm for computing $\tilde{\Omega}^{-1}$ involving matrix sub-block inversions. On each figure, the statistical evolution for the determinants associated with Z_0, W_k, ϕ_k, and $Z_k^{-1}(Z_{k-1}^{-1})$ are respectively indicated by labels det(Zo), det(Wk), det(Fik), and det(iZk,iZkm1), while the light-blue zone at bottom delimited by a red-dotted line exhibits the gap which marks the avoidance in rank-deficincy over the involved matrices. The zero-determinant value is marked with a black circle.

The next point of analysis for the behavior of the h^+ LPI-based iterative-recursive algorithm is complexity, which in essence will consist of a demand in matrix partitions (amount of matrix sub-blocks: PART) and arithmetic operations (amount of additions-subtractions: ADD-SUB; multiplications: MULT; and divisions: DIV). Let PART-mtx and ARITH-ops be the nomenclature for complexity cost related to matrix partitions and arithmetic operations, respectively. Without loss of generalization, define $\mathrm{C}[*]$ as the complexity in terms of the

[6] Matrix structure given in (21) is directly derived from applying Equation (6), and by the use of Lemma 1 as $\left(Z_{k-1} - Y_k W_k^{-1} X_k\right)^{-1} = Z_{k-1}^{-1} + Z_{k-1}^{-1} Y_k \left(W_k - X_k Z_{k-1}^{-1} Y_k\right)^{-1} X_k Z_{k-1}^{-1}$. See that this expansion is preferable instead of $\left(W_k - X_k Z_{k-1}^{-1} Y_k\right)^{-1} = W_k^{-1} + W_k^{-1} X_k \left(Z_{k-1} - Y_k W_k^{-1} X_k\right)^{-1} Y_k W_k^{-1}$, which is undesirable due to an unnecessary matrix operation overhead related to computing Z_{k-1}, e.g. inverting Z_{k-1}^{-1}, which comes preferably from the $Z_k^{-1}(Z_{k-1}^{-1})$ recursion.

costs PART-mtx and ARITH-ops belonging to operations involved in $*$. Henceforth, $C\left[h^{+}\right]=C\left[\tilde{\Omega}^{-1}\right]+C\left[\tilde{\Omega}^{-1}\bullet h^{T}\right]$ denotes the cost of computing h^{+} as the sum of the costs of inverting $\tilde{\Omega}$ and multiplying $\tilde{\Omega}^{-1}$ by h^{T}. It is evident that: a) $C\left[\tilde{\Omega}^{-1}\bullet h^{T}\right]$ implies PART=0 and ARITH-ops itemized into MULT=$8n_R n_T^2$, ADD-SUB=$4n_R n_T\left(2n_T-1\right)$, and DIV=0; b) $C\left[\tilde{\Omega}^{-1}\right]=C\left[h^{T}\bullet h\right]+C\left[(h^{T}h)^{-1}\right]$. Clearly, $C\left[h^{T}\bullet h\right]$ demands no partitions at all, but with a ARITH-ops cost of MULT=$8n_R n_T^2$, and ADD-SUB=$4\left(2n_R-1\right)n_T^2$. However, the principal complexity relies critically on $C\left[(h^{T}h)^{-1}\right]$, which is the backbone for h^{+}, as presented in [22]. Table 2 summerizes these complexity results. For this treatment, $C\left[(h^{T}h)^{-1}\right]$ consists of $3n_T-2$ partitions, MULT = $\sum_{k=1}^{n_T-1} C_k^{I}+6$, ADD-SUB = $\sum_{k=1}^{n_T-1} C_k^{II}+1$, and DIV = $\sum_{k=1}^{n_T-1} C_k^{III}+1$. The ARITH-ops cost depends on C_k^{I}, C_k^{II}, and C_k^{III}; the constant factors for each one of these items are proper of the complexity presented in $C\left[Z_0^{-1}\right]$. The remain of the complexities, i.e. C_k^{I}, C_k^{II}, and C_k^{III}, are calculated according to the iterative stpes defined in (18-20) and (21), particularly expressed in terms of

$$C\left[\phi_k^{-1}\right]+C\left[-\alpha_k\right]+C\left[-\theta_k Y_k W_k^{-1}\right]+C\left[\theta_k\right] \tag{22}$$

It can be checked out that: a) no PART-mtx cost is required; b) the ARITH-ops cost employs (22) for each item, yielding: $C_k^{I}=40k^2+24k+12$ (for MULT), $C_k^{II}=40k^2+2$ (for ADD_SUB), and $C_k^{III}=2$ (for DIV).

An illustrative application example is given next. It considers a MIMO channel matrix realization obeying statistical behavior according to (1) and a $\text{MIMO}(4,4)$ configuration:

$$H=\begin{bmatrix} -0.3059+j0.7543 & -0.8107+j0.2082 & 0.2314-j0.4892 & -0.416-j1.0189 \\ -1.1777+j0.0419 & 0.8421-j0.9448 & 0.1235+j0.6067 & 1.5437+j0.4039 \\ 0.0886-j0.0676 & 0.8409+j0.5051 & -0.132+j0.8867 & -0.0964-j0.2828 \\ 0.2034-j0.5886 & -0.0266+j1.148 & 0.5132-j1.1269 & 0.0806+j0.4879 \end{bmatrix}\in\mathbb{C}^{4\times4} \text{ with } \rho(H)=4.$$ As a consequence,

$$\tilde{\Omega}=\begin{bmatrix} 2.4516 & -1.2671 & 0.1362 & -2.7028 & 0 & -1.9448 & 0.6022 & -0.2002 \\ -1.2671 & 4.5832 & -1.7292 & 1.3776 & 1.9448 & 0 & -1.229 & -2.4168 \\ 0.1362 & -1.7292 & 3.0132 & 0.0913 & -0.6022 & 1.229 & 0 & 0.862 \\ -2.7028 & 1.3776 & 0.0913 & 4.0913 & 0.2002 & 2.4168 & -0.862 & 0 \\ 0 & 1.9448 & -0.6022 & 0.2002 & 2.4516 & -1.2671 & 0.1362 & -2.7028 \\ -1.9448 & 0 & 1.229 & 2.4168 & -1.2671 & 4.5832 & -1.7292 & 1.3776 \\ 0.6022 & -1.229 & 0 & -0.862 & 0.1362 & -1.7292 & 3.0132 & 0.0913 \\ -0.2002 & -2.4168 & 0.862 & 0 & -2.7028 & 1.3776 & 0.0913 & 4.0913 \end{bmatrix}\in\mathbb{R}^{8\times8} \text{ with } \rho(\tilde{\Omega})=8.$$

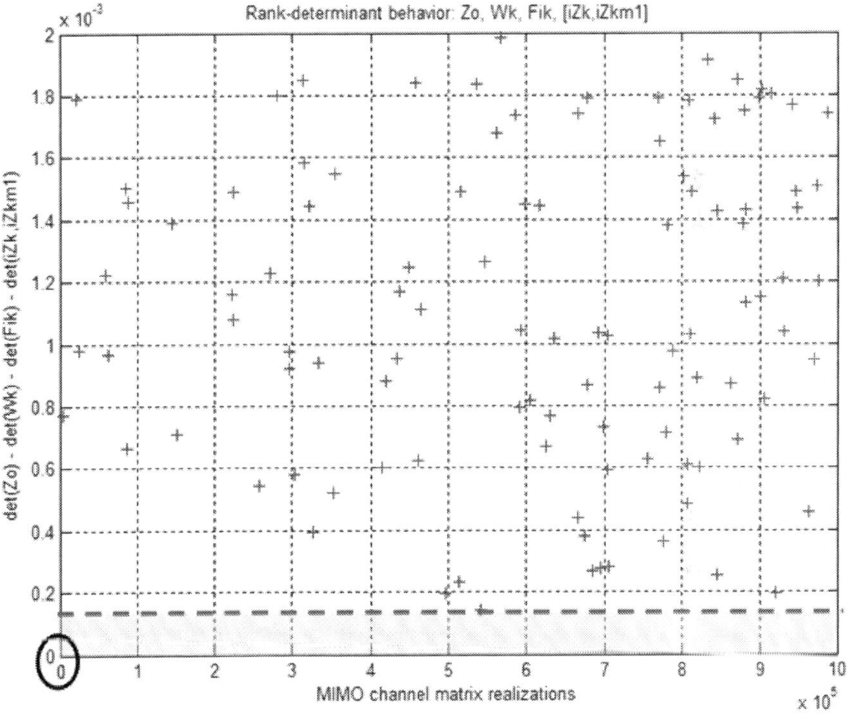

Figure 6. Statistical evolution of the rank-determinant behaviour concerning Z_0, W_k, ϕ_k, and $Z_k^{-1}(Z_{k-1}^{-1})$ for a MIMO(2,2) configuration.

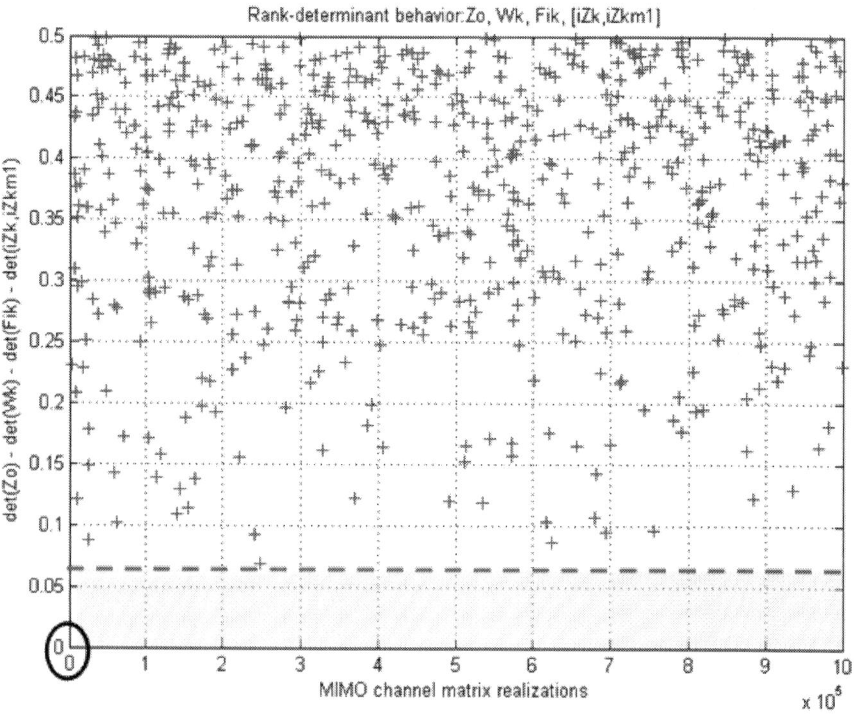

Figure 7. Statistical evolution of the rank-determinant behaviour concerning Z_0 , W_k , ϕ_k , and $Z_k^{-1}(Z_{k-1}^{-1})$ for a MIMO(4, 2) configuration.

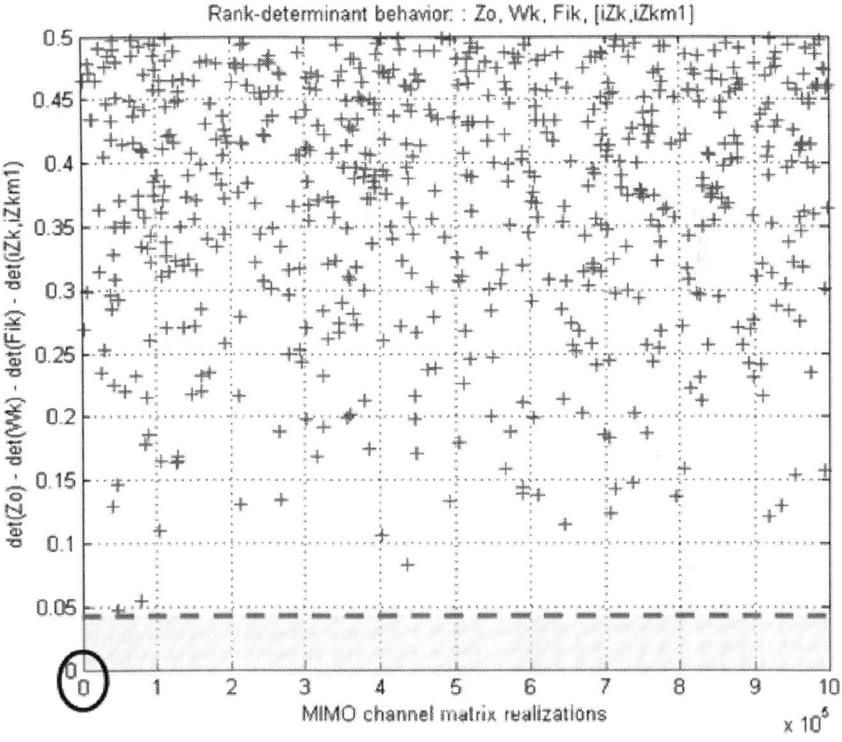

Figure 8. Statistical evolution of the rank-determinant behaviour concerning Z_0, W_k, ϕ_k, and $Z_k^{-1}(Z_{k-1}^{-1})$ for a MIMO$(4,4)$ configuration.

<table>
<tr><td colspan="3">$C\left[\mathrm{h}^{+}\right]$</td><td></td><td></td></tr>
<tr><td rowspan="2">$C\left[\tilde{\Omega}^{-1}\bullet\mathrm{h}^{T}\right]$</td><td colspan="2">$C\left[\tilde{\Omega}^{-1}\right]$</td><td></td><td></td></tr>
<tr><td>$C\left[\mathrm{h}^{T}\bullet\mathrm{h}\right]$</td><td>$C\left[\left(\mathrm{h}^{T}\mathrm{h}\right)^{-1}\right]$</td><td></td><td></td></tr>
<tr><td>0</td><td>0</td><td>$3n_T-2$</td><td>PART-mtx cost [in PART operations]</td><td></td></tr>
<tr><td>$8n_R n_T^2$</td><td>$8n_R n_T^2$</td><td>$\sum_{k=1}^{n_T-1} C_k^I+6$</td><td>MULT</td><td rowspan="3">ARITH-ops cost [in amount of itemized operations]</td></tr>
<tr><td>$4n_R n_T\left(2n_T-1\right)$</td><td>$4\left(2n_R-1\right)n_T^2$</td><td>$\sum_{k=1}^{n_T-1} C_k^{II}+1$</td><td>ADD-SUBB</td></tr>
<tr><td>0</td><td>0</td><td>$\sum_{k=1}^{n_T-1} C_k^{III}+1$</td><td>DIV</td></tr>
</table>

	Based on (22), and for		
	C_k^{I} :	C_k^{II} :	C_k^{III} :
$C\left[\phi_k^{-1}\right]$	$8k^2+8k+6$	$8k^2+4k+1$	1
$C\left[-\alpha_k\right]$	$8k^2+8k$	$8k^2$	0
$C\left[-\theta_k Y_k W_k^{-1}\right]$	$8k^2+8k+6$	$8k^2+1$	1
$C\left[\theta_k\right]$	$16k^2$	$16k^2-4k$	0

Table 2. Complexity cost results of the LPI-based iterative-recursive algorithm for h^{+}.

Applying partition criteria (13-16) and given $k=1:1:3$, the following matrix sub-blocks are generated:

$$W_1=\begin{bmatrix}2.4516 & -1.2671\\ -1.2671 & 4.5832\end{bmatrix},$$

$$X_1=\begin{bmatrix}0.1362 & -2.7028\\ -1.7292 & 1.3776\end{bmatrix},\ Y_1=\begin{bmatrix}0.1362 & -1.7292\\ -2.7028 & 1.3776\end{bmatrix},\ Z_0=\begin{bmatrix}3.0132 & 0.0913\\ 0.0913 & 4.0913\end{bmatrix},$$

$$W_2=\begin{bmatrix}3.0132 & 0.0913\\ 0.0913 & 4.0913\end{bmatrix},$$

$$X_2=\begin{bmatrix}-0.6022 & 1.2290 & 0 & 0.862\\ 0.2002 & 2.4168 & -0.862 & 0\end{bmatrix},\ Y_2=\begin{bmatrix}-0.6022 & 0.2002\\ 1.229 & 2.4168\\ 0 & -0.862\\ 0.862 & 0\end{bmatrix},$$

$$W_3 = \begin{bmatrix} 2.4516 & -1.2671 \\ -1.2671 & 4.5832 \end{bmatrix},\ X_3 = \begin{bmatrix} 0.1362 & -2.7028 & 0 & -1.9448 & 0.6022 & -0.2002 \\ -1.7292 & 1.3776 & 1.9448 & 0 & -1.229 & -2.4168 \end{bmatrix},$$

and $Y_3 = \begin{bmatrix} 0.1362 & -1.7292 \\ -2.7028 & 1.3776 \\ 0 & 1.9448 \\ -1.9448 & 0 \\ 0.6022 & -1.229 \\ -0.2002 & -2.4168 \end{bmatrix}$. Suggested by (18-20), iterative operations (23-25) are computed as:

$$\phi_1 = W_1 - X_1 Z_0^{-1} Y_1,\quad \alpha_1 = \phi_1^{-1} X_1 Z_0^{-1},\quad \theta_1 = Z_0^{-1} + Z_0^{-1} Y_1 \alpha_1 \tag{23}$$

$$\phi_2 = W_2 - X_2 Z_1^{-1} Y_2,\quad \alpha_2 = \phi_2^{-1} X_2 Z_1^{-1},\quad \theta_2 = Z_1^{-1} + Z_1^{-1} Y_2 \alpha_2 \tag{24}$$

$$\phi_3 = W_3 - X_3 Z_2^{-1} Y_3,\quad \alpha_3 = \phi_3^{-1} X_3 Z_2^{-1},\quad \theta_3 = Z_2^{-1} + Z_2^{-1} Y_3 \alpha_3 \tag{25}$$

From (21), the matrix assignments related to recursion $Z_k^{-1}(Z_{k-1}^{-1})$ produces the following intermediate blockwise matrix results:

$$Z_1^{-1}\left(Z_0^{-1}\right) = \tilde{\Omega}_1^{-1} = \begin{bmatrix} \phi_1^{-1} & -\alpha_1 \\ -\theta_1 Y_1 W_1^{-1} & \theta_1 \end{bmatrix} = \begin{bmatrix} 1.5765 & 0.1235 & -0.0307 & 1.0005 \\ 0.1235 & 0.3332 & 0.1867 & -0.0348 \\ -0.0307 & 0.1867 & 0.4432 & -0.093 \\ 1.0005 & -0.0348 & -0.093 & 0.9191 \end{bmatrix},$$

$$Z_2^{-1}\left(Z_1^{-1}\right) = \tilde{\Omega}_2^{-1} = \begin{bmatrix} \phi_2^{-1} & -\alpha_2 \\ -\theta_2 Y_2 W_2^{-1} & \theta_2 \end{bmatrix}$$

$$= \begin{bmatrix} 0.4098 & 0.0879 & -0.0829 & -0.1839 & -0.0743 & -0.0775 \\ 0.0879 & 0.4355 & -0.2847 & -0.3182 & -0.0422 & -0.0985 \\ -0.0829 & -0.2847 & 1.7642 & 0.3393 & 0.0012 & 1.0686 \\ -0.1839 & -0.3182 & 0.3393 & 0.6023 & 0.2376 & 0.0548 \\ -0.0743 & -0.0422 & 0.0012 & 0.2376 & 0.4583 & -0.0738 \\ -0.0775 & -0.0985 & 1.0686 & 0.0548 & -0.0738 & 0.9499 \end{bmatrix},\ Z_3^{-1}\left(Z_2^{-1}\right) = \tilde{\Omega}^{-1} = \begin{bmatrix} \phi_3^{-1} & -\alpha_3 \\ -\theta_3 Y_3 W_3^{-1} & \theta_3 \end{bmatrix}$$

$$= \begin{bmatrix} 1.9798 & 0.3808 & -0.1114 & 1.0224 & 0 & 0.3605 & 0.2524 & 0.2183 \\ 0.3808 & 0.6759 & 0.2619 & 0.0856 & -0.3605 & 0 & 0.2368 & 0.1193 \\ -0.1114 & 0.2619 & 0.5493 & -0.0218 & -0.2524 & -0.2368 & 0 & -0.0535 \\ 1.02224 & 0.0856 & -0.0218 & 0.9839 & -0.2183 & -0.1193 & 0.0535 & 0 \\ 0 & -0.3605 & -0.2524 & -0.2183 & 1.9798 & 0.3808 & -0.1114 & 1.0224 \\ 0.3605 & 0 & -0.2368 & -0.1193 & 0.3808 & 0.6759 & 0.2619 & 0.0856 \\ 0.2524 & 0.2368 & 0 & 0.0535 & -0.1114 & 0.2619 & 0.5493 & -0.0218 \\ 0.2183 & 0.1193 & -0.0535 & 0 & -0.1114 & 0.0856 & -0.0218 & 0.9839 \end{bmatrix}$$

. This last recursive outcome from $Z_k^{-1}(Z_{k-1}^{-1})$ corresponds to $\tilde{\Omega}^{-1}$, and is further used for calculating $h^{+} = \tilde{\Omega}^{-1} h^{T} \in \mathbb{R}^{8\times 8}$.

Moreover, notice that full-rank properties are always presented in matrices $Z_0, W_1, W_2, W_3, \phi_1, \phi_2, \phi_3, Z_1^{-1}, Z_2^{-1}$, and Z_3^{-1}.

7. VLSI implementation aspects

The arithmetic operations presented in the algorithm for computing h^+ can be implemented under a modular-iterative fashion towards a VLSI (Very Large Scale of Integration) design. The partition strategy comprised in (13-16) provides modularity, while (18-20) is naturally associated with iterativeness; recursion is just used for constructing matrix-blocks in (21). Several well-studied aspects aid to implement a further VLSI architecture [23-27] given the nature of the mathematical structure of the algorithm. For instance, systolic arrays [25-27] are a suitable choice for efficient, parallel-processing architectures concerning matrix multiplications-additions. Bidimensional processing arrays are typical architectural outcomes, whose design consist basically in interconnecting processing elements (PE) among different array layers. The configuration of each PE comes from projection or linear mapping techniques [25-27] derived from multiplications and additions presented in (18-20). Also, systolic arrays tend to concurrently perform arithmetic operations dealing with the matrix concatenated multiplications $X_k Z_{k-1}^{-1} Y_k, \phi_k^{-1} X_k Z_{k-1}^{-1}, Z_{k-1}^{-1} Y_k \alpha_k$, and $\theta_k Y_k W_k^{-1}$ presented in (18-20). Consecutive additions inside every PE can be favourably implemented via Carry-Save-Adder (CSA) architectures [23-24], while multiplications may recur to Booth multipliers [23-24] in order to reduce latencies caused by adding acummulated partial products. Divisions presented in W_k^{-1}, Z_0^{-1}, and ϕ_k^{-1} can be built through regular shift-and-subtract modules or classic serial-parallel subtractors [23-24]; in fact, CORDIC (Coordinate Rotate Digital Computer) processors [23] are also employed and configured in order to solve numerical divisions. The aforementioned architectural aspects offer an attractive and alternative framework for consolidating an ultimate VLSI design for implementing the h^+ algorithm without compromising the overall system data throughput (intrinsicly related to operation frequencies) for it.

8. Conclusions

This chapter presented the development of a novel iterative-recursive algorithm for computing a Left-Pseudoinverse (LPI) as a Generalized-Inverse for a MIMO channel matrix within a Rayleigh fading channel (RFC). The formulation of this algorithm consisted in the following step: i) first, structural properties for the MIMO channel matrix acquired permanent full-rank due to statistical properties of the RFC scenario; ii) second, Partition-Matrix Theory was applied allowing the generation of a block-matrix version of the MIMO channel matrix; iii) third, iterative addition-multiplication operations were applied at these matrix sub-blocks in order to construct blockwise sub-matrix inverses, and recursively reusing them for obtaining the LPI. For accomplishing this purpose, required mathematical background and MIMO systems concepts were provided for consolidating a solid scientific framework to understand the context of the problem this algorithm was attempting to solve.

Proper functionality for this approach was validated through simulation-driven experiments, as well as providing an example of this operation. As an additional remark, some VLSI aspects and architectures were outlined for basically implementing arithmetic operations within the proposed LPI-based algorithm.

Author details

P. Cervantes and L. F. González
Instituto Tecnológico y de Estudios Superiores de Monterrey, Campus Guadalajara, ITESM University, Mexico

F. J. Ortiz and A. D. García
Instituto Tecnológico y de Estudios Superiores de Monterrey, Campus Estado de México, ITESM University, Mexico

Acknowledgement

This work was supported by CONACYT (National Council of Science and Technology) under the supervision, revision, and sponsorship of ITESM University (Instituto Tecnológico y de Estudios Superiores de Monterrey).

9. References

[1] Abramovich YI, Johnson BA, and Spencer NK (2008) Two-Dimensional Multivariate Parametric Models for Radar Applications-Part I: Maximum-Entropy Extensions for Toeplitz-Block Matrices. IEEE Transactions on Signal Processing, vol. 56, no. 11. November 2008: 5509-5526.

[2] Bera TK, *et al* (2011) Improving the image reconstruction in Electrical Impedance Tomography (EIT) with block matrix-based Multiple Regularization (BMMR): A practical phantom study. World Congress on Information and Communication Technologies (WICT). 2011: 1346-1351.

[3] Kailath T (1980) Linear Systems. Prentice-Hall. 682 p.

[4] Spong MW (1998) Underactuated Mechanical Systems. Control Problems in Robotics and Automation, Lecture Notes in Control and Information Sciences, no. 230. Springer-Verlag: 135-150.

[5] Utkin V, Guldner J, and Shi JX (1992) Sliding Mode Control in Electro-Mechanical Systems. CRC Press. April 1999: 338 p.

[6] Juang J-N (1993) Applied System Identification. Prentice Hall. 400 p.

[7] Watt SM (2006) Pivot-Free Block Matrix Inversion. Proceedings of the Eighth International Symposium on Symbolic and Numeric Algorithms for Scientific Computing (SYNASC), IEEE Computer Society: 5 p.

[8] Tian Y, and Tanake Y (2009) The inverse of any two-by-two nonsingular partitioned matrix and three matrix inverse completion problems. Journal Computers & Mathematics with Applications, vol. 57, issue 8. April 2009: 12 p.

[9] Choi Y (2009) New Form of Block Matrix Inversion. International Conference on Advanced Intelligent Mechatronics. July 2009: 1952-1957.

[10] Choi Y, and Cheong J. (2009) New Expressions of 2X2 Block Matrix Inversion and Their Application. IEEE Transactions on Automatic Control, vol. 54, no. 11. November 2009: 2648-2653.

[11] Fontán FP, and Espiñera PM (2008) Modeling the Wireless Propagation Channel. Wiley. 268 p.

[12] El-Hajjar M, and Hanzo L (2010) Multifunctional MIMO Systems: A Combined Diversity and Multiplexing Design Perspective. IEEE Wireless Communications. April 2010: 73-79.

[13] Biglieri E, *et al* (2007) MIMO Wireless Communications. Cambridge University Press: United Kingdom. 344 p.

[14] Jankiraman M (2004) Space-Time Codes and MIMO Systems. Artech House: United States. 327 p.

[15] Biglieri E, Proakis J, and Shamai S (1998) Fading Channels: Information-Theoretic and Communications Aspects. IEEE Transactions on Information Theory, vol. 44, no. 6. October 1998: 2619-2692.

[16] Almers P, Bonek E, Burr A, *et al* (2007) Survey of Channel and Radio Propagation Models for Wireless MIMO Systems. EURASIP Journal on Wireless Communications and Networking, vol. 2011, issue 1. January 2007: 19 p.

[17] Golub GH, and Van Loan CF (1996) Matrix Computations. The Johns Hopkins University Press. 694 p.

[18] Serre D (2001) Matrices: Theory and Applications. Springer Verlag. 202 p.

[19] R&S®. Rohde & Schwarz GmbH & Co. KG. WLAN 802.11n: From SISO to MIMO. Application Note: 1MA179_9E. Available: www.rohde-schwarz.com: 59 p.

[20] © Agilent Technologies, Inc. (2008) Agilent MIMO Wireless LAN PHY Layer [RF] : Operation & Measurement: Application Note: 1509. Available: www.agilent.com: 48 p.

[21] Paul T, and Ogunfunmi T (2008) Wireless LAN Comes of Age : Understanding the IEEE 802.11n Amendment. IEEE Circuits and Systems Magazine. First Quarter 2008: 28-54.

[22] Cervantes P, González VM, and Mejía PA (2009) Left-Pseudoinverse MIMO Channel Matrix Computation. 19th International Conference on Electronics, Communications, and Computers (CONIELECOMP 2009). July 2009: 134-138.

[23] Milos E, and Tomas L (2004) Digital Arithmetic. Morgan Kauffmann Publishers. 709 p.

[24] Parhi KK (1999) VLSI Digital Signal Processing Systems: Design and Implementation. John Wiley & Sons. 784 p.

[25] Song SW (1994) Systolic Algorithms: Concepts, Synthesis, and Evolution. Institute of Mathematics, University of Sao Paulo, Brazil. Available: http://www.ime.usp.br/~song/papers/cimpa.pdf. DOI number: 10.1.1.160.4057: 40 p.

[26] Kung SY (1985) VLSI Array Processors. IEEE ASSP Magazine. July 1985: 4-22.

[27] Jagadish HV, Rao SK, and Kailath T (1987) Array Architectures for Iterative Algorithms. Proceedings of the IEEE, vol. 75, no. 9. September 1987: 1304-1321.

Chapter 8

Operator Means and Applications

Pattrawut Chansangiam

Additional information is available at the end of the chapter

http://dx.doi.org/10.5772/46479

1. Introduction

The theory of scalar means was developed since the ancient Greek by the Pythagoreans until the last century by many famous mathematicians. See the development of this subject in a survey article [24]. In Pythagorean school, various means are defined via the method of proportions (in fact, they are solutions of certain algebraic equations). The theory of matrix and operator means started from the presence of the notion of parallel sum as a tool for analyzing multi-port electrical networks in engineering; see [1]. Three classical means, namely, arithmetic mean, harmonic mean and geometric mean for matrices and operators are then considered, e.g., in [3, 4, 11, 12, 23]. These means play crucial roles in matrix and operator theory as tools for studying monotonicity and concavity of many interesting maps between algebras of operators; see the original idea in [3]. Another important mean in mathematics, namely the power mean, is considered in [6]. The parallel sum is characterized by certain properties in [22]. The parallel sum and these means share some common properties. This leads naturally to the definitions of the so-called connection and mean in a seminal paper [17]. This class of means cover many in-practice operator means. A major result of Kubo-Ando states that there are one-to-one correspondences between connections, operator monotone functions on the non-negative reals and finite Borel measures on the extended half-line. The mean theoretic approach has many applications in operator inequalities (see more information in Section 8), matrix and operator equations (see e.g. [2, 19]) and operator entropy. The concept of operator entropy plays an important role in mathematical physics. The *relative operator entropy* is defined in [13] for invertible positive operators A, B by

$$S(A|B) = A^{1/2} \log(A^{-1/2}BA^{-1/2})A^{1/2}. \tag{1}$$

In fact, this formula comes from the Kubo-Ando theory–$S(\cdot|\cdot)$ is the connection corresponds to the operator monotone function $t \mapsto \log t$. See more information in [7, Chapter IV] and its references.

In this chapter, we treat the theory of operator means by weakening the original definition of connection in such a way that the same theory is obtained. Moreover, there is a one-to-one correspondence between connections and finite Borel measures on the unit interval. Each connection can be regarded as a weighed series of weighed harmonic means. Hence, every mean in Kubo-Ando's sense corresponds to a probability Borel measure on the unit interval.

Various characterizations of means are obtained; one of them is a usual property of scalar mean, namely, the betweenness property. We provide some new properties of abstract operator connections, involving operator monotonicity and concavity, which include specific operator means as special cases.

For benefits of readers, we provide the development of the theory of operator means. In Section 2, we setup basic notations and state some background about operator monotone functions which play important roles in the theory of operator means. In Section 3, we consider the parallel sum together with its physical interpretation in electrical circuits. The arithmetic mean, the geometric mean and the harmonic mean of positive operators are investigated and characterized in Section 4. The original definition of connection is improved in Section 5 in such a way that the same theory is obtained. In Section 6, several characterizations and examples of Kubo-Ando means are given. We provide some new properties of general operator connections, related to operator monotonicity and concavity, in Section 7. Many operator versions of classical inequalities are obtained via the mean-theoretic approach in Section 8.

2. Preliminaries

Throughout, let $B(\mathcal{H})$ be the von Neumann algebra of bounded linear operators acting on a Hilbert space $\mathcal{H}$. Let $B(\mathcal{H})^{sa}$ be the real vector space of self-adjoint operators on $\mathcal{H}$. Equip $B(\mathcal{H})$ with a natural partial order as follows. For $A, B \in B(\mathcal{H})^{sa}$, we write $A \leqslant B$ if $B - A$ is a positive operator. The notation $T \in B(\mathcal{H})^+$ or $T \geqslant 0$ means that T is a positive operator. The case that $T \geqslant 0$ and T is invertible is denoted by $T > 0$ or $T \in B(\mathcal{H})^{++}$. Unless otherwise stated, every limit in $B(\mathcal{H})$ is taken in the strong-operator topology. Write $A_n \to A$ to indicate that A_n converges strongly to A. If A_n is a sequence in $B(\mathcal{H})^{sa}$, the expression $A_n \downarrow A$ means that A_n is a decreasing sequence and $A_n \to A$. Similarly, $A_n \uparrow A$ tells us that A_n is increasing and $A_n \to A$. We always reserve A, B, C, D for positive operators. The set of non-negative real numbers is denoted by $\mathbb{R}^+$.

Remark 0.1. It is important to note that if A_n is a decreasing sequence in $B(\mathcal{H})^{sa}$ such that $A_n \geqslant A$, then $A_n \to A$ if and only if $\langle A_n x, x\rangle \to \langle Ax, x\rangle$ for all $x \in \mathcal{H}$. Note first that this sequence is convergent by the order completeness of $B(\mathcal{H})$. For the sufficiency, if $x \in \mathcal{H}$, then

$$\|(A_n - A)^{1/2}x\|^2 = \langle (A_n - A)^{1/2}x, (A_n - A)^{1/2}x\rangle = \langle (A_n - A)x, x\rangle \to 0$$

and hence $\|(A_n - A)x\| \to 0$.

The spectrum of $T \in B(\mathcal{H})$ is defined by

$$\mathrm{Sp}(T) = \{\lambda \in \mathbb{C} : T - \lambda I \text{ is not invertible}\}.$$

Then $\mathrm{Sp}(T)$ is a nonempty compact Hausdorff space. Denote by $C(\mathrm{Sp}(T))$ the C*-algebra of continuous functions from $\mathrm{Sp}(T)$ to $\mathbb{C}$. Let $T \in B(\mathcal{H})$ be a normal operator and $z : \mathrm{Sp}(T) \to \mathbb{C}$ the inclusion. Then there exists a unique unital $*$-homomorphism $\phi : C(\mathrm{Sp}(T)) \to B(\mathcal{H})$ such that $\phi(z) = T$, i.e.,

- ϕ is linear
- $\phi(fg) = \phi(f)\phi(g)$ for all $f, g \in C(\mathrm{Sp}(T))$

- $\phi(\bar{f}) = (\phi(f))^*$ for all $f \in C(\mathrm{Sp}(T))$
- $\phi(1) = I$.

Moreover, ϕ is isometric. We call the unique isometric $*$-homomorphism which sends $f \in C(\mathrm{Sp}(T))$ to $\phi(f) \in B(\mathcal{H})$ the *continuous functional calculus* of T. We write $f(T)$ for $\phi(f)$.

Example 0.2. 1. If $f(t) = a_0 + a_1 t + \cdots + a_n t^n$, then $f(T) = a_0 I + a_1 T + \cdots + a_n T^n$.

2. If $f(t) = \bar{t}$, then $f(T) = \phi(f) = \phi(\bar{z}) = \phi(z)^* = T^*$
3. If $f(t) = t^{1/2}$ for $t \in \mathbb{R}^+$ and $T \geqslant 0$, then we define $T^{1/2} = f(T)$. Equivalently, $T^{1/2}$ is the unique positive square root of T.
4. If $f(t) = t^{-1/2}$ for $t > 0$ and $T > 0$, then we define $T^{-1/2} = f(T)$. Equivalently, $T^{-1/2} = (T^{1/2})^{-1} = (T^{-1})^{1/2}$.

A continuous real-valued function f on an interval I is called an *operator monotone function* if one of the following equivalent conditions holds:

(i) $A \leqslant B \implies f(A) \leqslant f(B)$ for all Hermitian matrices A, B of all orders whose spectrums are contained in I;

(ii) $A \leqslant B \implies f(A) \leqslant f(B)$ for all Hermitian operators $A, B \in B(\mathcal{H})$ whose spectrums are contained in I and for an infinite dimensional Hilbert space $\mathcal{H}$;

(iii) $A \leqslant B \implies f(A) \leqslant f(B)$ for all Hermitian operators $A, B \in B(\mathcal{H})$ whose spectrums are contained in I and for all Hilbert spaces $\mathcal{H}$.

This concept is introduced in [20]; see also [7, 10, 15, 16]. Every operator monotone function is always continuously differentiable and monotone increasing. Here are examples of operator monotone functions:

1) $t \mapsto \alpha t + \beta$ on $\mathbb{R}$, for $\alpha \geqslant 0$ and $\beta \in \mathbb{R}$,
2) $t \mapsto -t^{-1}$ on $(0, \infty)$,
3) $t \mapsto (c - t)^{-1}$ on (a, b), for $c \notin (a, b)$,
4) $t \mapsto \log t$ on $(0, \infty)$,
5) $t \mapsto (t-1)/\log t$ on $\mathbb{R}^+$, where $0 \mapsto 0$ and $1 \mapsto 1$.

The next result is called the Löwner-Heinz's inequality [20].

Theorem 0.3. *For $A, B \in B(\mathcal{H})^+$ and $r \in [0,1]$, if $A \leqslant B$, then $A^r \leqslant B^r$. That is the map $t \mapsto t^r$ is an operator monotone function on $\mathbb{R}^+$ for any $r \in [0,1]$.*

A key result about operator monotone functions is that there is a one-to-one correspondence between nonnegative operator monotone functions on $\mathbb{R}^+$ and finite Borel measures on $[0, \infty]$ via integral representations. We give a variation of this result in the next proposition.

Proposition 0.4. *A continuous function $f : \mathbb{R}^+ \to \mathbb{R}^+$ is operator monotone if and only if there exists a finite Borel measure μ on $[0,1]$ such that*

$$f(x) = \int_{[0,1]} 1 \,!_t\, x \, d\mu(t), \quad x \in \mathbb{R}^+. \tag{2}$$

Here, the weighed harmonic mean $!_t$ is defined for $a, b > 0$ by

$$a \,!_t\, b = [(1-t)a^{-1} + tb^{-1}]^{-1} \tag{3}$$

and extended to $a, b \geqslant 0$ by continuity. Moreover, the measure μ is unique. Hence, there is a one-to-one correspondence between operator monotone functions on the non-negative reals and finite Borel measures on the unit interval.

Proof. Recall that a continuous function $f : \mathbb{R}^+ \to \mathbb{R}^+$ is operator monotone if and only if there exists a unique finite Borel measure ν on $[0, \infty]$ such that

$$f(x) = \int_{[0,\infty]} \phi_x(\lambda)\, d\nu(\lambda), \quad x \in \mathbb{R}^+$$

where

$$\phi_x(\lambda) = \frac{x(\lambda+1)}{x+\lambda} \text{ for } \lambda > 0, \quad \phi_x(0) = 1, \quad \phi_x(\infty) = x.$$

Consider the Borel measurable function $\psi : [0,1] \to [0,\infty], t \mapsto \frac{t}{1-t}$. Then, for each $x \in \mathbb{R}^+$,

$$\begin{aligned} \int_{[0,\infty]} \phi_x(\lambda)\, d\nu(\lambda) &= \int_{[0,1]} \phi_x \circ \psi(t)\, d\nu\psi(t) \\ &= \int_{[0,1]} \frac{x}{x - xt + t}\, d\nu\psi(t) \\ &= \int_{[0,1]} 1 \,!_t\, x \, d\nu\psi(t). \end{aligned}$$

Now, set $\mu = \nu\psi$. Since ψ is bijective, there is a one-to-one corresponsence between the finite Borel measures on $[0, \infty]$ of the form ν and the finite Borel measures on $[0, 1]$ of the form $\nu\psi$. The map $f \mapsto \mu$ is clearly well-defined and bijective. □

3. Parallel sum: A notion from electrical networks

In connections with electrical engineering, Anderson and Duffin [1] defined the *parallel sum* of two positive definite matrices A and B by

$$A : B = (A^{-1} + B^{-1})^{-1}. \tag{4}$$

The impedance of an electrical network can be represented by a positive (semi)definite matrix. If A and B are impedance matrices of multi-port networks, then the parallel sum $A : B$ indicates the total impedance of two electrical networks connected in parallel. This notion plays a crucial role for analyzing multi-port electrical networks because many physical interpretations of electrical circuits can be viewed in a form involving parallel sums. This is a starting point of the study of matrix and operator means. This notion can be extended to invertible positive operators by the same formula.

Lemma 0.5. *Let $A, B, C, D, A_n, B_n \in B(\mathcal{H})^{++}$ for all $n \in \mathbb{N}$.*

(1) If $A_n \downarrow A$, then $A_n^{-1} \uparrow A^{-1}$. If $A_n \uparrow A$, then $A_n^{-1} \downarrow A^{-1}$.

(2) If $A \leqslant C$ and $B \leqslant D$, then $A : B \leqslant C : D$.

(3) If $A_n \downarrow A$ and $B_n \downarrow B$, then $A_n : B_n \downarrow A : B$.

(4) If $A_n \downarrow A$ and $B_n \downarrow B$, then $\lim A_n : B_n$ exists and does not depend on the choices of A_n, B_n.

Proof. (1) Assume $A_n \downarrow A$. Then A_n^{-1} is increasing and, for each $x \in \mathcal{H}$,

$$\langle (A_n^{-1} - A^{-1})x, x\rangle = \langle (A - A_n)A^{-1}x, A_n^{-1}x\rangle \leqslant \|(A - A_n)A^{-1}x\| \|A_n^{-1}\| \|x\| \to 0.$$

(2) Follow from (1).

(3) Let $A_n, B_n \in B(\mathcal{H})^{++}$ be such that $A_n \downarrow A$ and $B_n \downarrow A$ where $A, B > 0$. Then $A_n^{-1} \uparrow A^{-1}$ and $B_n^{-1} \uparrow B^{-1}$. So, $A_n^{-1} + B_n^{-1}$ is an increasing sequence in $B(\mathcal{H})^+$ such that

$$A_n^{-1} + B_n^{-1} \to A^{-1} + B^{-1},$$

i.e. $A_n^{-1} + B_n^{-1} \uparrow A^{-1} + B^{-1}$. By (1), we thus have $(A_n^{-1} + B_n^{-1})^{-1} \downarrow (A^{-1} + B^{-1})^{-1}$.

(4) Let $A_n, B_n \in B(\mathcal{H})^{++}$ be such that $A_n \downarrow A$ and $B_n \downarrow B$. Then, by (2), $A_n : B_n$ is a decreasing sequence of positive operators. The order completeness of $B(\mathcal{H})$ guaruntees the existence of the strong limit of $A_n : B_n$. Let A'_n and B'_n be another sequences such that $A'_n \downarrow A$ and $B'_n \downarrow B$. Note that for each $n, m \in \mathbb{N}$, we have $A_n \leqslant A_n + A'_m - A$ and $B_n \leqslant B_n + B'_m - B$. Then

$$A_n : B_n \leqslant (A_n + A'_m - A) : (B_n + B'_m - B).$$

Note that as $n \to \infty$, $A_n + A'_m - A \to A'_m$ and $B_n + B'_m - B \to B'_m$. We have that as $n \to \infty$,

$$(A_n + A'_m - A) : (B_n + B'_m - B) \to A'_m : B'_m.$$

Hence, $\lim_{n\to\infty} A_n : B_n \leqslant A'_m : B'_m$ and $\lim_{n\to\infty} A_n : B_n \leqslant \lim_{m\to\infty} A'_m : B'_m$. By symmetry, $\lim_{n\to\infty} A_n : B_n \geqslant \lim_{m\to\infty} A'_m : B'_m$. □

We define the *parallel sum* of $A, B \geqslant 0$ to be

$$A : B = \lim_{\epsilon \downarrow 0} (A + \epsilon I) : (B + \epsilon I) \tag{5}$$

where the limit is taken in the strong-operator topology.

Lemma 0.6. *For each $x \in \mathcal{H}$,*

$$\langle (A : B)x, x\rangle = \inf\{\langle Ay, y\rangle + \langle Bz, z\rangle : y, z \in \mathcal{H}, y + z = x\}. \tag{6}$$

Proof. First, assume that A, B are invertible. Then for all $x, y \in \mathcal{H}$,

$$\begin{aligned}
&\langle Ay, y\rangle + \langle B(x-y), x-y\rangle - \langle (A : B)x, x\rangle \\
&= \langle Ay, y\rangle + \langle Bx, x\rangle - 2Re\langle Bx, y\rangle + \langle By, y\rangle - \langle (B - B(A+B)^{-1}B)x, x\rangle \\
&= \langle (A+B)y, y\rangle - 2Re\langle Bx, y\rangle + \langle (A+B)^{-1}Bx, Bx\rangle \\
&= \|(A+B)^{1/2}y\|^2 - 2Re\langle Bx, y\rangle + \|(A+B)^{-1/2}Bx\|^2 \\
&\geqslant 0.
\end{aligned}$$

With $y = (A+B)^{-1}Bx$, we have

$$\langle Ay, y\rangle + \langle B(x-y), x-y\rangle - \langle (A:B)x, x\rangle = 0.$$

Hence, we have the claim for $A, B > 0$. For $A, B \geqslant 0$, consider $A + \epsilon I$ and $B + \epsilon I$ where $\epsilon \downarrow 0$. □

Remark 0.7. This lemma has a physical interpretation, called the *Maxwell's minimum power principle*. Recall that a positive operator represents the impedance of a electrical network while the power dissipation of network with impedance A and current x is the inner product $\langle Ax, x\rangle$. Consider two electrical networks connected in parallel. For a given current input x, the current will divide $x = y + z$, where y and z are currents of each network, in such a way that the power dissipation is minimum.

Theorem 0.8. *The parallel sum satisfies*

(1) monotonicity: $A_1 \leqslant A_2, B_1 \leqslant B_2 \Rightarrow A_1 : B_1 \leqslant A_2 : B_2$.

(2) transformer inequality: $S^*(A:B)S \leqslant (S^*AS) : (S^*BS)$ *for every* $S \in B(\mathcal{H})$.

(3) continuity from above: if $A_n \downarrow A$ *and* $B_n \downarrow B$*, then* $A_n : B_n \downarrow A : B$.

Proof. (1) The monotonicity follows from the formula (5) and Lemma 0.5(2).

(2) For each $x, y, z \in \mathcal{H}$ such that $x = y + z$, by the previous lemma,

$$\begin{aligned}\langle S^*(A:B)Sx, x\rangle &= \langle (A:B)Sx, Sx\rangle \\ &\leqslant \langle ASy, Sy\rangle + \langle S^*BSz, z\rangle \\ &= \langle S^*ASy, y\rangle + \langle S^*BSz, z\rangle.\end{aligned}$$

Again, the previous lemma assures $S^*(A:B)S \leqslant (S^*AS) : (S^*BS)$.

(3) Let A_n and B_n be decreasing sequences in $B(\mathcal{H})^+$ such that $A_n \downarrow A$ and $B_n \downarrow B$. Then $A_n : B_n$ is decreasing and $A : B \leqslant A_n : B_n$ for all $n \in \mathbb{N}$. We have that, by the joint monotonicity of parallel sum, for all $\epsilon > 0$

$$A_n : B_n \leqslant (A_n + \epsilon I) : (B_n + \epsilon I).$$

Since $A_n + \epsilon I \downarrow A + \epsilon I$ and $B_n + \epsilon I \downarrow B + \epsilon I$, by Lemma 3.1.4(3) we have $A_n : B_n \downarrow A : B$. □

Remark 0.9. The positive operator S^*AS represents the impedance of a network connected to a transformer. The transformer inequality means that the impedance of parallel connection with transformer first is greater than that with transformer last.

Proposition 0.10. *The set of positive operators on* $\mathcal{H}$ *is a partially ordered commutative semigroup with respect to the parallel sum.*

Proof. For $A, B, C > 0$, we have $(A:B):C = A:(B:C)$ and $A:B = B:A$. The continuity from above in Theorem 0.8 implies that $(A:B):C = A:(B:C)$ and $A:B = B:A$ for all $A, B, C \geqslant 0$. The monotonicity of the parallel sum means that the positive operators form a partially ordered semigroup. □

Theorem 0.11. *For $A, B, C, D \geqslant 0$, we have the series-parallel inequality*

$$(A + B) : (C + D) \geqslant A : C + B : D. \tag{7}$$

In other words, the parallel sum is concave.

Proof. For each $x, y, z \in \mathcal{H}$ such that $x = y + z$, we have by the previous lemma that

$$\begin{aligned}\langle (A : C + B : D)x, x\rangle &= \langle (A : C)x, x\rangle + \langle (B : D)x, x\rangle \\ &\leqslant \langle Ay, y\rangle + \langle Cz, z\rangle + \langle By, y\rangle + \langle Dz, z\rangle \\ &= \langle (A + B)y, y\rangle + \langle (C + D)z, z\rangle.\end{aligned}$$

Applying the previous lemma yields $(A + B) : (C + D) \geqslant A : C + B : D$. □

Remark 0.12. The ordinary sum of operators represents the total impedance of two networks with series connection while the parallel sum indicates the total impedance of two networks with parallel connection. So, the series-parallel inequality means that the impedance of a series-parallel connection is greater than that of a parallel-series connection.

4. Classical means: arithmetic, harmonic and geometric means

Some desired properties of any object that is called a "mean" M on $B(\mathcal{H})^+$ should have are given here.

(A1). *positivity*: $A, B \geqslant 0 \Rightarrow M(A, B) \geqslant 0$;

(A2). *monotonicity*: $A \geqslant A', B \geqslant B' \Rightarrow M(A, B) \geqslant M(A', B')$;

(A3). *positive homogeneity*: $M(kA, kB) = kM(A, B)$ for $k \in \mathbb{R}^+$;

(A4). *transformer inequality*: $X^*M(A, B)X \leqslant M(X^*AX, X^*BX)$ for $X \in B(\mathcal{H})$;

(A5). *congruence invariance*: $X^*M(A, B)X = M(X^*AX, X^*BX)$ for invertible $X \in B(\mathcal{H})$;

(A6). *concavity*: $M(tA + (1 - t)B, tA' + (1 - t)B') \geqslant tM(A, A') + (1 - t)M(B, B')$ for $t \in [0, 1]$;

(A7). *continuity from above*: if $A_n \downarrow A$ and $B_n \downarrow B$, then $M(A_n, B_n) \downarrow M(A, B)$;

(A8). *betweenness*: if $A \leqslant B$, then $A \leqslant M(A, B) \leqslant B$;

(A9). *fixed point property*: $M(A, A) = A$.

In order to study matrix or operator means in general, the first step is to consider three classical means in mathematics, namely, arithmetic, geometric and harmonic means.

The *arithmetic mean* of $A, B \in B(\mathcal{H})^+$ is defined by

$$A \triangledown B = \frac{1}{2}(A + B). \tag{8}$$

Then the arithmetic mean satisfies the properties (A1)–(A9). In fact, the properties (A5) and (A6) can be replaced by a stronger condition:

$$X^*M(A, B)X = M(X^*AX, X^*BX) \text{ for all } X \in B(\mathcal{H}).$$

Moreover, the arithmetic mean satisfies

affinity: $M(kA + C, kB + C) = kM(A, B) + C$ for $k \in \mathbb{R}^+$.

Define the *harmonic mean* of positive operators $A, B \in B(\mathcal{H})^+$ by

$$A \,!\, B = 2(A : B) = \lim_{\epsilon \downarrow 0} 2(A_\epsilon^{-1} + B_\epsilon^{-1})^{-1} \tag{9}$$

where $A_\epsilon \equiv A + \epsilon I$ and $B_\epsilon \equiv B + \epsilon I$. Then the harmonic mean satisfies the properties (A1)–(A9).

The geometric mean of matrices is defined in [23] and studied in details in [3]. A usage of congruence transformations for treating geometric means is given in [18]. For a given invertible operator $C \in B(\mathcal{H})$, define

$$\Gamma_C : B(\mathcal{H})^{sa} \to B(\mathcal{H})^{sa}, A \mapsto C^* AC.$$

Then each Γ_C is a linear isomorphism with inverse $\Gamma_{C^{-1}}$ and is called a *congruence transformation*. The set of congruence transformations is a group under multiplication. Each congruence transformation preserves positivity, invertibility and, hence, strictly positivity. In fact, Γ_C maps $B(\mathcal{H})^+$ and $B(\mathcal{H})^{++}$ onto themselves. Note also that Γ_C is order-preserving.

Define the *geometric mean* of $A, B > 0$ by

$$A \# B = A^{1/2}(A^{-1/2}BA^{-1/2})^{1/2}A^{1/2} = \Gamma_{A^{1/2}} \circ \Gamma^{1/2}_{A^{-1/2}}(B). \tag{10}$$

Then $A \# B > 0$ for $A, B > 0$. This formula comes from two natural requirements: This definition should coincide with the usual geometric mean in $\mathbb{R}^+$: $A \# B = (AB)^{1/2}$ provided that $AB = BA$. The second condition is that, for any invertible $T \in B(\mathcal{H})$,

$$T^*(A \# B)T = (T^*AT) \# (T^*BT). \tag{11}$$

The next theorem characterizes the geometric mean of A and B in term of the solution of a certain operator equation.

Theorem 0.13. *For each $A, B > 0$, the Riccati equation $\Gamma_X(A^{-1}) := XA^{-1}X = B$ has a unique positive solution, namely, $X = A \# B$.*

Proof. The direct computation shows that $(A \# B)A^{-1}(A \# B) = B$. Suppose there is another positive solution $Y \geqslant 0$. Then

$$(A^{-1/2}XA^{-1/2})^2 = A^{-1/2}XA^{-1}XA^{-1/2} = A^{-1/2}YA^{-1}YA^{-1/2} = (A^{-1/2}YA^{-1/2})^2.$$

The uniqueness of positive square roots implies that $A^{-1/2}XA^{-1/2} = A^{-1/2}YA^{-1/2}$, i.e., $X = Y$. □

Theorem 0.14 (Maximum property of geometric mean). *For $A, B > 0$,*

$$A \# B = \max\{X \geqslant 0 : XA^{-1}X \leqslant B\} \tag{12}$$

where the maximum is taken with respect to the positive semidefinite ordering.

Proof. If $XA^{-1}X \leqslant B$, then

$$(A^{-1/2}XA^{-1/2})^2 = A^{-1/2}XA^{-1}XA^{-1/2} \leqslant A^{-1/2}BA^{-1/2}$$

and $A^{-1/2}XA^{-1/2} \leqslant (A^{-1/2}BA^{-1/2})^{1/2}$ i.e. $X \leqslant A\# B$ by Theorem 0.3. □

Recall the fact that if $f : [a,b] \to \mathbb{C}$ is continuous and $A_n \to A$ with $\mathrm{Sp}(A_n) \subseteq [a,b]$ for all $n \in \mathbb{N}$, then $\mathrm{Sp}(A) \subseteq [a,b]$ and $f(A_n) \to f(A)$.

Lemma 0.15. *Let $A, B, C, D, A_n, B_n \in B(\mathcal{H})^{++}$ for all $n \in \mathbb{N}$.*

(1) If $A \leqslant C$ and $B \leqslant D$, then $A\# B \leqslant C\# D$.

(2) If $A_n \downarrow A$ and $B_n \downarrow B$, then $A_n \# B_n \downarrow A\# B$.

(3) If $A_n \downarrow A$ and $B_n \downarrow B$, then $\lim A_n \# B_n$ exists and does not depend on the choices of A_n, B_n.

Proof. (1) The extremal characterization allows us to prove only that $(A\# B)C^{-1}(A\# B) \leqslant D$. Indeed,

$$\begin{aligned}(A\# B)C^{-1}(A\# B) &= A^{1/2}(A^{-1/2}BA^{-1/2})^{1/2}A^{1/2}C^{-1}A^{1/2}(A^{-1/2}BA^{-1/2})^{1/2}A^{1/2} \\ &\leqslant A^{1/2}(A^{-1/2}BA^{-1/2})^{1/2}A^{1/2}A^{-1}A^{1/2}(A^{-1/2}BA^{-1/2})^{1/2}A^{1/2} \\ &= B \\ &\leqslant D.\end{aligned}$$

(2) Assume $A_n \downarrow A$ and $B_n \downarrow B$. Then $A_n \# B_n$ is a decreasing sequence of strictly positive operators which is bounded below by 0. The order completeness of $B(\mathcal{H})$ implies that this sequence converges strongly to a positive operator. Since $A_n^{-1} \leqslant A^{-1}$, the Löwner-Heinz's inequality assures that $A_n^{-1/2} \leqslant A^{-1/2}$ and hence $\|A_n^{-1/2}\| \leqslant \|A^{-1/2}\|$ for all $n \in \mathbb{N}$. Note also that $\|B_n\| \leqslant \|B_1\|$ for all $n \in \mathbb{N}$. Recall that the multiplication is jointly continuous in the strong-operator topology if the first variable is bounded in norm. So, $A_n^{-1/2}B_nA_n^{-1/2}$ converges strongly to $A^{-1/2}BA^{-1/2}$. It follows that

$$(A_n^{-1/2}B_nA_n^{-1/2})^{1/2} \to (A^{-1/2}BA^{-1/2})^{1/2}.$$

Since $A_n^{1/2}$ is norm-bounded by $\|A^{1/2}\|$ by Löwner-Heinz's inequality, we conclude that

$$A_n^{1/2}(A_n^{-1/2}B_nA_n^{-1/2})^{1/2}A_n^{1/2} \to A^{1/2}(A^{-1/2}BA^{-1/2})^{1/2}A^{1/2}.$$

The proof of (3) is just the same as the case of harmonic mean. □

We define the *geometric mean* of $A, B \geqslant 0$ by

$$A\# B = \lim_{\epsilon \downarrow 0}(A+\epsilon I)\#(B+\epsilon I). \tag{13}$$

Then $A\# B \geqslant 0$ for any $A, B \geqslant 0$.

Theorem 0.16. *The geometric mean enjoys the following properties*

(1) *monotonicity:* $A_1 \leqslant A_2, B_1 \leqslant B_2 \Rightarrow A_1 \# B_1 \leqslant A_2 \# B_2$.
(2) *continuity from above:* $A_n \downarrow A, B_n \downarrow B \Rightarrow A_n \# B_n \downarrow A \# B$.
(3) *fixed point property:* $A \# A = A$.
(4) *self-duality:* $(A \# B)^{-1} = A^{-1} \# B^{-1}$.
(5) *symmetry:* $A \# B = B \# A$.
(6) *congruence invariance:* $\Gamma_C(A) \# \Gamma_C(B) = \Gamma_C(A \# B)$ *for all invertible* C.

Proof. (1) Use the formula (13) and Lemma 0.15 (1).

(2) Follows from Lemma 0.15 and the definition of the geometric mean.

(3) The unique positive solution to the equation $XA^{-1}X = A$ is $X = A$.

(4) The unique positive solution to the equation $X^{-1}A^{-1}X^{-1} = B$ is $X^{-1} = A \# B$. But this equstion is equivalent to $XAX = B^{-1}$. So, $A^{-1} \# B^{-1} = X = (A \# B)^{-1}$.

(5) The equation $XA^{-1}X = B$ has the same solution to the equation $XB^{-1}X = A$ by taking inverse in both sides.

(6) We have

$$\begin{aligned}\Gamma_C(A \# B)(\Gamma_C(A))^{-1}\Gamma_C(A \# B) &= \Gamma_C(A \# B)\Gamma_{C^{-1}}(A^{-1})\Gamma_C(A \# B)\\ &= \Gamma_C((A \# B)A^{-1}(A \# B))\\ &= \Gamma_C(B).\end{aligned}$$

Then apply Theorem 0.13. □

The congruence invariance asserts that Γ_C is an isomorphism on $B(\mathcal{H})^{++}$ with respect to the operation of taking the geometric mean.

Lemma 0.17. *For* $A > 0$ *and* $B \geqslant 0$, *the operator*

$$\begin{pmatrix} A & C \\ C^* & B \end{pmatrix}$$

is positive if and only if $B - C^*A^{-1}C$ *is positive, i.e.,* $B \geqslant C^*A^{-1}C$.

Proof. By setting

$$X = \begin{pmatrix} I & -A^{-1}C \\ 0 & I \end{pmatrix},$$

we compute

$$\begin{aligned}\Gamma_X\begin{pmatrix} A & C \\ C^* & B \end{pmatrix} &= \begin{pmatrix} I & 0 \\ -C^*A^{-1} & I \end{pmatrix}\begin{pmatrix} A & C \\ C^* & B \end{pmatrix}\begin{pmatrix} I & -A^{-1}C \\ 0 & I \end{pmatrix}\\ &= \begin{pmatrix} A & 0 \\ 0 & B - C^*A^{-1}C \end{pmatrix}.\end{aligned}$$

Since Γ_G preserves positivity, we obtain the desired result. □

Theorem 0.18. *The geometric mean $A \# B$ of $A, B \in B(\mathcal{H})^+$ is the largest operator $X \in B(\mathcal{H})^{sa}$ for which the operator*

$$\begin{pmatrix} A & X \\ X^* & B \end{pmatrix} \tag{14}$$

is positive.

Proof. By continuity argumeny, we may assume that $A, B > 0$. If $X = A \# B$, then the operator (14) is positive by Lemma 0.17. Let $X \in B(\mathcal{H})^{sa}$ be such that the operator (14) is positive. Then Lemma 0.17 again implies that $XA^{-1}X \leqslant B$ and

$$(A^{-1/2}XA^{-1/2})^2 = A^{-1/2}XA^{-1}XA^{-1/2} \leqslant A^{-1/2}BA^{-1/2}.$$

The Löwner-Heinz's inequality forces $A^{-1/2}XA^{-1/2} \leqslant (A^{-1/2}BA^{-1/2})^{1/2}$. Now, applying $\Gamma_{A^{1/2}}$ yields $X \leqslant A \# B$. □

Remark 0.19. The arithmetic mean and the harmonic mean can be easily defined for multivariable positive operators. The case of geometric mean is not easy, even for the case of matrices. Many authors tried to defined geometric means for multivariable positive semidefinite matrices but there is no satisfactory definition until 2004 in [5].

5. Operator connections

We see that the arithmetic, harmonic and geometric means share the properties (A1)–(A9) in common. A mean in general should have algebraic, order and topological properties. Kubo and Ando [17] proposed the following definition:

Definition 0.20. A *connection* on $B(\mathcal{H})^+$ is a binary operation σ on $B(\mathcal{H})^+$ satisfying the following axioms for all $A, A', B, B', C \in B(\mathcal{H})^+$:

(M1) *monotonicity*: $A \leqslant A', B \leqslant B' \implies A \,\sigma\, B \leqslant A' \,\sigma\, B'$

(M2) *transformer inequality*: $C(A \,\sigma\, B)C \leqslant (CAC) \,\sigma\, (CBC)$

(M3) *joint continuity from above*: if $A_n, B_n \in B(\mathcal{H})^+$ satisfy $A_n \downarrow A$ and $B_n \downarrow B$, then $A_n \,\sigma\, B_n \downarrow A \,\sigma\, B$.

The term "connection" comes from the study of electrical network connections.

Example 0.21. The following are examples of connections:

1. the *left trivial mean* $(A, B) \mapsto A$ and the *right trivial mean* $(A, B) \mapsto B$
2. the sum $(A, B) \mapsto A + B$
3. the parallel sum
4. arithmetic, geometric and harmonic means
5. the *weighed arithmetic mean* with weight $\alpha \in [0, 1]$ which is defined for each $A, B \geqslant 0$ by $A \,\triangledown_\alpha\, B = (1 - \alpha)A + \alpha B$
6. the *weighed harmonic mean* with weight $\alpha \in [0, 1]$ which is defined for each $A, B > 0$ by $A \,!_\alpha\, B = [(1 - \alpha)A^{-1} + \alpha B^{-1}]^{-1}$ and extended to the case $A, B \geqslant 0$ by continuity.

From now on, assume $\dim \mathcal{H} = \infty$. Consider the following property:

(M3′) *separate continuity from above*: if $A_n, B_n \in B(\mathcal{H})^+$ satisfy $A_n \downarrow A$ and $B_n \downarrow B$, then $A_n \,\sigma\, B \downarrow A \,\sigma\, B$ and $A \,\sigma\, B_n \downarrow A \,\sigma\, B$.

The condition (M3′) is clearly weaker than (M3). The next theorem asserts that we can improve the definition of Kubo-Ando by replacing (M3) with (M3′) and still get the same theory. This theorem also provides an easier way for checking a binary opertion to be a connection.

Theorem 0.22. *If a binary operation σ on $B(\mathcal{H})^+$ satisfies (M1), (M2) and (M3′), then σ satisfies (M3), that is, σ is a connection.*

Denote by $OM(\mathbb{R}^+)$ the set of operator monotone functions from $\mathbb{R}^+$ to $\mathbb{R}^+$. If a binary operation σ has a property (A), we write $\sigma \in BO(A)$. The following properties for a binary operation σ and a function $f : \mathbb{R}^+ \to \mathbb{R}^+$ play important roles:

(P) : If a projection $P \in B(\mathcal{H})^+$ commutes with $A, B \in B(\mathcal{H})^+$, then

$$P(A \,\sigma\, B) = (PA) \,\sigma\, (PB) = (A \,\sigma\, B)P;$$

(F) : $f(t)I = I \,\sigma\, (tI)$ for any $t \in \mathbb{R}^+$.

Proposition 0.23. *The transformer inequality (M2) implies*

- Congruence invariance: *For $A, B \geqslant 0$ and $C > 0$, $C(A\sigma B)C = (CAC) \,\sigma\, (CBC)$;*
- Positive homogeneity: *For $A, B \geqslant 0$ and $\alpha \in (0, \infty)$, $\alpha(A \,\sigma\, B) = (\alpha A) \,\sigma\, (\alpha B)$.*

Proof. For $A, B \geqslant 0$ and $C > 0$, we have

$$C^{-1}[(CAC) \,\sigma\, (CBC)]C^{-1} \leqslant (C^{-1}CACC^{-1}) \,\sigma\, (C^{-1}CBCC^{-1}) = A \,\sigma\, B$$

and hence $(CAC) \,\sigma\, (CBC) \leqslant C(A \,\sigma\, B)C$. The positive homogeneity comes from the congruence invariance by setting $C = \sqrt{\alpha} I$. □

Lemma 0.24. *Let $f : \mathbb{R}^+ \to \mathbb{R}^+$ be an increasing function. If σ satisfies the positive homogeneity, (M3′) and (F), then f is continuous.*

Proof. To show that f is right continuous at each $t \in \mathbb{R}^+$, consider a sequence t_n in $\mathbb{R}^+$ such that $t_n \downarrow t$. Then by (M3′)

$$f(t_n)I = I \,\sigma\, t_n I \downarrow I \,\sigma\, tI = f(t)I,$$

i.e. $f(t_n) \downarrow f(t)$. To show that f is left continuous at each $t > 0$, consider a sequence $t_n > 0$ such that $t_n \uparrow t$. Then $t_n^{-1} \downarrow t^{-1}$ and

$$\begin{aligned} \lim t_n^{-1} f(t_n) I &= \lim t_n^{-1}(I \,\sigma\, t_n I) = \lim (t_n^{-1} I) \,\sigma\, I = (t^{-1} I) \,\sigma\, I \\ &= t^{-1}(I \,\sigma\, tI) = t^{-1} f(t) I \end{aligned}$$

Since f is increasing, $t_n^{-1} f(t_n)$ is decreasing. So, $t \mapsto t^{-1} f(t)$ and f are left continuous. □

Lemma 0.25. *Let σ be a binary operation on $B(\mathcal{H})^+$ satisfying (M3') and (P). If $f : \mathbb{R}^+ \to \mathbb{R}^+$ is an increasing continuous function such that σ and f satisfy (F), then $f(A) = I \sigma A$ for any $A \in B(\mathcal{H})^+$.*

Proof. First consider $A \in B(\mathcal{H})^+$ in the form $\sum_{i=1}^m \lambda_i P_i$ where $\{P_i\}_{i=1}^m$ is an orthogonal family of projections with sum I and $\lambda_i > 0$ for all $i = 1, \dots, m$. Since each P_i commutes with A, we have by the property (P) that

$$\begin{aligned} I \sigma A &= \sum P_i (I \sigma A) = \sum P_i \sigma P_i A = \sum P_i \sigma \lambda_i P_i \\ &= \sum P_i (I \sigma \lambda_i I) = \sum f(\lambda_i) P_i = f(A). \end{aligned}$$

Now, consider $A \in B(\mathcal{H})^+$. Then there is a sequence A_n of strictly positive operators in the above form such that $A_n \downarrow A$. Then $I \sigma A_n \downarrow I \sigma A$ and $f(A_n)$ converges strongly to $f(A)$. Hence, $I \sigma A = \lim I \sigma A_n = \lim f(A_n) = f(A)$. □

Proof of Theorem 0.22: Let $\sigma \in BO(M1, M2, M3')$. As in [17], the conditions (M1) and (M2) imply that σ satisfies (P) and there is a function $f : \mathbb{R}^+ \to \mathbb{R}^+$ subject to (F). If $0 \leqslant t_1 \leqslant t_2$, then by (M1)

$$f(t_1) I = I \sigma (t_1 I) \leqslant I \sigma (t_2 I) = f(t_2) I,$$

i.e. $f(t_1) \leqslant f(t_2)$. The assumption (M3') is enough to guarantee that f is continuous by Lemma 0.24. Then Lemma 0.25 results in $f(A) = I\sigma A$ for all $A \geqslant 0$. Now, (M1) and the fact that $\dim \mathcal{H} = \infty$ yield that f is operator monotone. If there is another $g \in OM(\mathbb{R}^+)$ satisfying (F), then $f(t)I = I \sigma tI = g(t)I$ for each $t \geqslant 0$, i.e. $f = g$. Thus, we establish a well-defined map $\sigma \in BO(M1, M2, M3') \mapsto f \in OM(\mathbb{R}^+)$ such that σ and f satisfy (F).

Now, given $f \in OM(\mathbb{R}^+)$, we construct σ from the integral representation (2) in Proposition 0.4. Define a binary operation $\sigma : B(\mathcal{H})^+ \times B(\mathcal{H})^+ \to B(\mathcal{H})^+$ by

$$A \sigma B = \int_{[0,1]} A \,!_t\, B \, d\mu(t) \tag{15}$$

where the integral is taken in the sense of Bochner. Consider $A, B \in B(\mathcal{H})^+$ and set $F_t = A \,!_t\, B$ for each $t \in [0,1]$. Since $A \leqslant \|A\| I$ and $B \leqslant \|B\| I$, we get

$$A \,!_t\, B \leqslant \|A\| I \,!_t\, \|B\| I = \frac{\|A\| \|B\|}{t\|A\| + (1-t)\|B\|} I.$$

By Banach-Steinhaus' theorem, there is an $M > 0$ such that $\|F_t\| \leqslant M$ for all $t \in [0,1]$. Hence,

$$\int_{[0,1]} \|F_t\| \, d\mu(t) \leqslant \int_{[0,1]} M \, d\mu(t) < \infty.$$

So, F_t is Bochner integrable. Since $F_t \geqslant 0$ for all $t \in [0,1]$, $\int_{[0,1]} F_t \, d\mu(t) \geqslant 0$. Thus, $A \sigma B$ is a well-defined element in $B(\mathcal{H})^+$. The monotonicity (M1) and the transformer inequality (M2) come from passing the monotonicity and the transformer inequality of the weighed harmonic mean through the Bochner integral. To show (M3'), let $A_n \downarrow A$ and $B_n \downarrow B$. Then $A_n \,!_t\, B \downarrow A \,!_t\, B$ for $t \in [0,1]$ by the monotonicity and the separate continuity from above of the weighed harmonic mean. Let $\xi \in H$. Define a bounded linear map $\Phi : B(\mathcal{H}) \to \mathbb{C}$ by $\Phi(T) = \langle T\xi, \xi \rangle$.

For each $n \in \mathbb{N}$, set $T_n(t) = A_n \,!_t\, B$ and put $T_\infty(t) = A \,!_t\, B$. Then for each $n \in \mathbb{N} \cup \{\infty\}$, $\Phi \circ T_n$ is Bochner integrable and

$$\langle \int T_n(t)\, d\mu(t)\xi, \xi \rangle = \Phi(\int T_n(t)\, d\mu(t)) = \int \Phi \circ T_n(t)\, d\mu(t).$$

Since $T_n(t) \downarrow T_\infty(t)$, we have that $\langle T_n(t)\xi, \xi\rangle \to \langle T_\infty(t)\xi, \xi\rangle$ as $n \to \infty$ for each $t \in [0,1]$. We obtain from the dominated convergence theorem that

$$\begin{aligned}
\lim_{n\to\infty} \langle (A_n \,\sigma\, B)\xi, \xi\rangle &= \lim_{n\to\infty} \langle \int T_n(t)\, d\mu(t)\xi, \xi\rangle \\
&= \lim_{n\to\infty} \int \langle T_n(t)\xi, \xi\rangle\, d\mu(t) \\
&= \int \langle T_\infty(t)\xi, \xi\rangle\, d\mu(t) \\
&= \langle \int T_\infty(t) d\mu(t)\xi, \xi\rangle \\
&= \langle (A \,\sigma\, B)\xi, \xi\rangle.
\end{aligned}$$

So, $A_n \,\sigma\, B \downarrow A \,\sigma\, B$. Similarly, $A \,\sigma\, B_n \downarrow A \,\sigma\, B$. Thus, σ satisfies (M3'). It is easy to see that $f(t)I = I \,\sigma\, (tI)$ for $t \geqslant 0$. This shows that the map $\sigma \mapsto f$ is surjective.

To show the injectivity of this map, let $\sigma_1, \sigma_2 \in BO(M1, M2, M3')$ be such that $\sigma_i \mapsto f$ where, for each $t \geqslant 0$, $I \,\sigma_i\, (tI) = f(t)I, i = 1,2$. Since σ_i satisfies the property (P), we have $I \,\sigma_i\, A = f(A)$ for $A \geqslant 0$ by Lemma 0.25. Since σ_i satisfies the congruence invariance, we have that for $A > 0$ and $B \geqslant 0$,

$$A \,\sigma_i\, B = A^{1/2}(I \,\sigma_i\, A^{-1/2}BA^{-1/2})A^{1/2} = A^{1/2} f(A^{-1/2}BA^{-1/2})A^{1/2}, \quad i = 1,2.$$

For each $A, B \geqslant 0$, we obtain by (M3') that

$$\begin{aligned}
A \,\sigma_1\, B &= \lim_{\epsilon\downarrow 0} A_\epsilon \,\sigma_1\, B \\
&= \lim_{\epsilon\downarrow 0} A_\epsilon^{1/2}(I \,\sigma_1\, A_\epsilon^{-1/2}BA_\epsilon^{-1/2})A_\epsilon^{1/2} \\
&= \lim_{\epsilon\downarrow 0} A_\epsilon^{1/2} f(A_\epsilon^{-1/2}BA_\epsilon^{-1/2})A_\epsilon^{1/2} \\
&= \lim_{\epsilon\downarrow 0} A_\epsilon^{1/2}(I \,\sigma_2\, A_\epsilon^{-1/2}BA_\epsilon^{-1/2})A_\epsilon^{1/2} \\
&= \lim_{\epsilon\downarrow 0} A_\epsilon \,\sigma_2\, B \\
&= A \,\sigma_2\, B,
\end{aligned}$$

where $A_\epsilon \equiv A + \epsilon I$. That is $\sigma_1 = \sigma_2$. Therefore, there is a bijection between $OM(\mathbb{R}^+)$ and $BO(M1, M2, M3')$. Every element in $BO(M1, M2, M3')$ admits an integral representation (15). Since the weighed harmonic mean possesses the joint continuity (M3), so is any element in $BO(M1, M2, M3')$. □

The next theorem is a fundamental result of [17].

Theorem 0.26. *There is a one-to-one correspondence between connections σ and operator monotone functions f on the non-negative reals satisfying*

$$f(t)I = I\sigma(tI), \quad t \in \mathbb{R}^+. \tag{16}$$

There is a one-to-one correspondence between connections σ and finite Borel measures ν on $[0,\infty]$ satisfying

$$A\sigma B = \int_{[0,\infty]} \frac{t+1}{t}(tA : B)\, d\nu(t), \quad A, B \geqslant 0. \tag{17}$$

Moreover, the map $\sigma \mapsto f$ is an affine order-isomorphism between connections and non-negative operator monotone functions on $\mathbb{R}^+$. Here, the order-isomorphism means that when $\sigma_i \mapsto f_i$ for $i = 1,2$, $A\sigma_1 B \leqslant A\sigma_2 B$ for all $A, B \in B(\mathcal{H})^+$ if and only if $f_1 \leqslant f_2$.

Each connection σ on $B(\mathcal{H})^+$ produces a unique scalar function on $\mathbb{R}^+$, denoted by the same notation, satisfying

$$(s\sigma t)I = (sI)\sigma(tI), \quad s,t \in \mathbb{R}^+. \tag{18}$$

Let $s, t \in \mathbb{R}^+$. If $s > 0$, then $s\sigma t = sf(t/s)$. If $t > 0$, then $s\sigma t = tf(s/t)$.

Theorem 0.27. *There is a one-to-one correspondence between connections and finite Borel measures on the unit interval. In fact, every connection takes the form*

$$A\sigma B = \int_{[0,1]} A\,!_t\, B\, d\mu(t), \quad A, B \geqslant 0 \tag{19}$$

for some finite Borel measure μ on $[0,1]$. Moreover, the map $\mu \mapsto \sigma$ is affine and order-preserving. Here, the order-presering means that when $\mu_i \mapsto \sigma_i$ (i=1,2), if $\mu_1(E) \leqslant \mu_2(E)$ for all Borel sets E in $[0,1]$, then $A\sigma_1 B \leqslant A\sigma_2 B$ for all $A, B \in B(\mathcal{H})^+$.

Proof. The proof of the first part is contained in the proof of Theorem 0.22. This map is affine because of the linearity of the map $\mu \mapsto \int f\, d\mu$ on the set of finite positive measures and the bijective correspondence between connections and Borel measures. It is straight forward to show that this map is order-preserving. □

Remark 0.28. Let us consider operator connections from electrical circuit viewpoint. A general connection represents a formulation of making a new impedance from two given impedances. The integral representation (19) shows that such a formulation can be described as a weighed series connection of (infinite) weighed harmonic means. From this point of view, the theory of operator connections can be regarded as a mathematical theory of electrical circuits.

Definition 0.29. Let σ be a connection. The operator monotone function f in (16) is called the *representing function* of σ. If μ is the measure corresponds to σ in Theorem 0.27, the measure $\mu\psi^{-1}$ that takes a Borel set E in $[0,\infty]$ to $\mu(\psi^{-1}(E))$ is called the *representing measure* of σ in the Kubo-Ando's theory. Here, $\psi : [0,1] \to [0,\infty]$ is a homeomorphism $t \mapsto t/(1-t)$.

Since every connection σ has an integral representation (19), properties of weighed harmonic means reflect properties of a general connection. Hence, every connection σ satisfies the following properties for all $A, B \geqslant 0, T \in B(\mathcal{H})$ and invertible $X \in B(\mathcal{H})$:

- *transformer inequality*: $T^*(A \,\sigma\, B)T \leqslant (T^*AT)\,\sigma\,(T^*BT)$;
- *congruence invariance*: $X^*(A \,\sigma\, B)X = (X^*AX)\,\sigma\,(X^*BX)$;
- *concavity*: $(tA + (1-t)B)\,\sigma\,(tA' + (1-t)B') \geqslant t(A\,\sigma\,A') + (1-t)(B\,\sigma\,B')$ for $t \in [0,1]$.

Moreover, if $A, B > 0$,

$$A \,\sigma\, B = A^{1/2} f(A^{-1/2} B A^{-1/2}) A^{1/2} \tag{20}$$

and, in general, for each $A, B \geqslant 0$,

$$A \,\sigma\, B = \lim_{\epsilon \downarrow 0} A_\epsilon \,\sigma\, B_\epsilon \tag{21}$$

where $A_\epsilon \equiv A + \epsilon I$ and $B_\epsilon \equiv B + \epsilon I$. These properties are useful tools for deriving operator inequalities involving connections. The formulas (20) and (21) give a way for computing the formula of connection from its representing function.

Example 0.30. 1. The left- and the right-trivial means have representing functions given by $t \mapsto 1$ and $t \mapsto t$, respectively. The representing measures of the left- and the right-trivial means are given respectively by δ_0 and δ_∞ where δ_x is the Dirac measure at x. So, the α-weighed arithmetic mean has the representing function $t \mapsto (1-\alpha) + \alpha t$ and it has $(1-\alpha)\delta_0 + \alpha\delta_\infty$ as the representing measure.

2. The geometric mean has the representing function $t \mapsto t^{1/2}$.
3. The harmonic mean has the representing function $t \mapsto 2t/(1+t)$ while $t \mapsto t/(1+t)$ corrsponds to the parallel sum.

Remark 0.31. The map $\sigma \mapsto \mu$, where μ is the representing measure of σ, is not order-preserving in general. Indeed, the representing measure of $\triangledown$ is given by $\mu = (\delta_0 + \delta_\infty)/2$ while the representing measure of ! is given by δ_1. We have $! \leqslant \triangledown$ but $\delta_1 \nleqslant \mu$.

6. Operator means

According to [24], a (scalar) mean is a binary operation M on $(0,\infty)$ such that $M(s,t)$ lies between s and t for any $s, t > 0$. For a connection, this property is equivalent to various properties in the next theorem.

Theorem 0.32. *The following are equivalent for a connection* σ *on* $B(\mathcal{H})^+$*:*

(i) σ *satisfies the* betweenness property, *i.e.* $A \leqslant B \Rightarrow A \leqslant A\,\sigma\,B \leqslant B$.
(ii) σ *satisfies the* fixed point property, *i.e.* $A\,\sigma\,A = A$ *for all* $A \in B(\mathcal{H})^+$.
(iii) σ *is normalized, i.e.* $I\,\sigma\,I = I$.
(iv) the representing function f *of* σ *is normalized, i.e.* $f(1) = 1$.
(v) the representing measure μ *of* σ *is normalized, i.e.* μ *is a probability measure.*

Proof. Clearly, (i) $\Rightarrow$ (iii) $\Rightarrow$ (iv). The implication (iii) $\Rightarrow$ (ii) follows from the congruence invariance and the continuity from above of σ. The monotonicity of σ is used to prove (ii) $\Rightarrow$ (i). Since

$$I\,\sigma\,I = \int_{[0,1]} I \,!_t\, I \, d\mu(t) = \mu([0,1])I,$$

we obtain that (iv) $\Rightarrow$ (v) $\Rightarrow$ (iii). □

Definition 0.33. A *mean* is a connection satisfying one, and thus all, of the properties in the previous theorem.

Hence, every mean in Kubo-Ando's sense satisfies the desired properties (A1)–(A9) in Section 3. As a consequence of Theorem 0.32, a convex combination of means is a mean.

Theorem 0.34. *Given a Hilbert space $\mathcal{H}$, there exist affine bijections between any pair of the following objects:*

(i) the means on $B(\mathcal{H})^+$,

(ii) the operator monotone functions $f : \mathbb{R}^+ \to \mathbb{R}^+$ such that $f(1) = 1$,

(iii) the probability Borel measures on $[0,1]$.

Moreover, these correspondences between (i) and (ii) are order isomorphic. Hence, there exists an affine order isomorphism between the means on the positive operators acting on different Hilbert spaces.

Proof. Follow from Theorems 0.27 and 0.32. □

Example 0.35. The left- and right-trivial means, weighed arithmetic means, the geometric mean and the harmonic mean are means. The parallel sum is not a mean since its representing function is not normalized.

Example 0.36. The function $t \mapsto t^{\alpha}$ is an operator monotone function on $\mathbb{R}^+$ for each $\alpha \in [0,1]$ by the Löwner-Heinz's inequality. So it produces a mean, denoted by $\#_{\alpha}$, on $B(\mathcal{H})^+$. By the direct computation,

$$s \,\#_{\alpha}\, t = s^{1-\alpha} t^{\alpha}, \tag{22}$$

i.e. $\#_{\alpha}$ is the α-weighed geometric mean on $\mathbb{R}^+$. So the α-weighed geometric mean on $\mathbb{R}^+$ is really a Kubo-Ando mean. The *α-weighed geometric mean* on $B(\mathcal{H})^+$ is defined to be the mean corresponding to that mean on $\mathbb{R}^+$. Since t^{α} has an integral expression

$$t^{\alpha} = \frac{\sin \alpha\pi}{\pi} \int_0^{\infty} \frac{t\lambda^{\alpha-1}}{t+\lambda}\, dm(\lambda) \tag{23}$$

(see [7]) where m denotes the Lebesgue measure, the representing measure of $\#_{\alpha}$ is given by

$$d\mu(\lambda) = \frac{\sin \alpha\pi}{\pi} \frac{\lambda^{\alpha-1}}{\lambda+1}\, dm(\lambda). \tag{24}$$

Example 0.37. Consider the operator monotone function

$$t \mapsto \frac{t}{(1-\alpha)t+\alpha}, \quad t \geqslant 0, \alpha \in [0,1].$$

The direct computation shows that

$$s \,!_{\alpha}\, t = \begin{cases} ((1-\alpha)s^{-1} + \alpha t^{-1})^{-1}, & s,t > 0; \\ 0, & \text{otherwise}, \end{cases} \tag{25}$$

which is the α-weighed harmonic mean. We define the *α-weighed harmonic mean* on $B(\mathcal{H})^+$ to be the mean corresponding to this operator monotone function.

Example 0.38. Consider the operator monotone function $f(t) = (t-1)/\log t$ for $t > 0, t \neq 1$, $f(0) \equiv 0$ and $f(1) \equiv 1$. Then it gives rise to a mean, denoted by λ, on $B(\mathcal{H})^+$. By the direct computation,

$$s \,\lambda\, t = \begin{cases} \frac{s-t}{\log s - \log t}, & s > 0, t > 0, s \neq t; \\ s, & s = t \\ 0, & \text{otherwise}, \end{cases} \tag{26}$$

i.e. λ is the logarithmic mean on $\mathbb{R}^+$. So the logarithmic mean on $\mathbb{R}^+$ is really a mean in Kubo-Ando's sense. The *logarithmic mean* on $B(\mathcal{H})^+$ is defined to be the mean corresponding to this operator monotone function.

Example 0.39. The map $t \mapsto (t^r + t^{1-r})/2$ is operator monotone for any $r \in [0,1]$. This function produces a mean on $B(\mathcal{H})^+$. The computation shows that

$$(s,t) \mapsto \frac{s^r t^{1-r} + s^{1-r} t^r}{2}.$$

However, the corresponding mean on $B(\mathcal{H})^+$ is not given by the formula

$$(A,B) \mapsto \frac{A^r B^{1-r} + A^{1-r} B^r}{2} \tag{27}$$

since it is not a binary operation on $B(\mathcal{H})^+$. In fact, the formula (27) is considered in [8], called the *Heinz mean* of A and B.

Example 0.40. For each $p \in [-1,1]$ and $\alpha \in [0,1]$, the map

$$t \mapsto [(1-\alpha) + \alpha t^p]^{1/p}$$

is an operator monotone function on $\mathbb{R}^+$. Here, the case $p = 0$ is understood that we take limit as $p \to 0$. Then

$$s \,\#_{p,\alpha}\, t = [(1-\alpha)s^p + \alpha t^p]^{1/p}. \tag{28}$$

The corresponding mean on $B(\mathcal{H})^+$ is called the *quasi-arithmetic power mean* with parameter (p,α), defined for $A > 0$ and $B \geqslant 0$ by

$$A \,\#_{p,\alpha}\, B \;=\; A^{1/2}[(1-\alpha)I + \alpha(A^{-1/2}BA^{-1/2})^p]^{1/p} A^{1/2}. \tag{29}$$

The class of quasi-arithmetic power means contain many kinds of means: The mean $\#_{1,\alpha}$ is the α-weighed arithmetic mean. The case $\#_{0,\alpha}$ is the α-weighed geometric mean. The case $\#_{-1,\alpha}$ is the α-weighed harmonic mean. The mean $\#_{p,1/2}$ is the *power mean* or *binomial mean* of order p. These means satisfy the property that

$$A \,\#_{p,\alpha}\, B \;=\; B \,\#_{p,1-\alpha}\, A. \tag{30}$$

Moreover, they are interpolated in the sense that for all $p, q, \alpha \in [0,1]$,

$$(A \,\#_{r,p}\, B) \,\#_{r,\alpha}\, (A \,\#_{r,q}\, B) \;=\; A \,\#_{r,(1-\alpha)p+\alpha q}\, B. \tag{31}$$

Example 0.41. If σ_1, σ_2 are means such that $\sigma_1 \leqslant \sigma_2$, then there is a family of means that interpolates between σ_1 and σ_2, namely, $(1-\alpha)\sigma_1 + \alpha\sigma_2$ for all $\alpha \in [0,1]$. Note that the map $\alpha \mapsto (1-\alpha)\sigma_1 + \alpha\sigma_2$ is increasing. For instance, the *Heron mean* with weight $\alpha \in [0,1]$ is defined to be $h_\alpha = (1-\alpha)\,\#\, + \alpha\,\nabla$. This family is the linear interpolations between the geometric mean and the arithmetic mean. The representing function of h_α is given by

$$t \mapsto (1-\alpha)t^{1/2} + \frac{\alpha}{2}(1+t).$$

The case $\alpha = 2/3$ is called the *Heronian mean* in the literature.

7. Applications to operator monotonicity and concavity

In this section, we generalize the matrix and operator monotonicity and concavity in the literature (see e.g. [3, 9]) in such a way that the geometric mean, the harmonic mean or specific operator means are replaced by general connections. Recall the following terminology. A continuous function $f : I \to \mathbb{R}$ is called an *operator concave function* if

$$f(tA + (1-t)B) \geqslant tf(A) + (1-t)f(B)$$

for any $t \in [0,1]$ and Hermitian operators $A, B \in B(\mathcal{H})$ whose spectrums are contained in the interval I and for all Hilbert spaces $\mathcal{H}$. A well-known result is that a continuous function $f : \mathbb{R}^+ \to \mathbb{R}^+$ is operator monotone if and only if it is operator concave. Hence, the maps $t \mapsto t^r$ and $t \mapsto \log t$ are operator concave for $r \in [0,1]$. Let $\mathcal{H}$ and $\mathcal{K}$ be Hilbert spaces. A map $\Phi : B(\mathcal{H}) \to B(\mathcal{K})$ is said to be *positive* if $\Phi(A) \geqslant 0$ whenever $A \geqslant 0$. It is called *unital* if $\Phi(I) = I$. We say that a positive map Φ is *strictly positive* if $\Phi(A) > 0$ when $A > 0$. A map Ψ from a convex subset $\mathcal{C}$ of $B(\mathcal{H})^{sa}$ to $B(\mathcal{K})^{sa}$ is called *concave* if for each $A, B \in \mathcal{C}$ and $t \in [0,1]$,

$$\Psi(tA + (1-t)B) \geqslant t\Psi(A) + (1-t)\Psi(B).$$

A map $\Psi : B(\mathcal{H})^{sa} \to B(\mathcal{K})^{sa}$ is called *monotone* if $A \leqslant B$ assures $\Psi(A) \leqslant \Psi(B)$. So, in particular, the map $A \mapsto A^r$ is monotone and concave on $B(\mathcal{H})^+$ for each $r \in [0,1]$. The map $A \mapsto \log A$ is monotone and concave on $B(\mathcal{H})^{++}$.

Note first that, from the previous section, the quasi-arithmetic power mean $(A, B) \mapsto A \#_{p,\alpha} B$ is monotone and concave for any $p \in [-1,1]$ and $\alpha \in [0,1]$. In particular, the following are monotone and concave:

(i) any weighed arithmetic mean,

(ii) any weighed geometric mean,

(iii) any weighed harmonic mean,

(iv) the logarithmic mean,

(v) any weighed power mean of order $p \in [-1,1]$.

Recall the following lemma from [9].

Lemma 0.42 (Choi's inequality). *If $\Phi : B(\mathcal{H}) \to B(\mathcal{K})$ is linear, strictly positive and unital, then for every $A > 0$, $\Phi(A)^{-1} \leqslant \Phi(A^{-1})$.*

Proposition 0.43. *If $\Phi : B(\mathcal{H}) \to B(\mathcal{K})$ is linear and strictly positive, then for any $A, B > 0$*

$$\Phi(A)\Phi(B)^{-1}\Phi(A) \leqslant \Phi(AB^{-1}A). \tag{32}$$

Proof. For each $X \in B(\mathcal{H})$, set $\Psi(X) = \Phi(A)^{-1/2}\Phi(A^{1/2}XA^{1/2})\Phi(A)^{-1/2}$. Then Ψ is a unital strictly positive linear map. So, by Choi's inequality, $\Psi(A)^{-1} \leqslant \Psi(A^{-1})$ for all $A > 0$. For each $A, B > 0$, we have by Lemma 0.42 that

$$\begin{aligned}\Phi(A)^{1/2}\Phi(B)^{-1}\Phi(A)^{1/2} &= \Psi(A^{-1/2}BA^{-1/2})^{-1} \\ &\leqslant \Psi\left((A^{-1/2}BA^{-1/2})^{-1}\right) \\ &= \Phi(A)^{-1/2}\Phi(AB^{-1}A)\Phi(A)^{-1/2}.\end{aligned}$$

So, we have the claim. □

Theorem 0.44. *If $\Phi : B(\mathcal{H}) \to B(\mathcal{K})$ is a positive linear map which is norm-continuous, then for any connection σ on $B(\mathcal{K})^+$ and for each $A, B > 0$,*

$$\Phi(A \,\sigma\, B) \leqslant \Phi(A) \,\sigma\, \Phi(B). \tag{33}$$

If, addition, Φ is strongly continuous, then (33) *holds for any $A, B \geqslant 0$.*

Proof. First, consider $A, B > 0$. Assume that Φ is strictly positive. For each $X \in B(\mathcal{H})$, set

$$\Psi(X) = \Phi(B)^{-1/2}\Phi(B^{1/2}XB^{1/2})\Phi(B)^{-1/2}.$$

Then Ψ is a unital strictly positive linear map. So, by Choi's inequality, $\Psi(C)^{-1} \leqslant \Psi(C^{-1})$ for all $C > 0$. For each $t \in [0,1]$, put $X_t = B^{-1/2}(A \,!_t\, B)B^{-1/2} > 0$. We obtain from the previous proposition that

$$\begin{aligned}\Phi(A \,!_t\, B) &= \Phi(B)^{1/2}\Psi(X_t)\Phi(B)^{1/2} \\ &\leqslant \Phi(B)^{1/2}[\Psi(X_t^{-1})]^{-1}\Phi(B)^{1/2} \\ &= \Phi(B)[\Phi(B((1-t)A^{-1} + tB^{-1})B)]^{-1}\Phi(B) \\ &= \Phi(B)[(1-t)\Phi(BA^{-1}B) + t\Phi(B)]^{-1}\Phi(B) \\ &\leqslant \Phi(B)[(1-t)\Phi(B)\Phi(A)^{-1}\Phi(B) + t\Phi(B)]^{-1}\Phi(B) \\ &= \Phi(A) \,!_t\, \Phi(B).\end{aligned}$$

For general case of Φ, consider the family $\Phi_\epsilon(A) = \Phi(A) + \epsilon I$ where $\epsilon > 0$. Since the map $(A, B) \mapsto A \,!_t\, B = [(1-t)A^{-1} + tB^{-1}]^{-1}$ is norm-continuous, we arrive at

$$\Phi(A \,!_t\, B) \leqslant \Phi(A) \,!_t\, \Phi(B).$$

For each connection σ, since Φ is a bounded linear operator, we have

$$\begin{aligned}\Phi(A\,\sigma\,B) &= \Phi\Big(\int_{[0,1]} A\,!_t\,B\,d\mu(t)\Big) = \int_{[0,1]} \Phi(A\,!_t\,B)\,d\mu(t) \\ &\leqslant \int_{[0,1]} \Phi(A)\,!_t\,\Phi(B)\,d\mu(t) = \Phi(A)\,\sigma\,\Phi(B).\end{aligned}$$

Suppose further that Φ is strongly continuous. Then, for each $A, B \geqslant 0$,

$$\begin{aligned}\Phi(A\,\sigma\,B) &= \Phi(\lim_{\epsilon\downarrow 0}(A+\epsilon I)\,\sigma\,(B+\epsilon I)) = \lim_{\epsilon\downarrow 0}\Phi((A+\epsilon I)\,\sigma\,(B+\epsilon I)) \\ &\leqslant \lim_{\epsilon\downarrow 0}\Phi(A+\epsilon I)\,\sigma\,\Phi(B+\epsilon I) = \Phi(A)\,\sigma\,\Phi(B).\end{aligned}$$

The proof is complete. $\square$

As a special case, if $\Phi : M_n(\mathbb{C}) \to M_n(\mathbb{C})$ is a positive linear map, then for any connection σ and for any positive semidefinite matrices $A, B \in M_n(\mathbb{C})$, we have

$$\Phi(A\sigma B) \leqslant \Phi(A)\,\sigma\,\Phi(B).$$

In particular, $\Phi(A)\,\#_{p,\alpha}\,\Phi(B) \leqslant \Phi(A)\,\#_{p,\alpha}\,\Phi(B)$ for any $p \in [-1,1]$ and $\alpha \in [0,1]$.

Theorem 0.45. *If $\Phi_1, \Phi_2 : B(\mathcal{H})^+ \to B(\mathcal{K})^+$ are concave, then the map*

$$(A_1, A_2) \mapsto \Phi_1(A_1)\,\sigma\,\Phi_2(A_2) \tag{34}$$

is concave for any connection σ on $B(\mathcal{K})^+$.

Proof. Let $A_1, A_1', A_2, A_2' \geqslant 0$ and $t \in [0,1]$. The concavity of Φ_1 and Φ_2 means that for $i = 1, 2$

$$\Phi_i(tA_i + (1-t)A_i') \geqslant t\Phi_i(A_i) + (1-t)\Phi_i(A_i').$$

It follows from the monotonicity and concavity of σ that

$$\begin{aligned}&\Phi_1(tA_1 + (1-t)A_1')\,\sigma\,\Phi_2(tA_2 + (1-t)A_2') \\ &\qquad\geqslant [t\Phi_1(A_1) + (1-t)\Phi_1(A_1')]\,\sigma\,[t\Phi_2(A_2) + (1-t)\Phi_2(A_2')] \\ &\qquad\geqslant t[\Phi_1(A_1)\,\sigma\,\Phi_2(A_2)] + (1-t)[\Phi_1(A_1)\,\sigma\,\Phi_2(A_2)].\end{aligned}$$

This shows the concavity of the map $(A_1, A_2) \mapsto \Phi_1(A_1)\,\sigma\,\Phi_2(A_2)$. $\square$

In particular, if Φ_1 and Φ_2 are concave, then so is $(A, B) \mapsto \Phi_1(A)\,\#_{p,\alpha}\Phi_2(B)$ for $p \in [-1,1]$ and $\alpha \in [0,1]$.

Corollary 0.46. *Let σ be a connection. Then, for any operator monotone functions $f, g : \mathbb{R}^+ \to \mathbb{R}^+$, the map $(A, B) \mapsto f(A)\,\sigma\,g(B)$ is concave. In particular,*

(1) the map $(A, B) \mapsto A^r\,\sigma\,B^s$ is concave on $B(\mathcal{H})^+$ for any $r, s \in [0,1]$,

(2) the map $(A, B) \mapsto (\log A)\,\sigma\,(\log B)$ is concave on $B(\mathcal{H})^{++}$.

Theorem 0.47. *If $\Phi_1, \Phi_2 : B(\mathcal{H})^+ \to B(\mathcal{K})^+$ are monotone, then the map*

$$(A_1, A_2) \mapsto \Phi_1(A_1) \,\sigma\, \Phi_2(A_2) \tag{35}$$

is monotone for any connection σ on $B(\mathcal{K})^+$.

Proof. Let $A_1 \leqslant A_1'$ and $A_2 \leqslant A_2'$. Then $\Phi_1(A_1) \leqslant \Phi_1(A_1')$ and $\Phi_2(A_2) \leqslant \Phi_2(A_2')$ by the monotonicity of Φ_1 and Φ_2. Now, the monotonicity of σ forces $\Phi_1(A_1) \,\sigma\, \Phi_2(A_2) \leqslant \Phi_1(A_1') \,\sigma\, \Phi_2(A_2')$. □

In particular, if Φ_1 and Φ_2 are monotone, then so is $(A, B) \mapsto \Phi_1(A) \,\#_{p,\alpha} \Phi_2(B)$ for $p \in [-1, 1]$ and $\alpha \in [0, 1]$.

Corollary 0.48. *Let σ be a connection. Then, for any operator monotone functions $f, g : \mathbb{R}^+ \to \mathbb{R}^+$, the map $(A, B) \mapsto f(A) \,\sigma\, g(B)$ is monotone. In particular,*

(1) *the map $(A, B) \mapsto A^r \,\sigma\, B^s$ is monotone on $B(\mathcal{H})^+$ for any $r, s \in [0, 1]$,*

(2) *the map $(A, B) \mapsto (\log A) \,\sigma\, (\log B)$ is monotone on $B(\mathcal{H})^{++}$.*

Corollary 0.49. *Let σ be a connection on $B(\mathcal{H})^+$. If $\Phi_1, \Phi_2 : B(\mathcal{H})^+ \to B(\mathcal{H})^+$ is monotone and strongly continuous, then the map*

$$(A, B) \mapsto \Phi_1(A) \,\sigma\, \Phi_2(B) \tag{36}$$

is a connection on $B(\mathcal{H})^+$. Hence, the map

$$(A, B) \mapsto f(A) \,\sigma\, g(B) \tag{37}$$

is a connection for any operator monotone functions $f, g : \mathbb{R}^+ \to \mathbb{R}^+$.

Proof. The monotonicity of this map follows from the previous result. It is easy to see that this map satisfies the transformer inequality. Since Φ_1 and Φ_2 strongly continuous, this binary operation satisfies the (separate or joint) continuity from above. The last statement follows from the fact that if $A_n \downarrow A$, then $\mathrm{Sp}(A_n) \subseteq [0, \|A_1\|]$ for all n and hence $f(A_n) \to f(A)$. □

8. Applications to operator inequalities

In this section, we apply Kubo-Ando's theory in order to get simple proofs of many classical inequalities in the context of operators.

Theorem 0.50 (AM-LM-GM-HM inequalities). *For $A, B \geqslant 0$, we have*

$$A \,!\, B \leqslant A \,\#\, B \leqslant A \,\lambda\, B \leqslant A \,\triangledown\, B. \tag{38}$$

Proof. It is easy to see that, for each $t > 0, t \neq 1$,

$$\frac{2t}{1+t} \leqslant t^{1/2} \leqslant \frac{t-1}{\log t} \leqslant \frac{1+t}{2}.$$

Now, we apply the order isomorphism which converts inequalities of operator monotone functions to inequalities of the associated operator connections. □

Theorem 0.51 (Weighed AM-GM-HM inequalities). *For $A, B \geqslant 0$ and $\alpha \in [0,1]$, we have*

$$A \,!_\alpha\, B \leqslant A \,\#_\alpha\, B \leqslant A \,\nabla_\alpha\, B. \tag{39}$$

Proof. Apply the order isomorphism to the following inequalities:

$$\frac{t}{(1-\alpha)t+\alpha} \leqslant t^\alpha \leqslant 1-\alpha+\alpha t, \quad t \geqslant 0.$$

□

The next two theorems are given in [21].

Theorem 0.52. *For each $i = 1, \cdots, n$, let $A_i, B_i \in B(\mathcal{H})^+$. Then for each connection σ*

$$\sum_{i=1}^n (A_i \,\sigma\, B_i) \leqslant \sum_{i=1}^n A_i \,\sigma\, \sum_{i=1}^n B_i. \tag{40}$$

Proof. Use the concavity of σ together with the induction. □

By replacing σ with appropriate connections, we get some interesting inequalities.

(1) Cauchy-Schwarz's inequality: For $A_i, B_i \in B(\mathcal{H})^{sa}$,

$$\sum_{i=1}^n A_i^2 \,\#\, B_i^2 \leqslant \sum_{i=1}^n A_i^2 \,\#\, \sum_{i=1}^n B_i^2. \tag{41}$$

(2) Hölder's inequality: For $A_i, B_i \in B(\mathcal{H})^+$ and $p, q > 0$ such that $1/p + 1/q = 1$,

$$\sum_{i=1}^n A_i^p \,\#_{1/p}\, B_i^q \leqslant \sum_{i=1}^n A_i^p \,\#_{1/p}\, \sum_{i=1}^n B_i^q. \tag{42}$$

(3) Minkowski's inequality: For $A_i, B_i \in B(\mathcal{H})^{++}$,

$$\left(\sum_{i=1}^n (A_i + B_i)^{-1}\right)^{-1} \geqslant \left(\sum_{i=1}^n A_i^{-1}\right)^{-1} + \left(\sum_{i=1}^n B_i^{-1}\right)^{-1}. \tag{43}$$

Theorem 0.53. *Let $A_i, B_i \in B(\mathcal{H})^+, i = 1, \cdots, n$, be such that*

$$A_1 - A_2 - \cdots - A_n \geqslant 0 \quad \text{and} \quad B_1 - B_2 - \cdots - B_n \geqslant 0.$$

Then

$$A_1 \,\sigma\, B_1 - \sum_{i=2}^n A_i \,\sigma\, B_i \geqslant \left(A_1 - \sum_{i=2}^n A_i\right) \sigma \left(B_1 - \sum_{i=2}^n B_i\right). \tag{44}$$

Proof. Substitute A_1 to $A_1 - A_2 - \cdots - A_n$ and B_1 to $B_1 - B_2 - \cdots - B_n$ in (40). □

Here are consequences.

(1) Aczél's inequality: For $A_i, B_i \in B(\mathcal{H})^{sa}$, if

$$A_1^2 - A_2^2 - \cdots - A_n^2 \geqslant 0 \quad \text{and} \quad B_1^2 - B_2^2 - \cdots - B_n^2 \geqslant 0,$$

then

$$A_1^2 \# B_1^2 - \sum_{i=2}^{n} A_i^2 \# B_i^2 \geqslant \left(A_1^2 - \sum_{i=2}^{n} A_i^2\right) \# \left(B_1^2 - \sum_{i=2}^{n} B_i^2\right). \tag{45}$$

(2) Popoviciu's inequality: For $A_i, B_i \in B(\mathcal{H})^+$ and $p, q > 0$ such that $1/p + 1/q = 1$, if $p, q > 0$ are such that $1/p + 1/q = 1$ and

$$A_1^p - A_2^p - \cdots - A_n^p \geqslant 0 \quad \text{and} \quad B_1^q - B_2^q - \cdots - B_n^q \geqslant 0,$$

then

$$A_1^p \#_{1/p} B_1^q - \sum_{i=2}^{n} A_i^p \#_{1/p} B_i^q \geqslant \left(A_1^p - \sum_{i=2}^{n} A_i^p\right) \#_{1/p} \left(B_1^q - \sum_{i=2}^{n} B_i^q\right). \tag{46}$$

(3) Bellman's inequality: For $A_i, B_i \in B(\mathcal{H})^{++}$, if

$$A_1^{-1} - A_2^{-1} - \cdots - A_n^{-1} > 0 \quad \text{and} \quad B_1^{-1} - B_2^{-1} - \cdots - B_n^{-1} > 0,$$

then

$$\left[(A_1^{-1} + B_1^{-1}) - \sum_{i=2}^{n} (A_i + B_i)^{-1}\right]^{-1} \leqslant \left(A_1^{-1} - \sum_{i=2}^{n} A_i^{-1}\right)^{-1} + \left(B_1^{-1} - \sum_{i=2}^{n} B_i^{-1}\right)^{-1}. \tag{47}$$

The mean-theoretic approach can be used to prove the famous Furuta's inequality as follows. We cite [14] for the proof.

Theorem 0.54 (Furuta's inequality). *For $A \geqslant B \geqslant 0$, we have*

$$(B^r A^p B^r)^{1/q} \geqslant B^{(p+2r)/q} \tag{48}$$

$$A^{(p+2r)/q} \geqslant (A^r B^p A^r)^{1/q} \tag{49}$$

where $r \geqslant 0, p \geqslant 0, q \geqslant 1$ and $(1+2r)q \geqslant p + 2r$.

Proof. By the continuity argument, assume that $A, B > 0$. Note that (48) and (49) are equivalent. Indeed, if (48) holds, then (49) comes from applying (48) to $A^{-1} \leqslant B^{-1}$ and taking inverse on both sides. To prove (48), first consider the case $0 \leqslant p \leqslant 1$. We have $B^{p+2r} = B^r B^p B^r \leqslant B^r A^p B^r$ and the Löwner-Heinz's inequality (LH) implies the desired result. Now, consider the case $p \geqslant 1$ and $q = (p+2r)/(1+2r)$, since (48) for $q > (p+2r)/(1+2r)$ can be obtained by (LH). Let $f(t) = t^{1/q}$ and let σ be the associated connection (in fact,

$\sigma = \#_{1/q}$). Must show that, for any $r \geqslant 0$,

$$B^{-2r} \sigma A^p \geqslant B. \qquad (50)$$

For $0 \leqslant r \leqslant \frac{1}{2}$, we have by (LH) that $A^{2r} \geqslant B^{2r}$ and

$$B^{-2r} \sigma A^p \geqslant A^{-2r} \sigma A^p = A^{-2r(1-1/q)} A^{p/q} = A \geqslant B = B^{-2r} \sigma B^p.$$

Now, set $s = 2r + \frac{1}{2}$ and $q_1 = (p + 2s)/(1 + 2s) \geqslant 1$. Let $f_1(t) = t^{1/q_1}$ and consider the associated connection σ_1. The previous step, the monotonicity and the congruence invariance of connections imply that

$$\begin{aligned} B^{-2s} \sigma_1 A^p &= B^{-r}[B^{-(2r+1)} \sigma_1 (B^r A^p B^r)]B^{-r} \\ &\geqslant B^{-r}[(B^r A^p B^r)^{-1/q_1} \sigma_1 (B^r A^p B^r)]B^{-r} \\ &= B^{-r}(B^r A^p B^r)^{1/q} B^{-r} \\ &\geqslant B^{-r} B^{1+2r} B^{-r} \\ &= B. \end{aligned}$$

Note that the above result holds for $A, B \geqslant 0$ via the continuity of a connection. The desired equation (50) holds for all $r \geqslant 0$ by repeating this process. □

Acknowledgement

The author thanks referees for article processing.

Author details

Pattrawut Chansangiam
King Mongkut's Institute of Technology Ladkrabang, Thailand

9. References

[1] Anderson, W. & Duffin, R. (1969). Series and parallel addition of matrices, *Journal of Mathematical Analysis and Applications*, Vol. 26, 576–594.
[2] Anderson, W. & Morley, T. & Trapp, G. (1990). Positive solutions to $X = A - BX^{-1}B$, *Linear Algebra and its Applications*, Vol. 134, 53–62.
[3] Ando, T. (1979). Concavity of certain maps on positive definite matrices and applications to Hadamard products, *Linear Algebra and its Applications*, Vol. 26, 203–241.
[4] Ando, T. (1983). On the arithmetic-geometric-harmonic mean inequality for positive definite matrices, *Linear Algebra and its Applications*, Vol. 52-53, 31–37.
[5] Ando, T. & Li, C. & Mathias, R. (2004). Geometric Means, *Linear Algebra and its Applications*, Vol. 385, 305–334.
[6] Bhagwat, K. & Subramanian, A. (1978). Inequalities between means of positive operators, *Mathematical Proceedings of the Cambridge Philosophical Society*, Vol. 83, 393–401.
[7] Bhatia, R. (1997). *Matrix Analysis*, Springer-Verlag New York Inc., New York.
[8] Bhatia, R. (2006). Interpolating the arithmetic-geometric mean inequality and its operator version, *Linear Algebra and its Applications*, Vol. 413, 355–363.

[9] Bhatia, R. (2007). *Positive Definite Matrices*, Princeton University Press, New Jersey.

[10] Donoghue, W. (1974). *Monotone matrix functions and analytic continuation*, Springer-Verlag New York Inc., New York.

[11] Fujii, J. (1978). Arithmetico-geometric mean of operators, *Mathematica Japonica*, Vol. 23, 667–669.

[12] Fujii, J. (1979). On geometric and harmonic means of positive operators, *Mathematica Japonica*, Vol. 24, No. 2, 203–207.

[13] Fujii, J. (1992). Operator means and the relative operator entropy, *Operator Theory: Advances and Applications*, Vol. 59, 161–172.

[14] Furuta, T. (1989). A proof via operator means of an order preserving inequality, *Linear Algebra and its Applications*, Vol. 113, 129–130.

[15] Hiai, F. (2010). Matrix analysis: matrix monotone functions, matrix means, and majorizations, *Interdisciplinary Information Sciences*, Vol. 16, No. 2, 139–248,

[16] Hiai, F. & Yanagi, K. (1995). *Hilbert spaces and linear operators*, Makino Pub. Ltd.

[17] Kubo, F. & Ando, T. (1980). Means of positive linear operators, *Mathematische Annalen*, Vol. 246, 205–224.

[18] Lawson, J. & Lim, Y. (2001). The geometric mean, matrices, metrices and more, *The American Mathematical Monthly*, Vol. 108, 797–812.

[19] Lim, Y. (2008). On Ando–Li–Mathias geometric mean equations, *Linear Algebra and its Applications*, Vol. 428, 1767–1777.

[20] Löwner, C. (1934). Über monotone matrix funktionen. *Mathematische Zeitschrift*, Vol. 38, 177–216.

[21] Mond, B. & Pečarić, J. & Šunde, J. & Varošanec, S. (1997). Operator versions of some classical inequalities, *Linear Algebra and its Applications*, Vol. 264, 117–126.

[22] Nishio, K. & Ando, T. (1976). Characterizations of operations derived from network connections. *Journal of Mathematical Analysis and Applications*, Vol. 53, 539–549.

[23] Pusz, W. & Woronowicz, S. (1975). Functional calculus for sesquilinear forms and the purification map, *Reports on Mathematical Physics*, Vol. 8, 159–170.

[24] Toader, G. & Toader, S. (2005). Greek means and the arithmetic-geometric mean, RGMIA Monographs, Victoria University, (ONLINE: http://rgmia.vu.edu.au/monographs).

Chapter 9

Recent Research on Jensen's Inequality for Operators

Jadranka Mićić and Josip Pečarić

Additional information is available at the end of the chapter

http://dx.doi.org/10.5772/48468

1. Introduction

The self-adjoint operators on Hilbert spaces with their numerous applications play an important part in the operator theory. The bounds research for self-adjoint operators is a very useful area of this theory. There is no better inequality in bounds examination than Jensen's inequality. It is an extensively used inequality in various fields of mathematics.

Let I be a real interval of any type. A continuous function $f : I \to \mathbb{R}$ is said to be operator convex if

$$f(\lambda x + (1-\lambda)y) \leq \lambda f(x) + (1-\lambda) f(y) \tag{1}$$

holds for each $\lambda \in [0,1]$ and every pair of self-adjoint operators x and y (acting) on an infinite dimensional Hilbert space H with spectra in I (the ordering is defined by setting $x \leq y$ if $y - x$ is positive semi-definite).

Let f be an operator convex function defined on an interval I. Ch. Davis [1] proved[1] a Schwarz inequality

$$f(\phi(x)) \leq \phi(f(x)) \tag{2}$$

where $\phi\colon \mathcal{A} \to B(K)$ is a unital completely positive linear mapping from a C^*-algebra $\mathcal{A}$ to linear operators on a Hilbert space K, and x is a self-adjoint element in $\mathcal{A}$ with spectrum in I. Subsequently M. D. Choi [2] noted that it is enough to assume that ϕ is unital and positive. In fact, the restriction of ϕ to the commutative C^*-algebra generated by x is automatically completely positive by a theorem of Stinespring.

F. Hansen and G. K. Pedersen [3] proved a Jensen type inequality

$$f\left(\sum_{i=1}^{n} a_i^* x_i a_i\right) \leq \sum_{i=1}^{n} a_i^* f(x_i) a_i \tag{3}$$

[1] There is small typo in the proof. Davis states that ϕ by Stinespring's theorem can be written on the form $\phi(x) = P\rho(x)P$ where ρ is a $*$-homomorphism to $B(H)$ and P is a projection on H. In fact, H may be embedded in a Hilbert space K on which ρ and P acts. The theorem then follows by the calculation $f(\phi(x)) = f(P\rho(x)P) \leq Pf(\rho(x))P = P\rho(f(x)P = \phi(f(x))$, where the pinching inequality, proved by Davis in the same paper, is applied.

for operator convex functions f defined on an interval $I = [0, \alpha)$ (with $\alpha \leq \infty$ and $f(0) \leq 0$) and self-adjoint operators $x_1, \ldots, x_n$ with spectra in I assuming that $\sum_{i=1}^{n} a_i^* a_i = \mathbf{1}$. The restriction on the interval and the requirement $f(0) \leq 0$ was subsequently removed by B. Mond and J. Pečarić in [4], cf. also [5].

The inequality (3) is in fact just a reformulation of (2) although this was not noticed at the time. It is nevertheless important to note that the proof given in [3] and thus the statement of the theorem, when restricted to $n \times n$ matrices, holds for the much richer class of $2n \times 2n$ matrix convex functions. Hansen and Pedersen used (3) to obtain elementary operations on functions, which leave invariant the class of operator monotone functions. These results then served as the basis for a new proof of Löwner's theorem applying convexity theory and Krein-Milman's theorem.

B. Mond and J. Pečarić [6] proved the inequality

$$f\left(\sum_{i=1}^{n} w_i \phi_i(x_i)\right) \leq \sum_{i=1}^{n} w_i \phi_i(f(x_i)) \tag{4}$$

for operator convex functions f defined on an interval I, where $\phi_i : B(H) \to B(K)$ are unital positive linear mappings, $x_1, \ldots, x_n$ are self-adjoint operators with spectra in I and $w_1, \ldots, w_n$ are are non-negative real numbers with sum one.

Also, B. Mond, J. Pečarić, T. Furuta et al. [6–11] observed conversed of some special case of Jensen's inequality. So in [10] presented the following generalized converse of a Schwarz inequality (2)

$$F\left[\phi\left(f(A)\right), g\left(\phi(A)\right)\right] \leq \max_{m \leq t \leq M} F\left[f(m) + \frac{f(M) - f(m)}{M - m}(t - m), g(t)\right] 1_{\tilde{n}} \tag{5}$$

for convex functions f defined on an interval $[m, M]$, $m < M$, where g is a real valued continuous function on $[m, M]$, $F(u, v)$ is a real valued function defined on $U \times V$, matrix non-decreasing in u, $U \supset f[m, M]$, $V \supset g[m, M]$, $\phi : H_n \to H_{\tilde{n}}$ is a unital positive linear mapping and A is a Hermitian matrix with spectrum contained in $[m, M]$.

There are a lot of new research on the classical Jensen inequality (4) and its reverse inequalities. For example, J.I. Fujii et all. in [12, 13] expressed these inequalities by externally dividing points.

2. Classic results

In this section we present a form of Jensen's inequality which contains (2), (3) and (4) as special cases. Since the inequality in (4) was the motivating step for obtaining converses of Jensen's inequality using the so-called Mond-Pečarić method, we also give some results pertaining to converse inequalities in the new formulation.

We recall some definitions. Let T be a locally compact Hausdorff space and let $\mathcal{A}$ be a C^*-algebra of operators on some Hilbert space H. We say that a field $(x_t)_{t \in T}$ of operators in $\mathcal{A}$ is continuous if the function $t \mapsto x_t$ is norm continuous on T. If in addition μ is a Radon measure on T and the function $t \mapsto \|x_t\|$ is integrable, then we can form *the Bochner integral* $\int_T x_t \, d\mu(t)$, which is the unique element in $\mathcal{A}$ such that

$$\varphi\left(\int_T x_t \, d\mu(t)\right) = \int_T \varphi(x_t) \, d\mu(t)$$

for every linear functional φ in the norm dual $\mathcal{A}^*$.

Assume furthermore that there is a field $(\phi_t)_{t\in T}$ of positive linear mappings $\phi_t : \mathcal{A} \to \mathcal{B}$ from $\mathcal{A}$ to another C^*-algebra $\mathcal{B}$ of operators on a Hilbert space K. We recall that a linear mapping $\phi_t : \mathcal{A} \to \mathcal{B}$ is said to be a positive mapping if $\phi_t(x_t) \geq 0$ for all $x_t \geq 0$. We say that such a field is continuous if the function $t \mapsto \phi_t(x)$ is continuous for every $x \in \mathcal{A}$. Let the C^*-algebras include the identity operators and the function $t \mapsto \phi_t(1_H)$ be integrable with $\int_T \phi_t(1_H) \, d\mu(t) = k1_K$ for some positive scalar k. Specially, if $\int_T \phi_t(1_H) \, d\mu(t) = 1_K$, we say that a *field* $(\phi_t)_{t\in T}$ is *unital*.

Let $B(H)$ be the C^*-algebra of all bounded linear operators on a Hilbert space H. We define bounds of an operator $x \in B(H)$ by

$$m_x = \inf_{\|\xi\|=1} \langle x\xi, \xi\rangle \quad \text{and} \quad M_x = \sup_{\|\xi\|=1} \langle x\xi, \xi\rangle \tag{6}$$

for $\xi \in H$. If $\mathsf{Sp}(x)$ denotes the spectrum of x, then $\mathsf{Sp}(x) \subseteq [m_x, M_x]$.

For an operator $x \in B(H)$ we define operators $|x|, x^+, x^-$ by

$$|x| = (x^*x)^{1/2}, \qquad x^+ = (|x| + x)/2, \qquad x^- = (|x| - x)/2$$

Obviously, if x is self-adjoint, then $|x| = (x^2)^{1/2}$ and $x^+, x^- \geq 0$ (called positive and negative parts of $x = x^+ - x^-$).

2.1. Jensen's inequality with operator convexity

Firstly, we give a general formulation of Jensen's operator inequality for a unital field of positive linear mappings (see [14]).

Theorem 1. *Let $f : I \to \mathbb{R}$ be an operator convex function defined on an interval I and let $\mathcal{A}$ and $\mathcal{B}$ be unital C^*-algebras acting on a Hilbert space H and K respectively. If $(\phi_t)_{t\in T}$ is a unital field of positive linear mappings $\phi_t : \mathcal{A} \to \mathcal{B}$ defined on a locally compact Hausdorff space T with a bounded Radon measure μ, then the inequality*

$$f\left(\int_T \phi_t(x_t) \, d\mu(t)\right) \leq \int_T \phi_t(f(x_t)) \, d\mu(t) \tag{7}$$

holds for every bounded continuous field $(x_t)_{t\in T}$ of self-adjoint elements in $\mathcal{A}$ with spectra contained in I.

Proof. We first note that the function $t \mapsto \phi_t(x_t) \in \mathcal{B}$ is continuous and bounded, hence integrable with respect to the bounded Radon measure μ. Furthermore, the integral is an element in the multiplier algebra $M(\mathcal{B})$ acting on K. We may organize the set $CB(T, \mathcal{A})$ of bounded continuous functions on T with values in $\mathcal{A}$ as a normed involutive algebra by applying the point-wise operations and setting

$$\|(y_t)_{t\in T}\| = \sup_{t\in T}\|y_t\| \qquad (y_t)_{t\in T} \in CB(T,\mathcal{A})$$

and it is not difficult to verify that the norm is already complete and satisfy the C^*-identity. In fact, this is a standard construction in C^*-algebra theory. It follows that $f((x_t)_{t\in T}) = (f(x_t))_{t\in T}$. We then consider the mapping

$$\pi\colon CB(T,\mathcal{A}) \to M(\mathcal{B}) \subseteq B(K)$$

defined by setting

$$\pi\left((x_t)_{t\in T}\right) = \int_T \phi_t(x_t)\, d\mu(t)$$

and note that it is a unital positive linear map. Setting $x = (x_t)_{t\in T} \in CB(T,\mathcal{A})$, we use inequality (2) to obtain

$$f\left(\pi\left((x_t)_{t\in T}\right)\right) = f(\pi(x)) \le \pi(f(x)) = \pi\left(f\left((x_t)_{t\in T}\right)\right) = \pi\left(\left(f(x_t)\right)_{t\in T}\right)$$

but this is just the statement of the theorem. □

2.2. Converses of Jensen's inequality

In the present context we may obtain results of the Li-Mathias type cf. [15, Chapter 3] and [16, 17].

Theorem 2. *Let T be a locally compact Hausdorff space equipped with a bounded Radon measure μ. Let $(x_t)_{t\in T}$ be a bounded continuous field of self-adjoint elements in a unital C^*-algebra $\mathcal{A}$ with spectra in $[m,M]$, $m < M$. Furthermore, let $(\phi_t)_{t\in T}$ be a field of positive linear mappings $\phi_t : \mathcal{A} \to \mathcal{B}$ from $\mathcal{A}$ to another unital C^*-algebra $\mathcal{B}$, such that the function $t \mapsto \phi_t(1_H)$ is integrable with $\int_T \phi_t(1_H)\, d\mu(t) = k1_K$ for some positive scalar k. Let m_x and M_x, $m_x \le M_x$, be the bounds of the self-adjoint operator $x = \int_T \phi_t(x_t)\, d\mu(t)$ and $f : [m,M] \to \mathbb{R}$, $g : [m_x, M_x] \to \mathbb{R}$, $F : U \times V \to \mathbb{R}$ be functions such that $(kf)([m,M]) \subset U$, $g([m_x,M_x]) \subset V$ and F is bounded. If F is operator monotone in the first variable, then*

$$\begin{aligned}\inf_{m_x\le z\le M_x} F\left[k\cdot h_1\left(\frac{1}{k}z\right), g(z)\right] 1_K &\le F\left[\int_T \phi_t\left(f(x_t)\right) d\mu(t), g\left(\int_T \phi_t(x_t) d\mu(t)\right)\right] \\ &\le \sup_{m_x\le z\le M_x} F\left[k\cdot h_2\left(\frac{1}{k}z\right), g(z)\right] 1_K\end{aligned} \tag{8}$$

holds for every operator convex function h_1 on $[m,M]$ such that $h_1 \le f$ and for every operator concave function h_2 on $[m,M]$ such that $h_2 \ge f$.

Proof. We prove only RHS of (8). Let h_2 be operator concave function on $[m,M]$ such that $f(z) \le h_2(z)$ for every $z \in [m,M]$. By using the functional calculus, it follows that $f(x_t) \le h_2(x_t)$ for every $t \in T$. Applying the positive linear mappings ϕ_t and integrating, we obtain

$$\int_T \phi_t\left(f(x_t)\right) d\mu(t) \le \int_T \phi_t\left(h_2(x_t)\right) d\mu(t)$$

Furthermore, replacing ϕ_t by $\frac{1}{k}\phi_t$ in Theorem 1, we obtain $\frac{1}{k}\int_T \phi_t\left(h_2(x_t)\right) d\mu(t) \le h_2\left(\frac{1}{k}\int_T \phi_t(x_t)\, d\mu(t)\right)$, which gives $\int_T \phi_t\left(f(x_t)\right) d\mu(t) \le k\cdot h_2\left(\frac{1}{k}\int_T \phi_t(x_t)\, d\mu(t)\right)$. Since $m_x 1_K \le \int_T \phi_t(x_t) d\mu(t) \le M_x 1_K$, then using operator monotonicity of $F(\cdot, v)$ we obtain

$$
\begin{aligned}
&F\left[\int_T \phi_t\left(f(x_t)\right) d\mu(t), g\left(\int_T \phi_t(x_t) d\mu(t)\right)\right] && (9)\\
&\le F\left[k\cdot h_2\left(\frac{1}{k}\int_T \phi_t(x_t)\, d\mu(t)\right), g\left(\int_T \phi_t(x_t) d\mu(t)\right)\right] \le \sup_{m_x\le z\le M_x} F\left[k\cdot h_2\left(\frac{1}{k}z\right), g(z)\right] 1_K
\end{aligned}
$$

□

Applying RHS of (8) for a convex function f (or LHS of (8) for a concave function f) we obtain the following generalization of (5).

Theorem 3. *Let $(x_t)_{t\in T}$, m_x, M_x and $(\phi_t)_{t\in T}$ be as in Theorem 2. Let $f : [m, M] \to \mathbb{R}$, $g : [m_x, M_x] \to \mathbb{R}$, $F : U \times V \to \mathbb{R}$ be functions such that $(kf)([m, M]) \subset U$, $g([m_x, M_x]) \subset V$ and F is bounded. If F is operator monotone in the first variable and f is convex on the interval $[m, M]$, then*

$$
\begin{aligned}
&F\left[\int_T \phi_t\left(f(x_t)\right) d\mu(t), g\left(\int_T \phi_t(x_t) d\mu(t)\right)\right]\\
&\le \sup_{m_x\le z\le M_x} F\left[\frac{Mk-z}{M-m} f(m) + \frac{z-km}{M-m} f(M), g(z)\right] 1_K
\end{aligned} \tag{10}
$$

In the dual case (when f is concave) the opposite inequalities hold in (10) *with* inf *instead of* sup.

Proof. We prove only the convex case. For convex f the inequality $f(z) \le \frac{M-z}{M-m} f(m) + \frac{z-m}{M-m} f(M)$ holds for every $z \in [m, M]$. Thus, by putting $h_2(z) = \frac{M-z}{M-m} f(m) + \frac{z-m}{M-m} f(M)$ in (9) we obtain (10). □

Numerous applications of the previous theorem can be given (see [15]). Applying Theorem 3 for the function $F(u, v) = u - \alpha v$ and $k = 1$, we obtain the following generalization of [15, Theorem 2.4].

Corollary 4. *Let $(x_t)_{t\in T}$, m_x, M_x be as in Theorem 2 and $(\phi_t)_{t\in T}$ be a unital field of positive linear mappings $\phi_t : \mathcal{A} \to \mathcal{B}$. If $f : [m, M] \to \mathbb{R}$ is convex on the interval $[m, M]$, $m < M$, and $g : [m, M] \to \mathbb{R}$, then for any $\alpha \in \mathbb{R}$*

$$
\int_T \phi_t\left(f(x_t)\right) d\mu(t) \le \alpha\, g\left(\int_T \phi_t(x_t) d\mu(t)\right) + C 1_K \tag{11}
$$

where

$$
\begin{aligned}
C &= \max_{m_x\le z\le M_x}\left\{\frac{M-z}{M-m} f(m) + \frac{z-m}{M-m} f(M) - \alpha g(z)\right\}\\
&\le \max_{m\le z\le M}\left\{\frac{M-z}{M-m} f(m) + \frac{z-m}{M-m} f(M) - \alpha g(z)\right\}
\end{aligned}
$$

If furthermore αg is strictly convex differentiable, then the constant $C \equiv C(m, M, f, g, \alpha)$ can be written more precisely as

$$C = \frac{M - z_0}{M - m} f(m) + \frac{z_0 - m}{M - m} f(M) - \alpha g(z_0)$$

where

$$z_0 = \begin{cases} g'^{-1}\left(\frac{f(M)-f(m)}{\alpha(M-m)}\right) & \text{if } \alpha g'(m_x) \leq \frac{f(M)-f(m)}{M-m} \leq \alpha g'(M_x) \\ m_x & \text{if } \alpha g'(m_x) \geq \frac{f(M)-f(m)}{M-m} \\ M_x & \text{if } \alpha g'(M_x) \leq \frac{f(M)-f(m)}{M-m} \end{cases}$$

In the dual case (when f is concave and αg is strictly concave differentiable) the opposite inequalities hold in (11) *with* min *instead of* max *with the opposite condition while determining z_0.*

3. Inequalities with conditions on spectra

In this section we present Jensens's operator inequality for real valued continuous convex functions with conditions on the spectra of the operators. A discrete version of this result is given in [18]. Also, we obtain generalized converses of Jensen's inequality under the same conditions.

Operator convexity plays an essential role in (2). In fact, the inequality (2) will be false if we replace an operator convex function by a general convex function. For example, M.D. Choi in [2, Remark 2.6] considered the function $f(t) = t^4$ which is convex but not operator convex. He demonstrated that it is sufficient to put $\dim H = 3$, so we have the matrix case as follows. Let $\Phi : M_3(\mathbb{C}) \to M_2(\mathbb{C})$ be the contraction mapping $\Phi((a_{ij})_{1\leq i,j\leq 3}) = (a_{ij})_{1\leq i,j\leq 2}$. If $A = \begin{pmatrix} 1 & 0 & 1 \\ 0 & 0 & 1 \\ 1 & 1 & 1 \end{pmatrix}$, then $\Phi(A)^4 = \begin{pmatrix} 1 & 0 \\ 0 & 0 \end{pmatrix} \not\leq \begin{pmatrix} 9 & 5 \\ 5 & 3 \end{pmatrix} = \Phi(A^4)$ and no relation between $\Phi(A)^4$ and $\Phi(A^4)$ under the operator order.

Example 5. *It appears that the inequality* (7) *will be false if we replace the operator convex function by a general convex function. We give a small example for the matrix cases and $T = \{1, 2\}$. We define mappings $\Phi_1, \Phi_2 : M_3(\mathbb{C}) \to M_2(\mathbb{C})$ by $\Phi_1((a_{ij})_{1\leq i,j\leq 3}) = \frac{1}{2}(a_{ij})_{1\leq i,j\leq 2}$, $\Phi_2 = \Phi_1$. Then $\Phi_1(I_3) + \Phi_2(I_3) = I_2$.*

I) If

$$X_1 = 2\begin{pmatrix} 1 & 0 & 1 \\ 0 & 0 & 1 \\ 1 & 1 & 1 \end{pmatrix} \quad \text{and} \quad X_2 = 2\begin{pmatrix} 1 & 0 & 0 \\ 0 & 0 & 0 \\ 0 & 0 & 0 \end{pmatrix}$$

then

$$(\Phi_1(X_1) + \Phi_2(X_2))^4 = \begin{pmatrix} 16 & 0 \\ 0 & 0 \end{pmatrix} \not\leq \begin{pmatrix} 80 & 40 \\ 40 & 24 \end{pmatrix} = \Phi_1\left(X_1^4\right) + \Phi_2\left(X_2^4\right)$$

Given the above, there is no relation between $(\Phi_1(X_1) + \Phi_2(X_2))^4$ and $\Phi_1\left(X_1^4\right) + \Phi_2\left(X_2^4\right)$ under the operator order. We observe that in the above case the following stands $X = \Phi_1(X_1) + \Phi_2(X_2) = \begin{pmatrix} 2 & 0 \\ 0 & 0 \end{pmatrix}$ and $[m_x, M_x] = [0, 2]$, $[m_1, M_1] \subset [-1.60388, 4.49396]$, $[m_2, M_2] = [0, 2]$, i.e.

$$(m_x, M_x) \subset [m_1, M_1] \cup [m_2, M_2]$$

(see Fig. 1.a).

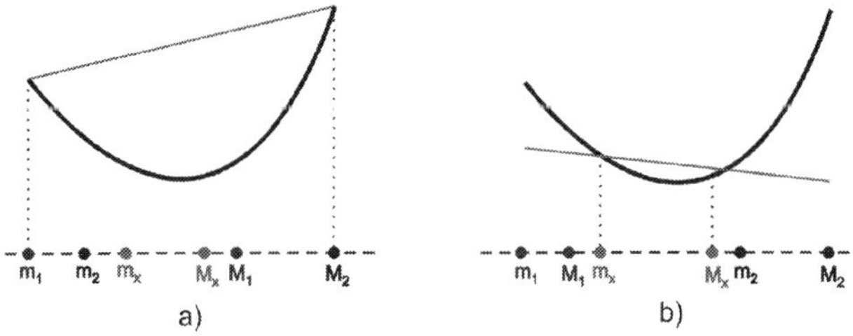

Figure 1. Spectral conditions for a convex function f

II) If

$$X_1 = \begin{pmatrix} -14 & 0 & 1 \\ 0 & -2 & -1 \\ 1 & -1 & -1 \end{pmatrix} \quad \textit{and} \quad X_2 = \begin{pmatrix} 15 & 0 & 0 \\ 0 & 2 & 0 \\ 0 & 0 & 15 \end{pmatrix}$$

then

$$(\Phi_1(X_1) + \Phi_2(X_2))^4 = \begin{pmatrix} \frac{1}{16} & 0 \\ 0 & 0 \end{pmatrix} < \begin{pmatrix} 89660 & -247 \\ -247 & 51 \end{pmatrix} = \Phi_1\left(X_1^4\right) + \Phi_2\left(X_2^4\right)$$

So we have that an inequality of type (7) *now is valid. In the above case the following stands* $X = \Phi_1(X_1) + \Phi_2(X_2) = \begin{pmatrix} \frac{1}{2} & 0 \\ 0 & 0 \end{pmatrix}$ *and* $[m_x, M_x] = [0, 0.5]$, $[m_1, M_1] \subset [-14.077, -0.328566]$, $[m_2, M_2] = [2, 15]$, *i.e.*

$$(m_x, M_x) \cap [m_1, M_1] = \emptyset \quad \textit{and} \quad (m_x, M_x) \cap [m_2, M_2] = \emptyset$$

(see Fig. 1.b).

3.1. Jensen's inequality without operator convexity

It is no coincidence that the inequality (7) is valid in Example 18-II). In the following theorem we prove a general result when Jensen's operator inequality (7) holds for convex functions.

Theorem 6. *Let $(x_t)_{t\in T}$ be a bounded continuous field of self-adjoint elements in a unital C^*-algebra $\mathcal{A}$ defined on a locally compact Hausdorff space T equipped with a bounded Radon measure μ. Let m_t and M_t, $m_t \leq M_t$, be the bounds of x_t, $t \in T$. Let $(\phi_t)_{t\in T}$ be a unital field of positive linear mappings $\phi_t : \mathcal{A} \to \mathcal{B}$ from $\mathcal{A}$ to another unital C^*-algebra $\mathcal{B}$. If*

$$(m_x, M_x) \cap [m_t, M_t] = \emptyset, \qquad t \in T$$

where m_x and M_x, $m_x \leq M_x$, are the bounds of the self-adjoint operator $x = \int_T \phi_t(x_t)\, d\mu(t)$, then

$$f\left(\int_T \phi_t(x_t)\, d\mu(t)\right) \leq \int_T \phi_t(f(x_t))\, d\mu(t) \tag{12}$$

holds for every continuous convex function $f : I \to \mathbb{R}$ provided that the interval I contains all m_t, M_t.

If $f : I \to \mathbb{R}$ is concave, then the reverse inequality is valid in (12).

Proof. We prove only the case when f is a convex function. If we denote $m = \inf_{t\in T}\{m_t\}$ and $M = \sup_{t\in T}\{M_t\}$, then $[m,M] \subseteq I$ and $m1_H \le A_t \le M1_H$, $t \in T$. It follows $m1_K \le \int_T \phi_t(x_t)\,d\mu(t) \le M1_K$. Therefore $[m_x, M_x] \subseteq [m,M] \subseteq I$.

a) Let $m_x < M_x$. Since f is convex on $[m_x, M_x]$, then

$$f(z) \le \frac{M_x - z}{M_x - m_x} f(m_x) + \frac{z - m_x}{M_x - m_x} f(M_x), \quad z \in [m_x, M_x] \tag{13}$$

but since f is convex on $[m_t, M_t]$ and since $(m_x, M_x) \cap [m_t, M_t] = \emptyset$, then

$$f(z) \ge \frac{M_x - z}{M_x - m_x} f(m_x) + \frac{z - m_x}{M_x - m_x} f(M_x), \quad z \in [m_t, M_t], \quad t \in T \tag{14}$$

Since $m_x 1_K \le \int_T \phi_t(x_t)\,d\mu(t) \le M_x 1_K$, then by using functional calculus, it follows from (13)

$$f\left(\int_T \phi_t(x_t)\,d\mu(t)\right) \le \frac{M_x 1_K - \int_T \phi_t(x_t)\,d\mu(t)}{M_x - m_x} f(m_x) + \frac{\int_T \phi_t(x_t)\,d\mu(t) - m_x 1_K}{M_x - m_x} f(M_x) \tag{15}$$

On the other hand, since $m_t 1_H \le x_t \le M_t 1_H$, $t \in T$, then by using functional calculus, it follows from (14)

$$f(x_t) \ge \frac{M_x 1_H - x_t}{M_x - m_x} f(m_x) + \frac{x_t - m_x 1_H}{M_x - m_x} f(M_x), \qquad t \in T$$

Applying a positive linear mapping ϕ_t and summing, we obtain

$$\int_T \phi_t\left(f(x_t)\right) d\mu(t) \ge \frac{M_x 1_K - \int_T \phi_t(x_t)\,d\mu(t)}{M_x - m_x} f(m_x) + \frac{\int_T \phi_t(x_t)\,d\mu(t) - m_x 1_K}{M_x - m_x} f(M_x) \tag{16}$$

since $\int_T \phi_t(1_H)\,d\mu(t) = 1_K$. Combining the two inequalities (15) and (16), we have the desired inequality (12).

b) Let $m_x = M_x$. Since f is convex on $[m,M]$, we have

$$f(z) \ge f(m_x) + l(m_x)(z - m_x) \quad \text{for every } z \in [m,M] \tag{17}$$

where l is the subdifferential of f. Since $m1_H \le x_t \le M1_H$, $t \in T$, then by using functional calculus, applying a positive linear mapping ϕ_t and summing, we obtain from (17)

$$\int_T \phi_t\left(f(x_t)\right) d\mu(t) \ge f(m_x)1_K + l(m_x)\left(\int_T \phi_t(x_t)\,d\mu(t) - m_x 1_K\right)$$

Since $m_x 1_K = \int_T \phi_t(x_t)\,d\mu(t)$, it follows

$$\int_T \phi_t\left(f(x_t)\right) d\mu(t) \ge f(m_x)1_K = f\left(\int_T \phi_t(x_t)\,d\mu(t)\right)$$

which is the desired inequality (12). □

Putting $\phi_t(y) = a_t y$ for every $y \in \mathcal{A}$, where $a_t \ge 0$ is a real number, we obtain the following obvious corollary of Theorem 6.

Corollary 7. *Let $(x_t)_{t\in T}$ be a bounded continuous field of self-adjoint elements in a unital C^*-algebra $\mathcal{A}$ defined on a locally compact Hausdorff space T equipped with a bounded Radon measure μ. Let m_t and M_t, $m_t \le M_t$, be the bounds of x_t, $t \in T$. Let $(a_t)_{t\in T}$ be a continuous field of nonnegative real numbers such that $\int_T a_t \, d\mu(t) = 1$. If*

$$(m_x, M_x) \cap [m_t, M_t] = \emptyset, \qquad t \in T$$

where m_x and M_x, $m_x \le M_x$, are the bounds of the self-adjoint operator $x = \int_T a_t x_t \, d\mu(t)$, then

$$f\left(\int_T a_t x_t \, d\mu(t)\right) \le \int_T a_t f(x_t) \, d\mu(t) \tag{18}$$

holds for every continuous convex function $f : I \to \mathbb{R}$ provided that the interval I contains all m_t, M_t.

3.2. Converses of Jensen's inequality with conditions on spectra

Using the condition on spectra we obtain the following extension of Theorem 3.

Theorem 8. *Let $(x_t)_{t\in T}$ be a bounded continuous field of self-adjoint elements in a unital C^*-algebra $\mathcal{A}$ defined on a locally compact Hausdorff space T equipped with a bounded Radon measure μ. Furthermore, let $(\phi_t)_{t\in T}$ be a field of positive linear mappings $\phi_t : \mathcal{A} \to \mathcal{B}$ from $\mathcal{A}$ to another unital C^*-algebra $\mathcal{B}$, such that the function $t \mapsto \phi_t(1_H)$ is integrable with $\int_T \phi_t(1_H) \, d\mu(t) = k1_K$ for some positive scalar k. Let m_t and M_t, $m_t \le M_t$, be the bounds of x_t, $t \in T$, $m = \inf_{t\in T}\{m_t\}$, $M = \sup_{t\in T}\{M_t\}$, and m_x and M_x, $m_x < M_x$, be the bounds of $x = \int_T \phi_t(x_t) \, d\mu(t)$. If*

$$(m_x, M_x) \cap [m_t, M_t] = \emptyset, \qquad t \in T$$

and $f : [m, M] \to \mathbb{R}$, $g : [m_x, M_x] \to \mathbb{R}$, $F : U \times V \to \mathbb{R}$ are functions such that $(kf)([m, M]) \subset U$, $g([m_x, M_x]) \subset V$, f is convex, F is bounded and operator monotone in the first variable, then

$$\begin{aligned}
\inf_{m_x \le z \le M_x} F\left[\frac{M_x k - z}{M_x - m_x} f(m_x) + \frac{z - k m_x}{M_x - m_x} f(M_x), g(z)\right] 1_K \\
F\left[\int_T \phi_t\left(f(x_t)\right) d\mu(t), g\left(\int_T \phi_t(x_t) d\mu(t)\right)\right] \\
\le \sup_{m_x \le z \le M_x} F\left[\frac{Mk - z}{M - m} f(m) + \frac{z - km}{M - m} f(M), g(z)\right] 1_K
\end{aligned} \tag{19}$$

In the dual case (when f is concave) the opposite inequalities hold in (19) by replacing inf *and* sup *with* sup *and* inf, *respectively.*

Proof. We prove only LHS of (19). It follows from (14) (compare it to (16))

$$\int_T \phi_t\left(f(x_t)\right) d\mu(t) \ge \frac{M_x k 1_K - \int_T \phi_t(x_t) \, d\mu(t)}{M_x - m_x} f(m_x) + \frac{\int_T \phi_t(x_t) \, d\mu(t) - m_x k 1_K}{M_x - m_x} f(M_x)$$

since $\int_T \phi_t(1_H) \, d\mu(t) = k1_K$. By using operator monotonicity of $F(\cdot, v)$ we obtain

$$\begin{aligned}
& F\left[\int_T \phi_t\left(f(x_t)\right) d\mu(t), g\left(\int_T \phi_t(x_t) \, d\mu(t)\right)\right] \\
\ge\ & F\left[\tfrac{M_x k 1_K - \int_T \phi_t(x_t) \, d\mu(t)}{M_x - m_x} f(m_x) + \tfrac{\int_T \phi_t(x_t) \, d\mu(t) - m_x k 1_K}{M_x - m_x} f(M_x), g\left(\int_T \phi_t(x_t) \, d\mu(t)\right)\right] \\
\ge\ & \inf_{m_x \le z \le M_x} F\left[\frac{M_x k - z}{M_x - m_x} f(m_x) + \frac{z - k m_x}{M_x - m_x} f(M_x), g(z)\right] 1_K
\end{aligned}$$

□

Putting $F(u,v) = u - \alpha v$ or $F(u,v) = v^{-1/2}uv^{-1/2}$ in Theorem 8, we obtain the next corollary.

Corollary 9. *Let $(x_t)_{t\in T}$, m_t, M_t, m_x, M_x, m, M, $(\phi_t)_{t\in T}$ be as in Theorem 8 and $f:[m,M]\to\mathbb{R}$, $g:[m_x,M_x]\to\mathbb{R}$ be continuous functions. If*

$$(m_x, M_x)\cap[m_t,M_t]=\emptyset, \qquad t\in T$$

and f is convex, then for any $\alpha\in\mathbb{R}$

$$\begin{aligned}\min_{m_x\le z\le M_x}&\left\{\frac{M_xk-z}{M_x-m_x}f(m_x)+\frac{z-km_x}{M_x-m_x}f(M_x)-g(z)\right\}1_K+\alpha g\left(\int_T\phi_t(x_t)d\mu(t)\right)\\ &\le\int_T\phi_t\left(f(x_t)\right)d\mu(t)\\ \le\alpha g\left(\int_T\phi_t(x_t)d\mu(t)\right)+&\max_{m_x\le z\le M_x}\left\{\frac{Mk-z}{M-m}f(m)+\frac{z-km}{M-m}f(M)-g(z)\right\}1_K\end{aligned}\tag{20}$$

If additionally $g>0$ on $[m_x,M_x]$, then

$$\begin{aligned}&\min_{m_x\le z\le M_x}\left\{\frac{\frac{M_xk-z}{M_x-m_x}f(m_x)+\frac{z-km_x}{M_x-m_x}f(M_x)}{g(z)}\right\}g\left(\int_T\phi_t(x_t)d\mu(t)\right)\\ \le\int_T\phi_t\left(f(x_t)\right)d\mu(t)\le&\max_{m_x\le z\le M_x}\left\{\frac{\frac{Mk-z}{M-m}f(m)+\frac{z-km}{M-m}f(M)}{g(z)}\right\}g\left(\int_T\phi_t(x_t)d\mu(t)\right)\end{aligned}\tag{21}$$

In the dual case (when f is concave) the opposite inequalities hold in (20) *by replacing* min *and* max *with* max *and* min, *respectively. If additionally $g>0$ on $[m_x,M_x]$, then the opposite inequalities also hold in* (21) *by replacing* min *and* max *with* max *and* min, *respectively.*

4. Refined Jensen's inequality

In this section we present a refinement of Jensen's inequality for real valued continuous convex functions given in Theorem 6. A discrete version of this result is given in [19].

To obtain our result we need the following two lemmas.

Lemma 10. *Let f be a convex function on an interval I, $m,M\in I$ and $p_1,p_2\in[0,1]$ such that $p_1+p_2=1$. Then*

$$\min\{p_1,p_2\}\left[f(m)+f(M)-2f\left(\frac{m+M}{2}\right)\right]\le p_1f(m)+p_2f(M)-f(p_1m+p_2M)\tag{22}$$

Proof. These results follows from [20, Theorem 1, p. 717]. □

Lemma 11. *Let x be a bounded self-adjoint elements in a unital C^*-algebra $\mathcal{A}$ of operators on some Hilbert space H. If the spectrum of x is in $[m,M]$, for some scalars $m<M$, then*

$$f(x)\le\frac{M1_H-x}{M-m}f(m)+\frac{x-m1_H}{M-m}f(M)-\delta_f\widetilde{x}\tag{23}$$

$$\left(\textit{resp.}\quad f(x)\ge\frac{M1_H-x}{M-m}f(m)+\frac{x-m1_H}{M-m}f(M)+\delta_f\widetilde{x}\quad\right)$$

holds for every continuous convex (resp. concave) function $f : [m, M] \to \mathbb{R}$, *where*

$$\delta_f = f(m) + f(M) - 2f\left(\tfrac{m+M}{2}\right) \quad (\textit{resp. } \delta_f = 2f\left(\tfrac{m+M}{2}\right) - f(m) - f(M))$$

$$\textit{and} \quad \widetilde{x} = \tfrac{1}{2}1_H - \tfrac{1}{M-m}\left|x - \tfrac{m+M}{2}1_H\right|$$

Proof. We prove only the convex case. It follows from (22) that

$$\begin{aligned} f\left(p_1 m + p_2 M\right) &\leq p_1 f(m) + p_2 f(M) \\ &\quad - \min\{p_1, p_2\}\left(f(m) + f(M) - 2f\left(\tfrac{m+M}{2}\right)\right) \end{aligned} \tag{24}$$

for every $p_1, p_2 \in [0,1]$ such that $p_1 + p_2 = 1$. For any $z \in [m, M]$ we can write

$$f(z) = f\left(\frac{M-z}{M-m}m + \frac{z-m}{M-m}M\right)$$

Then by using (24) for $p_1 = \frac{M-z}{M-m}$ and $p_2 = \frac{z-m}{M-m}$ we obtain

$$\begin{aligned} f(z) &\leq \frac{M-z}{M-m}f(m) + \frac{z-m}{M-m}f(M) \\ &- \left(\frac{1}{2} - \frac{1}{M-m}\left|z - \frac{m+M}{2}\right|\right)\left(f(m) + f(M) - 2f\left(\frac{m+M}{2}\right)\right) \end{aligned} \tag{25}$$

since

$$\min\left\{\frac{M-z}{M-m}, \frac{z-m}{M-m}\right\} = \frac{1}{2} - \frac{1}{M-m}\left|z - \frac{m+M}{2}\right|$$

Finally we use the continuous functional calculus for a self-adjoint operator x: $f, g \in C(I), Sp(x) \subseteq I$ and $f \leq g$ on I implies $f(x) \leq g(x)$; and $h(z) = |z|$ implies $h(x) = |x|$. Then by using (25) we obtain the desired inequality (23). □

Theorem 12. *Let* $(x_t)_{t\in T}$ *be a bounded continuous field of self-adjoint elements in a unital* C^**-algebra* $\mathcal{A}$ *defined on a locally compact Hausdorff space* T *equipped with a bounded Radon measure* μ. *Let* m_t *and* M_t, $m_t \leq M_t$, *be the bounds of* x_t, $t \in T$. *Let* $(\phi_t)_{t\in T}$ *be a unital field of positive linear mappings* $\phi_t : \mathcal{A} \to \mathcal{B}$ *from* $\mathcal{A}$ *to another unital* C^*-*algebra* $\mathcal{B}$. *Let*

$$(m_x, M_x) \cap [m_t, M_t] = \emptyset, \quad t \subset T, \qquad \textit{and} \qquad m < M$$

where m_x *and* M_x, $m_x \leq M_x$, *be the bounds of the operator* $x = \int_T \phi_t(x_t)\, d\mu(t)$ *and*

$$m = \sup\left\{M_t\colon M_t \leq m_x, t \in T\right\}, \; M = \inf\left\{m_t\colon m_t \geq M_x, t \in T\right\}$$

If $f : I \to \mathbb{R}$ *is a continuous convex (resp. concave) function provided that the interval* I *contains all* m_t, M_t, *then*

$$f\left(\int_T \phi_t(x_t)\, d\mu(t)\right) \leq \int_T \phi_t(f(x_t))\, d\mu(t) - \delta_f \widetilde{x} \leq \int_T \phi_t(f(x_t))\, d\mu(t) \tag{26}$$

(resp.

$$f\left(\int_T \phi_t(x_t)\,d\mu(t)\right) \geq \int_T \phi_t(f(x_t))\,d\mu(t) - \delta_f \widetilde{x} \geq \int_T \phi_t(f(x_t))\,d\mu(t)\ ) \tag{27}$$

holds, where

$$\begin{aligned} \delta_f &\equiv \delta_f(\bar{m},\bar{M}) = f(\bar{m}) + f(\bar{M}) - 2f\left(\tfrac{\bar{m}+\bar{M}}{2}\right) \\ (\textit{resp.}\quad \delta_f &\equiv \delta_f(\bar{m},\bar{M}) = 2f\left(\tfrac{\bar{m}+\bar{M}}{2}\right) - f(\bar{m}) - f(\bar{M})\ ) \\ \widetilde{x} &\equiv \widetilde{x}_x(\bar{m},\bar{M}) = \tfrac{1}{2}1_K - \tfrac{1}{\bar{M}-\bar{m}}\left|x - \tfrac{\bar{m}+\bar{M}}{2}1_K\right| \end{aligned} \tag{28}$$

and $\bar{m} \in [m, m_A]$, $\bar{M} \in [M_A, M]$, $\bar{m} < \bar{M}$, *are arbitrary numbers.*

Proof. We prove only the convex case. Since $x = \int_T \phi_t(x_t)\,d\mu(t) \in \mathcal{B}$ is the self-adjoint elements such that $\bar{m}1_K \leq m_x 1_K \leq \int_T \phi_t(x_t)\,d\mu(t) \leq M_x 1_K \leq \bar{M}1_K$ and f is convex on $[\bar{m},\bar{M}] \subseteq I$, then by Lemma 11 we obtain

$$f\left(\int_T \phi_t(x_t)\,d\mu(t)\right) \leq \frac{\bar{M}1_K - \int_T \phi_t(x_t)\,d\mu(t)}{\bar{M}-\bar{m}} f(\bar{m}) + \frac{\int_T \phi_t(x_t)\,d\mu(t) - \bar{m}1_K}{\bar{M}-\bar{m}} f(\bar{M}) - \delta_f \widetilde{x} \tag{29}$$

where δ_f and $\widetilde{x}$ are defined by (28).

But since f is convex on $[m_t, M_t]$ and $(m_x, M_x) \cap [m_t, M_t] = \emptyset$ implies $(\bar{m},\bar{M}) \cap [m_t, M_t] = \emptyset$, then

$$f(x_t) \geq \frac{\bar{M}1_H - x_t}{\bar{M}-\bar{m}} f(\bar{m}) + \frac{x_t - \bar{m}1_H}{\bar{M}-\bar{m}} f(\bar{M}), \quad t \in T$$

Applying a positive linear mapping ϕ_t, integrating and adding $-\delta_f \widetilde{x}$, we obtain

$$\int_T \phi_t(f(x_t))\,d\mu(t) - \delta_f \widetilde{x} \geq \frac{\bar{M}1_K - \int_T \phi_t(x_t)\,d\mu(t)}{\bar{M}-\bar{m}} f(\bar{m}) + \frac{\int_T \phi_t(x_t)\,d\mu(t) - \bar{m}1_K}{\bar{M}-\bar{m}} f(\bar{M}) - \delta_f \widetilde{x} \tag{30}$$

since $\int_T \phi_t(1_H)\,d\mu(t) = 1_K$. Combining the two inequalities (29) and (30), we have LHS of (26). Since $\delta_f \geq 0$ and $\widetilde{x} \geq 0$, then we have RHS of (26). □

If $m < M$ and $m_x = M_x$, then the inequality (26) holds, but $\delta_f(m_x, M_x)\,\widetilde{x}(m_x, M_x)$ is not defined (see Example 13 I) and II)).

Example 13. *We give examples for the matrix cases and* $T = \{1,2\}$. *Then we have refined inequalities given in Fig. 2. We put* $f(t) = t^4$ *which is convex but not operator convex in* (26). *Also, we define mappings* $\Phi_1, \Phi_2 : M_3(\mathbb{C}) \to M_2(\mathbb{C})$ *as follows:* $\Phi_1((a_{ij})_{1\leq i,j\leq 3}) = \frac{1}{2}(a_{ij})_{1\leq i,j\leq 2}$, $\Phi_2 = \Phi_1$ *(then* $\Phi_1(I_3) + \Phi_2(I_3) = I_2$*).*

I) *First, we observe an example when* $\delta_f \widetilde{X}$ *is equal to the difference RHS and LHS of Jensen's inequality. If* $X_1 = -3I_3$ *and* $X_2 = 2I_3$, *then* $X = \Phi_1(X_1) + \Phi_2(X_2) = -0.5I_2$, *so* $m = -3$, $M = 2$. *We also put* $\bar{m} = -3$ *and* $\bar{M} = 2$. *We obtain*

$$(\Phi_1(X_1) + \Phi_2(X_2))^4 = 0.0625I_2 < 48.5I_2 = \Phi_1\left(X_1^4\right) + \Phi_2\left(X_2^4\right)$$

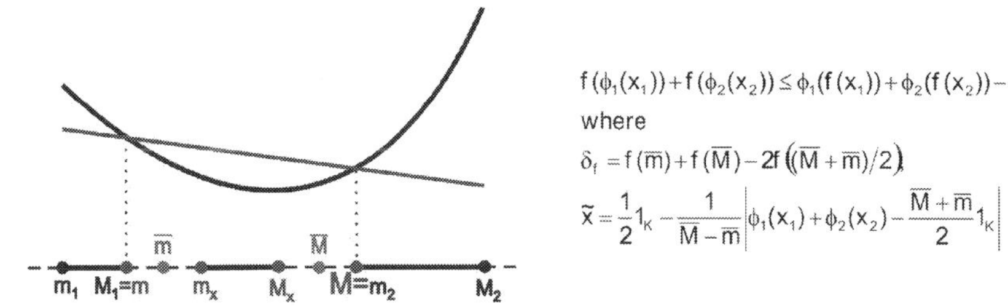

Figure 2. Refinement for two operators and a convex function f

and its improvement

$$(\Phi_1(X_1) + \Phi_2(X_2))^4 = 0.0625 I_2 = \Phi_1\left(X_1^4\right) + \Phi_2\left(X_2^4\right) - 48.4375 I_2$$

since $\delta_f = 96.875$, $\widetilde{X} = 0.5 I_2$. *We remark that in this case* $m_x = M_x = -1/2$ *and* $\widetilde{X}(m_x, M_x)$ *is not defined.*

II) *Next, we observe an example when* $\delta_f \widetilde{X}$ *is not equal to the difference RHS and LHS of Jensen's inequality and* $m_x = M_x$. *If*

$$X_1 = \begin{pmatrix} -1 & 0 & 0 \\ 0 & -2 & 0 \\ 0 & 0 & -1 \end{pmatrix}, \; X_2 = \begin{pmatrix} 2 & 0 & 0 \\ 0 & 3 & 0 \\ 0 & 0 & 4 \end{pmatrix}, \text{ then } X = \frac{1}{2}\begin{pmatrix} 1 & 0 \\ 0 & 1 \end{pmatrix} \text{ and } m = -1, \; M = 2$$

In this case $\widetilde{x}(m_x, M_x)$ *is not defined, since* $m_x = M_x = 1/2$. *We have*

$$(\Phi_1(X_1) + \Phi_2(X_2))^4 = \frac{1}{16}\begin{pmatrix} 1 & 0 \\ 0 & 1 \end{pmatrix} < \begin{pmatrix} \frac{17}{2} & 0 \\ 0 & \frac{97}{2} \end{pmatrix} = \Phi_1\left(X_1^4\right) + \Phi_2\left(X_2^4\right)$$

and putting $\bar{m} = -1$, $\bar{M} = 2$ *we obtain* $\delta_f = 135/8$, $\widetilde{X} = I_2/2$ *which give the following improvement*

$$(\Phi_1(X_1) + \Phi_2(X_2))^4 = \frac{1}{16}\begin{pmatrix} 1 & 0 \\ 0 & 1 \end{pmatrix} < \frac{1}{16}\begin{pmatrix} 1 & 0 \\ 0 & 641 \end{pmatrix} = \Phi_1\left(X_1^4\right) + \Phi_2\left(X_2^4\right) - \frac{135}{16}\begin{pmatrix} 1 & 0 \\ 0 & 1 \end{pmatrix}$$

III) *Next, we observe an example with matrices that are not special. If*

$$X_1 = \begin{pmatrix} -4 & 1 & 1 \\ 1 & -2 & -1 \\ 1 & -1 & -1 \end{pmatrix} \quad \text{and} \quad X_2 = \begin{pmatrix} 5 & -1 & -1 \\ -1 & 2 & 1 \\ -1 & 1 & 3 \end{pmatrix}, \quad \text{then} \quad X = \frac{1}{2}\begin{pmatrix} 1 & 0 \\ 0 & 0 \end{pmatrix}$$

so $m_1 = -4.8662$, $M_1 = -0.3446$, $m_2 = 1.3446$, $M_2 = 5.8662$, $m = -0.3446$, $M = 1.3446$ *and we put* $\bar{m} = m$, $\bar{M} = M$ *(rounded to four decimal places). We have*

$$(\Phi_1(X_1) + \Phi_2(X_2))^4 = \frac{1}{16}\begin{pmatrix} 1 & 0 \\ 0 & 0 \end{pmatrix} < \begin{pmatrix} \frac{1283}{2} & -255 \\ -255 & \frac{237}{2} \end{pmatrix} = \Phi_1\left(X_1^4\right) + \Phi_2\left(X_2^4\right)$$

and its improvement

$$(\Phi_1(X_1) + \Phi_2(X_2))^4 = \frac{1}{16}\begin{pmatrix} 1 & 0 \\ 0 & 0 \end{pmatrix}$$
$$< \begin{pmatrix} 639.9213 & -255 \\ -255 & 117.8559 \end{pmatrix} = \Phi_1\left(X_1^4\right) + \Phi_2\left(X_2^4\right) - \begin{pmatrix} 1.5787 & 0 \\ 0 & 0.6441 \end{pmatrix}$$

(rounded to four decimal places), since $\delta_f = 3.1574$, $\widetilde{X} = \begin{pmatrix} 0.5 & 0 \\ 0 & 0.2040 \end{pmatrix}$. *But, if we put* $\bar{m} = m_x = 0$, $\bar{M} = M_x = 0.5$, *then* $\widetilde{X} = \mathbf{0}$, *so we do not have an improvement of Jensen's inequality. Also, if we put* $\bar{m} = 0$, $\bar{M} = 1$, *then* $\widetilde{X} = 0.5\begin{pmatrix} 1 & 0 \\ 0 & 1 \end{pmatrix}$, $\delta_f = 7/8$ *and* $\delta_f\widetilde{X} = 0.4375\begin{pmatrix} 1 & 0 \\ 0 & 1 \end{pmatrix}$, *which is worse than the above improvement.*

Putting $\Phi_t(y) = a_t y$ for every $y \in \mathcal{A}$, where $a_t \geq 0$ is a real number, we obtain the following obvious corollary of Theorem 12.

Corollary 14. *Let* $(x_t)_{t\in T}$ *be a bounded continuous field of self-adjoint elements in a unital* C^**-algebra* $\mathcal{A}$ *defined on a locally compact Hausdorff space* T *equipped with a bounded Radon measure* μ. *Let* m_t *and* M_t, $m_t \leq M_t$, *be the bounds of* x_t, $t \in T$. *Let* $(a_t)_{t\in T}$ *be a continuous field of nonnegative real numbers such that* $\int_T a_t \, d\mu(t) = 1$. *Let*

$$(m_x, M_x) \cap [m_t, M_t] = \emptyset, \quad t \in T, \qquad \text{and} \qquad m < M$$

where m_x *and* M_x, $m_x \leq M_x$, *are the bounds of the operator* $x = \int_T \phi_t(x_t) \, d\mu(t)$ *and*

$$m = \sup\{M_t \colon M_t \leq m_x, t \in T\}, \; M = \inf\{m_t \colon m_t \geq M_x, t \in T\}$$

If $f : I \to \mathbb{R}$ *is a continuous convex (resp. concave) function provided that the interval* I *contains all* m_t, M_t, *then*

$$f\left(\int_T a_t x_t \, d\mu(t)\right) \leq \int_T a_t f(x_t) \, d\mu(t) - \delta_f \tilde{\tilde{x}} \leq \int_T a_t f(x_t) \, d\mu(t)$$
$$(\text{resp.} \;\; f\left(\int_T a_t x_t \, d\mu(t)\right) \geq \int_T a_t f(x_t) \, d\mu(t) + \delta_f \tilde{\tilde{x}} \geq \int_T a_t f(x_t) \, d\mu(t) \;)$$

holds, where δ_f *is defined by* (28), $\tilde{\tilde{x}} = \frac{1}{2}1_H - \frac{1}{\bar{M}-\bar{m}}\left|\int_T a_t x_t \, d\mu(t) - \frac{\bar{m}+\bar{M}}{2}1_H\right|$ *and* $\bar{m} \in [m, m_A]$, $\bar{M} \in [M_A, M]$, $\bar{m} < \bar{M}$, *are arbitrary numbers.*

5. Extension Jensen's inequality

In this section we present an extension of Jensen's operator inequality for $n-$tuples of self-adjoint operators, unital $n-$tuples of positive linear mappings and real valued continuous convex functions with conditions on the spectra of the operators.

In a discrete version of Theorem 6 we prove that Jensen's operator inequality holds for every continuous convex function and for every $n-$tuple of self-adjoint operators $(A_1, \ldots, A_n)$, for every $n-$tuple of positive linear mappings $(\Phi_1, \ldots, \Phi_n)$ in the case when the interval with bounds of the operator $A = \sum_{i=1}^n \Phi_i(A_i)$ has no intersection points with the interval with bounds of the operator A_i for each $i = 1, \ldots, n$, i.e. when $(m_A, M_A) \cap [m_i, M_i] = \emptyset$

for $i = 1, \ldots, n$, where m_A and M_A, $m_A \le M_A$, are the bounds of A, and m_i and M_i, $m_i \le M_i$, are the bounds of A_i, $i = 1, \ldots, n$. It is interesting to consider the case when $(m_A, M_A) \cap [m_i, M_i] = \emptyset$ is valid for several $i \in \{1, \ldots, n\}$, but not for all $i = 1, \ldots, n$. We study it in the following theorem (see [21]).

Theorem 15. *Let $(A_1, \ldots, A_n)$ be an $n-$tuple of self-adjoint operators $A_i \in B(H)$ with the bounds m_i and M_i, $m_i \le M_i$, $i = 1, \ldots, n$. Let $(\Phi_1, \ldots, \Phi_n)$ be an $n-$tuple of positive linear mappings $\Phi_i : B(H) \to B(K)$, such that $\sum_{i=1}^n \Phi_i(1_H) = 1_K$. For $1 \le n_1 < n$, we denote $m = \min\{m_1, \ldots, m_{n_1}\}$, $M = \max\{M_1, \ldots, M_{n_1}\}$ and $\sum_{i=1}^{n_1} \Phi_i(1_H) = \alpha\, 1_K$, $\sum_{i=n_1+1}^{n} \Phi_i(1_H) = \beta\, 1_K$, where $\alpha, \beta > 0$, $\alpha + \beta = 1$. If*

$$(m, M) \cap [m_i, M_i] = \emptyset, \qquad i = n_1 + 1, \ldots, n$$

and one of two equalities

$$\frac{1}{\alpha} \sum_{i=1}^{n_1} \Phi_i(A_i) = \frac{1}{\beta} \sum_{i=n_1+1}^{n} \Phi_i(A_i) = \sum_{i=1}^{n} \Phi_i(A_i)$$

is valid, then

$$\frac{1}{\alpha} \sum_{i=1}^{n_1} \Phi_i(f(A_i)) \le \sum_{i=1}^{n} \Phi_i(f(A_i)) \le \frac{1}{\beta} \sum_{i=n_1+1}^{n} \Phi_i(f(A_i)) \tag{31}$$

holds for every continuous convex function $f : I \to \mathbb{R}$ provided that the interval I contains all m_i, M_i, $i = 1, \ldots, n$. If $f : I \to \mathbb{R}$ is concave, then the reverse inequality is valid in (31).

Proof. We prove only the case when f is a convex function. Let us denote

$$A = \frac{1}{\alpha} \sum_{i=1}^{n_1} \Phi_i(A_i), \qquad B = \frac{1}{\beta} \sum_{i=n_1+1}^{n} \Phi_i(A_i), \qquad C = \sum_{i=1}^{n} \Phi_i(A_i)$$

It is easy to verify that $A = B$ or $B = C$ or $A = C$ implies $A = B = C$.
a) Let $m < M$. Since f is convex on $[m, M]$ and $[m_i, M_i] \subseteq [m, M]$ for $i = 1, \ldots, n_1$, then

$$f(z) \le \frac{M - z}{M - m} f(m) + \frac{z - m}{M - m} f(M), \quad z \in [m_i, M_i] \text{ for } i = 1, \ldots, n_1 \tag{32}$$

but since f is convex on all $[m_i, M_i]$ and $(m, M) \cap [m_i, M_i] = \emptyset$ for $i = n_1 + 1, \ldots, n$, then

$$f(z) \ge \frac{M - z}{M - m} f(m) + \frac{z - m}{M - m} f(M), \quad z \in [m_i, M_i] \text{ for } i = n_1 + 1, \ldots, n \tag{33}$$

Since $m_i 1_H \le A_i \le M_i 1_H$, $i = 1, \ldots, n_1$, it follows from (32)

$$f(A_i) \le \frac{M 1_H - A_i}{M - m} f(m) + \frac{A_i - m 1_H}{M - m} f(M), \qquad i = 1, \ldots, n_1$$

Applying a positive linear mapping Φ_i and summing, we obtain

$$\sum_{i=1}^{n_1} \Phi_i(f(A_i)) \le \frac{M\alpha 1_K - \sum_{i=1}^{n_1} \Phi_i(A_i)}{M - m} f(m) + \frac{\sum_{i=1}^{n_1} \Phi_i(A_i) - m\alpha 1_K}{M - m} f(M)$$

since $\sum_{i=1}^{n_1} \Phi_i(1_H) = \alpha 1_K$. It follows

$$\frac{1}{\alpha}\sum_{i=1}^{n_1} \Phi_i\left(f(A_i)\right) \le \frac{M1_K - A}{M-m} f(m) + \frac{A - m1_K}{M-m} f(M) \tag{34}$$

Similarly to (34) in the case $m_i 1_H \le A_i \le M_i 1_H$, $i = n_1+1, \dots, n$, it follows from (33)

$$\frac{1}{\beta}\sum_{i=n_1+1}^{n} \Phi_i\left(f(A_i)\right) \ge \frac{M1_K - B}{M-m} f(m) + \frac{B - m1_K}{M-m} f(M) \tag{35}$$

Combining (34) and (35) and taking into account that $A = B$, we obtain

$$\frac{1}{\alpha}\sum_{i=1}^{n_1} \Phi_i\left(f(A_i)\right) \le \frac{1}{\beta}\sum_{i=n_1+1}^{n} \Phi_i\left(f(A_i)\right) \tag{36}$$

It follows

$$\begin{aligned}
\frac{1}{\alpha}\sum_{i=1}^{n_1} \Phi_i(f(A_i)) &= \sum_{i=1}^{n_1} \Phi_i(f(A_i)) + \frac{\beta}{\alpha}\sum_{i=1}^{n_1} \Phi_i(f(A_i)) && (\text{by } \alpha+\beta=1)\\
&\le \sum_{i=1}^{n_1} \Phi_i(f(A_i)) + \sum_{i=n_1+1}^{n} \Phi_i(f(A_i)) && (\text{by (36)})\\
&= \sum_{i=1}^{n} \Phi_i(f(A_i))\\
&\le \frac{\alpha}{\beta}\sum_{i=n_1+1}^{n} \Phi_i(f(A_i)) + \sum_{i=n_1+1}^{n} \Phi_i(f(A_i)) && (\text{by (36)})\\
&= \frac{1}{\beta}\sum_{i=n_1+1}^{n} \Phi_i(f(A_i)) && (\text{by } \alpha+\beta=1)
\end{aligned}$$

which gives the desired double inequality (31).
b) Let $m = M$. Since $[m_i, M_i] \subseteq [m, M]$ for $i = 1, \dots, n_1$, then $A_i = m1_H$ and $f(A_i) = f(m)1_H$ for $i = 1, \dots, n_1$. It follows

$$\frac{1}{\alpha}\sum_{i=1}^{n_1} \Phi_i(A_i) = m1_K \quad \text{and} \quad \frac{1}{\alpha}\sum_{i=1}^{n_1} \Phi_i\left(f(A_i)\right) = f(m)1_K \tag{37}$$

On the other hand, since f is convex on I, we have

$$f(z) \ge f(m) + l(m)(z-m) \quad \text{for every } z \in I \tag{38}$$

where l is the subdifferential of f. Replacing z by A_i for $i = n_1+1, \dots, n$, applying Φ_i and summing, we obtain from (38) and (37)

$$\begin{aligned}
\frac{1}{\beta}\sum_{i=n_1+1}^{n} \Phi_i\left(f(A_i)\right) &\ge f(m)1_K + l(m)\left(\frac{1}{\beta}\sum_{i=n_1+1}^{n} \Phi_i(A_i) - m1_K\right)\\
&= f(m)1_K = \frac{1}{\alpha}\sum_{i=1}^{n_1} \Phi_i\left(f(A_i)\right)
\end{aligned}$$

So (36) holds again. The remaining part of the proof is the same as in the case a). □

Remark 16. *We obtain the equivalent inequality to the one in Theorem 15 in the case when* $\sum_{i=1}^{n}\Phi_i(1_H)=\gamma 1_K$, *for some positive scalar* γ. *If* $\alpha+\beta=\gamma$ *and one of two equalities*

$$\frac{1}{\alpha}\sum_{i=1}^{n_1}\Phi_i(A_i)=\frac{1}{\beta}\sum_{i=n_1+1}^{n}\Phi_i(A_i)=\frac{1}{\gamma}\sum_{i=1}^{n}\Phi_i(A_i)$$

is valid, then

$$\frac{1}{\alpha}\sum_{i=1}^{n_1}\Phi_i(f(A_i))\leq\frac{1}{\gamma}\sum_{i=1}^{n}\Phi_i(f(A_i))\leq\frac{1}{\beta}\sum_{i=n_1+1}^{n}\Phi_i(f(A_i))$$

holds for every continuous convex function f.

Remark 17. *Let the assumptions of Theorem 15 be valid.*

1. *We observe that the following inequality*

$$f\left(\frac{1}{\beta}\sum_{i=n_1+1}^{n}\Phi_i(A_i)\right)\leq\frac{1}{\beta}\sum_{i=n_1+1}^{n}\Phi_i(f(A_i))$$

holds for every continuous convex function $f: I\to\mathbb{R}$.

Indeed, by the assumptions of Theorem 15 we have

$$m\alpha 1_H\leq\sum_{i=1}^{n_1}\Phi_i(f(A_i))\leq M\alpha 1_H \quad and \quad \frac{1}{\alpha}\sum_{i=1}^{n_1}\Phi_i(A_i)=\frac{1}{\beta}\sum_{i=n_1+1}^{n}\Phi_i(A_i)$$

which implies

$$m1_H\leq\sum_{i=n_1+1}^{n}\frac{1}{\beta}\Phi_i(f(A_i))\leq M1_H$$

Also $(m,M)\cap[m_i,M_i]=\emptyset$ *for* $i=n_1+1,\dots,n$ *and* $\sum_{i=n_1+1}^{n}\frac{1}{\beta}\Phi_i(1_H)=1_K$ *hold. So we can apply Theorem 6 on operators* $A_{n_1+1},\dots,A_n$ *and mappings* $\frac{1}{\beta}\Phi_i$ *and obtain the desired inequality.*

2. *We denote by* m_C *and* M_C *the bounds of* $C=\sum_{i=1}^{n}\Phi_i(A_i)$. *If* $(m_C,M_C)\cap[m_i,M_i]=\emptyset$, $i=1,\dots,n_1$ *or* f *is an operator convex function on* $[m,M]$, *then the double inequality (31) can be extended from the left side if we use Jensen's operator inequality (see [16, Theorem 2.1])*

$$\begin{aligned}f\left(\sum_{i=1}^{n}\Phi_i(A_i)\right)&=f\left(\frac{1}{\alpha}\sum_{i=1}^{n_1}\Phi_i(A_i)\right)\\&\leq\frac{1}{\alpha}\sum_{i=1}^{n_1}\Phi_i(f(A_i))\leq\sum_{i=1}^{n}\Phi_i(f(A_i))\leq\frac{1}{\beta}\sum_{i=n_1+1}^{n}\Phi_i(f(A_i))\end{aligned}$$

Example 18. *If neither assumptions* $(m_C,M_C)\cap[m_i,M_i]=\emptyset$, $i=1,\dots,n_1$, *nor* f *is operator convex in Remark 17 - 2. is satisfied and if* $1<n_1<n$, *then (31) can not be extended by Jensen's operator inequality, since it is not valid. Indeed, for* $n_1=2$ *we define mappings* $\Phi_1,\Phi_2: M_3(\mathbb{C})\to M_2(\mathbb{C})$ *by* $\Phi_1((a_{ij})_{1\leq i,j\leq 3})=\frac{\alpha}{2}(a_{ij})_{1\leq i,j\leq 2}$, $\Phi_2=\Phi_1$. *Then* $\Phi_1(I_3)+\Phi_2(I_3)=\alpha I_2$. *If*

$$A_1=2\begin{pmatrix}1&0&1\\0&0&1\\1&1&1\end{pmatrix} \quad and \quad A_2=2\begin{pmatrix}1&0&0\\0&0&0\\0&0&0\end{pmatrix}$$

then

$$\left(\frac{1}{\alpha}\Phi_1(A_1)+\frac{1}{\alpha}\Phi_2(A_2)\right)^4=\frac{1}{\alpha^4}\begin{pmatrix}16 & 0\\ 0 & 0\end{pmatrix}\not\leq\frac{1}{\alpha}\begin{pmatrix}80 & 40\\ 40 & 24\end{pmatrix}=\frac{1}{\alpha}\Phi_1\left(A_1^4\right)+\frac{1}{\alpha}\Phi_2\left(A_2^4\right)$$

for every $\alpha \in (0,1)$. *We observe that* $f(t)=t^4$ *is not operator convex and* $(m_C, M_C)\cap[m_i, M_i]\neq\emptyset$, *since* $C = A = \frac{1}{\alpha}\Phi_1(A_1)+\frac{1}{\alpha}\Phi_2(A_2) = \frac{1}{\alpha}\begin{pmatrix}2 & 0\\ 0 & 0\end{pmatrix}$, $[m_C, M_C] = [0, 2/\alpha]$, $[m_1, M_1] \subset [-1.60388, 4.49396]$ *and* $[m_2, M_2] = [0, 2]$.

With respect to Remark 16, we obtain the following obvious corollary of Theorem 15.

Corollary 19. *Let* $(A_1,\ldots,A_n)$ *be an* $n-$*tuple of self-adjoint operators* $A_i \in B(H)$ *with the bounds* m_i *and* M_i, $m_i \leq M_i$, $i=1,\ldots,n$. *For some* $1 \leq n_1 < n$, *we denote* $m = \min\{m_1,\ldots,m_{n_1}\}$, $M = \max\{M_1,\ldots,M_{n_1}\}$. *Let* $(p_1,\ldots,p_n)$ *be an* $n-$*tuple of non-negative numbers, such that* $0 < \sum_{i=1}^{n_1} p_i = \mathbf{p_{n_1}} < \mathbf{p_n} = \sum_{i=1}^{n} p_i$. *If*

$$(m, M)\cap[m_i, M_i]=\emptyset, \qquad i=n_1+1,\ldots,n$$

and one of two equalities

$$\frac{1}{\mathbf{p_{n_1}}}\sum_{i=1}^{n_1} p_i A_i = \frac{1}{\mathbf{p_n}}\sum_{i=1}^{n} p_i A_i = \frac{1}{\mathbf{p_n}-\mathbf{p_{n_1}}}\sum_{i=n_1+1}^{n} p_i A_i$$

is valid, then

$$\frac{1}{\mathbf{p_{n_1}}}\sum_{i=1}^{n_1} p_i f(A_i) \leq \frac{1}{\mathbf{p_n}}\sum_{i=1}^{n} p_i f(A_i) \leq \frac{1}{\mathbf{p_n}-\mathbf{p_{n_1}}}\sum_{i=n_1+1}^{n} p_i f(A_i) \tag{39}$$

holds for every continuous convex function $f: I \to \mathbb{R}$ *provided that the interval* I *contains all* m_i, M_i, $i=1,\ldots,n$.

If $f: I \to \mathbb{R}$ *is concave, then the reverse inequality is valid in* (39).

As a special case of Corollary 19 we can obtain a discrete version of Corollary 7 as follows.

Corollary 20 (Discrete version of Corollary 7). *Let* $(A_1,\ldots,A_n)$ *be an* $n-$*tuple of self-adjoint operators* $A_i \in B(H)$ *with the bounds* m_i *and* M_i, $m_i \leq M_i$, $i=1,\ldots,n$. *Let* $(\alpha_1,\ldots,\alpha_n)$ *be an* $n-$*tuple of nonnegative real numbers such that* $\sum_{i=1}^{n}\alpha_i = 1$. *If*

$$(m_A, M_A)\cap[m_i, M_i]=\emptyset, \qquad i=1,\ldots,n \tag{40}$$

where m_A *and* M_A, $m_A \leq M_A$, *are the bounds of* $A=\sum_{i=1}^{n}\alpha_i A_i$, *then*

$$f\left(\sum_{i=1}^{n}\alpha_i A_i\right) \leq \sum_{i=1}^{n}\alpha_i f(A_i) \tag{41}$$

holds for every continuous convex function $f: I \to \mathbb{R}$ *provided that the interval* I *contains all* m_i, M_i.

Proof. We prove only the convex case. We define $(n+1)-$tuple of operators $(B_1,\ldots,B_{n+1})$, $B_i \in B(H)$, by $B_1 = A = \sum_{i=1}^{n}\alpha_i A_i$ and $B_i = A_{i-1}$, $i = 2,\ldots,n+1$. Then $m_{B_1} = m_A$, $M_{B_1} = M_A$ are the bounds of B_1 and $m_{B_i} = m_{i-1}$, $M_{B_i} = M_{i-1}$ are the ones of B_i, $i = 2,\ldots,n+1$. Also, we define $(n+1)-$tuple of non-negative numbers $(p_1,\ldots,p_{n+1})$ by $p_1 = 1$ and $p_i = \alpha_{i-1}$, $i = 2,\ldots,n+1$. Then $\sum_{i=1}^{n+1} p_i = 2$ and by using (40) we have

$$(m_{B_1}, M_{B_1}) \cap [m_{B_i}, M_{B_i}] = \emptyset, \qquad i = 2,\ldots,n+1 \tag{42}$$

Since

$$\sum_{i=1}^{n+1} p_i B_i = B_1 + \sum_{i=2}^{n+1} p_i B_i = \sum_{i=1}^{n}\alpha_i A_i + \sum_{i=1}^{n}\alpha_i A_i = 2B_1$$

then

$$p_1 B_1 = \frac{1}{2}\sum_{i=1}^{n+1} p_i B_i = \sum_{i=2}^{n+1} p_i B_i \tag{43}$$

Taking into account (42) and (43), we can apply Corollary 19 for $n_1 = 1$ and B_i, p_i as above, and we get

$$p_1 f(B_1) \le \frac{1}{2}\sum_{i=1}^{n+1} p_i f(B_i) \le \sum_{i=2}^{n+1} p_i f(B_i)$$

which gives the desired inequality (41). □

6. Extension of the refined Jensen's inequality

There is an extensive literature devoted to Jensen's inequality concerning different refinements and extensive results, see, for example [22–29].

In this section we present an extension of the refined Jensen's inequality obtained in Section 4 and a refinement of the same inequality obtained in Section 5.

Theorem 21. *Let $(A_1,\ldots,A_n)$ be an $n-$tuple of self-adjoint operators $A_i \in B(H)$ with the bounds m_i and M_i, $m_i \le M_i$, $i = 1,\ldots,n$. Let $(\Phi_1,\ldots,\Phi_n)$ be an $n-$tuple of positive linear mappings $\Phi_i : B(H) \to B(K)$, such that $\sum_{i=1}^{n_1}\Phi_i(1_H) = \alpha\, 1_K$, $\sum_{i=n_1+1}^{n}\Phi_i(1_H) = \beta\, 1_K$, where $1 \le n_1 < n$, $\alpha, \beta > 0$ and $\alpha + \beta = 1$. Let $m_L = \min\{m_1,\ldots,m_{n_1}\}$, $M_R = \max\{M_1,\ldots,M_{n_1}\}$ and*

$$\begin{aligned} m &= \max\{M_i\colon M_i \le m_L, i \in \{n_1+1,\ldots,n\}\} \\ M &= \min\{m_i\colon m_i \ge M_R, i \in \{n_1+1,\ldots,n\}\} \end{aligned}$$

If

$$(m_L, M_R) \cap [m_i, M_i] = \emptyset, \quad i = n_1+1,\ldots,n, \qquad \text{and} \qquad m < M$$

and one of two equalities

$$\frac{1}{\alpha}\sum_{i=1}^{n_1}\Phi_i(A_i) = \sum_{i=1}^{n}\Phi_i(A_i) = \frac{1}{\beta}\sum_{i=n_1+1}^{n}\Phi_i(A_i)$$

is valid, then

$$\begin{aligned} \frac{1}{\alpha}\sum_{i=1}^{n_1}\Phi_i(f(A_i)) &\le \frac{1}{\alpha}\sum_{i=1}^{n_1}\Phi_i(f(A_i)) + \beta\delta_f\widetilde{A} \le \sum_{i=1}^{n}\Phi_i(f(A_i)) \\ &\le \frac{1}{\beta}\sum_{i=n_1+1}^{n}\Phi_i(f(A_i)) - \alpha\delta_f\widetilde{A} \le \frac{1}{\beta}\sum_{i=n_1+1}^{n}\Phi_i(f(A_i)) \end{aligned} \tag{44}$$

holds for every continuous convex function $f : I \to \mathbb{R}$ *provided that the interval* I *contains all* m_i, M_i, $i = 1, \ldots, n$, *where*

$$\begin{aligned} \delta_f &\equiv \delta_f(\bar{m}, \bar{M}) = f(\bar{m}) + f(\bar{M}) - 2f\left(\frac{\bar{m}+\bar{M}}{2}\right) \\ \widetilde{A} &\equiv \widetilde{A}_{A,\Phi,n_1,\alpha}(\bar{m}, \bar{M}) = \frac{1}{2}1_K - \frac{1}{\alpha(\bar{M}-\bar{m})}\sum_{i=1}^{n_1}\Phi_i\left(\left|A_i - \frac{\bar{m}+\bar{M}}{2}1_H\right|\right) \end{aligned} \tag{45}$$

and $\bar{m} \in [m, m_L]$, $\bar{M} \in [M_R, M]$, $\bar{m} < \bar{M}$, *are arbitrary numbers. If* $f : I \to \mathbb{R}$ *is concave, then the reverse inequality is valid in* (44).

Proof. We prove only the convex case. Let us denote

$$A = \frac{1}{\alpha}\sum_{i=1}^{n_1}\Phi_i(A_i), \qquad B = \frac{1}{\beta}\sum_{i=n_1+1}^{n}\Phi_i(A_i), \qquad C = \sum_{i=1}^{n}\Phi_i(A_i)$$

It is easy to verify that $A = B$ or $B = C$ or $A = C$ implies $A = B = C$.
Since f is convex on $[\bar{m}, \bar{M}]$ and $\mathsf{Sp}(A_i) \subseteq [m_i, M_i] \subseteq [\bar{m}, \bar{M}]$ for $i = 1, \ldots, n_1$, it follows from Lemma 11 that

$$f(A_i) \le \frac{\bar{M}1_H - A_i}{\bar{M}-\bar{m}}f(\bar{m}) + \frac{A_i - \bar{m}1_H}{\bar{M}-\bar{m}}f(\bar{M}) - \delta_f\widetilde{A}_i, \qquad i = 1, \ldots, n_1$$

holds, where $\delta_f = f(\bar{m}) + f(\bar{M}) - 2f\left(\frac{\bar{m}+\bar{M}}{2}\right)$ and $\widetilde{A}_i = \frac{1}{2}1_H - \frac{1}{\bar{M}-\bar{m}}\left|A_i - \frac{\bar{m}+\bar{M}}{2}1_H\right|$.
Applying a positive linear mapping Φ_i and summing, we obtain

$$\begin{aligned} \textstyle\sum_{i=1}^{n_1}\Phi_i(f(A_i)) \le\ & \frac{\bar{M}\alpha 1_K - \sum_{i=1}^{n_1}\Phi_i(A_i)}{\bar{M}-\bar{m}}f(\bar{m}) + \frac{\sum_{i=1}^{n_1}\Phi_i(A_i) - \bar{m}\alpha 1_K}{\bar{M}-\bar{m}}f(\bar{M}) \\ & - \delta_f\left(\frac{\alpha}{2}1_K - \frac{1}{\bar{M}-\bar{m}}\sum_{i=1}^{n_1}\Phi_i\left(\left|A_i - \frac{\bar{m}+\bar{M}}{2}1_H\right|\right)\right) \end{aligned}$$

since $\sum_{i=1}^{n_1}\Phi_i(1_H) = \alpha 1_K$. It follows that

$$\frac{1}{\alpha}\sum_{i=1}^{n_1}\Phi_i(f(A_i)) \le \frac{\bar{M}1_K - A}{\bar{M}-\bar{m}}f(\bar{m}) + \frac{A - \bar{m}1_K}{\bar{M}-\bar{m}}f(\bar{M}) - \delta_f\widetilde{A} \tag{46}$$

where $\widetilde{A} = \frac{1}{2}1_K - \frac{1}{\alpha(\bar{M}-\bar{m})}\sum_{i=1}^{n_1}\Phi_i\left(\left|A_i - \frac{\bar{m}+\bar{M}}{2}1_H\right|\right)$.

Additionally, since f is convex on all $[m_i, M_i]$ and $(\bar{m}, \bar{M}) \cap [m_i, M_i] = \emptyset$, $i = n_1+1, \ldots, n$, then

$$f(A_i) \ge \frac{\bar{M}1_H - A_i}{\bar{M}-\bar{m}}f(\bar{m}) + \frac{A_i - \bar{m}1_H}{\bar{M}-\bar{m}}f(\bar{M}), \qquad i = n_1+1, \ldots, n$$

It follows

$$\frac{1}{\beta}\sum_{i=n_1+1}^{n}\Phi_i(f(A_i)) - \delta_f\widetilde{A} \ge \frac{\bar{M}1_K - B}{\bar{M}-\bar{m}}f(\bar{m}) + \frac{B - \bar{m}1_K}{\bar{M}-\bar{m}}f(\bar{M}) - \delta_f\widetilde{A} \tag{47}$$

Combining (46) and (47) and taking into account that $A = B$, we obtain

$$\frac{1}{\alpha}\sum_{i=1}^{n_1}\Phi_i\left(f(A_i)\right) \le \frac{1}{\beta}\sum_{i=n_1+1}^{n}\Phi_i\left(f(A_i)\right) - \delta_f\widetilde{A} \tag{48}$$

Next, we obtain

$$\begin{aligned}
&\frac{1}{\alpha}\sum_{i=1}^{n_1}\Phi_i(f(A_i)) \\
&= \sum_{i=1}^{n_1}\Phi_i(f(A_i)) + \frac{\beta}{\alpha}\sum_{i=1}^{n_1}\Phi_i(f(A_i)) \quad (\text{by } \alpha+\beta=1) \\
&\le \sum_{i=1}^{n_1}\Phi_i(f(A_i)) + \sum_{i=n_1+1}^{n}\Phi_i(f(A_i)) - \beta\delta_f\widetilde{A} \qquad (\text{by (48)}) \\
&\le \frac{\alpha}{\beta}\sum_{i=n_1+1}^{n}\Phi_i(f(A_i)) - \alpha\delta_f\widetilde{A} + \sum_{i=n_1+1}^{n}\Phi_i(f(A_i)) - \beta\delta_f\widetilde{A} \qquad (\text{by (48)}) \\
&= \frac{1}{\beta}\sum_{i=n_1+1}^{n}\Phi_i(f(A_i)) - \delta_f\widetilde{A} \quad (\text{by } \alpha+\beta=1)
\end{aligned}$$

which gives the following double inequality

$$\frac{1}{\alpha}\sum_{i=1}^{n_1}\Phi_i(f(A_i)) \le \sum_{i=1}^{n}\Phi_i(f(A_i)) - \beta\delta_f\widetilde{A} \le \frac{1}{\beta}\sum_{i=n_1+1}^{n}\Phi_i(f(A_i)) - \delta_f\widetilde{A}$$

Adding $\beta\delta_f\widetilde{A}$ in the above inequalities, we get

$$\frac{1}{\alpha}\sum_{i=1}^{n_1}\Phi_i(f(A_i)) + \beta\delta_f\widetilde{A} \le \sum_{i=1}^{n}\Phi_i(f(A_i)) \le \frac{1}{\beta}\sum_{i=n_1+1}^{n}\Phi_i(f(A_i)) - \alpha\delta_f\widetilde{A} \tag{49}$$

Now, we remark that $\delta_f > 0$ and $\widetilde{A} \ge 0$. (Indeed, since f is convex, then $f\left((\bar{m}+\bar{M})/2\right) \le (f(\bar{m})+f(\bar{M}))/2$, which implies that $\delta_f \ge 0$. Also, since

$$\mathsf{Sp}(A_i) \subseteq [\bar{m}, \bar{M}] \quad \Rightarrow \quad \left|A_i - \frac{\bar{M}+\bar{m}}{2}1_H\right| \le \frac{\bar{M}-\bar{m}}{2}1_H, \qquad i = 1,\ldots,n_1$$

then

$$\sum_{i=1}^{n_1}\Phi_i\left(\left|A_i - \frac{\bar{M}+\bar{m}}{2}1_H\right|\right) \le \frac{\bar{M}-\bar{m}}{2}\alpha 1_K$$

which gives

$$0 \le \frac{1}{2}1_K - \frac{1}{\alpha(\bar{M}-\bar{m})}\sum_{i=1}^{n_1}\Phi_i\left(\left|A_i - \frac{\bar{M}+\bar{m}}{2}1_H\right|\right) = \widetilde{A}\ )$$

Consequently, the following inequalities

$$\frac{1}{\alpha}\sum_{i=1}^{n_1}\Phi_i(f(A_i)) \le \frac{1}{\alpha}\sum_{i=1}^{n_1}\Phi_i(f(A_i)) + \beta\delta_f\widetilde{A}$$
$$\frac{1}{\beta}\sum_{i=n_1+1}^{n}\Phi_i(f(A_i)) - \alpha\delta_f\widetilde{A} \le \frac{1}{\beta}\sum_{i=n_1+1}^{n}\Phi_i(f(A_i))$$

hold, which with (49) proves the desired series inequalities (44). □

Example 22. *We observe the matrix case of Theorem 21 for $f(t) = t^4$, which is the convex function but not operator convex, $n = 4$, $n_1 = 2$ and the bounds of matrices as in Fig. 3. We show an example*

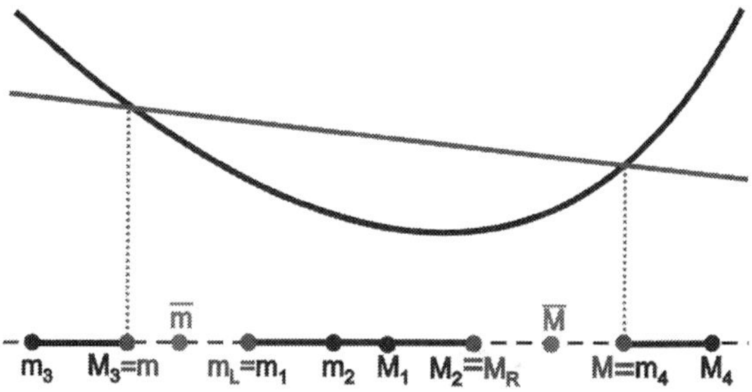

Figure 3. An example a convex function and the bounds of four operators

such that

$$\begin{aligned}
&\tfrac{1}{\alpha}\left(\Phi_1(A_1^4)+\Phi_2(A_2^4)\right) < \tfrac{1}{\alpha}\left(\Phi_1(A_1^4)+\Phi_2(A_2^4)\right) + \beta\delta_f\widetilde{A}\\
&\quad < \Phi_1(A_1^4)+\Phi_2(A_2^4)+\Phi_3(A_3^4)+\Phi_4(A_4^4)\\
&< \tfrac{1}{\beta}\left(\Phi_3(A_3^4)+\Phi_4(A_4^4)\right) - \alpha\delta_f\widetilde{A} < \tfrac{1}{\beta}\left(\Phi_3(A_3^4)+\Phi_4(A_4^4)\right)
\end{aligned} \tag{50}$$

holds, where $\delta_f = \bar{M}^4 + \bar{m}^4 - (\bar{M}+\bar{m})^4/8$ and

$$\widetilde{A} = \frac{1}{2}I_2 - \frac{1}{\alpha(\bar{M}-\bar{m})}\left(\Phi_1\left(|A_1 - \frac{\bar{M}+\bar{m}}{2}I_h|\right) + \Phi_2\left(|A_2 - \frac{\bar{M}+\bar{m}}{2}I_3|\right)\right)$$

We define mappings $\Phi_i : M_3(\mathbb{C}) \to M_2(\mathbb{C})$ as follows: $\Phi_i((a_{jk})_{1\le j,k\le 3}) = \frac{1}{4}(a_{jk})_{1\le j,k\le 2}$, $i = 1,\dots,4$. Then $\sum_{i=1}^{4}\Phi_i(I_3) = I_2$ and $\alpha = \beta = \frac{1}{2}$.

Let

$$A_1 = 2\begin{pmatrix} 2 & 9/8 & 1\\ 9/8 & 2 & 0\\ 1 & 0 & 3\end{pmatrix}, A_2 = 3\begin{pmatrix} 2 & 9/8 & 0\\ 9/8 & 1 & 0\\ 0 & 0 & 2\end{pmatrix}, A_3 = -3\begin{pmatrix} 4 & 1/2 & 1\\ 1/2 & 4 & 0\\ 1 & 0 & 2\end{pmatrix}, A_4 = 12\begin{pmatrix} 5/3 & 1/2 & 0\\ 1/2 & 3/2 & 0\\ 0 & 0 & 3\end{pmatrix}$$

Then $m_1 = 1.28607$, $M_1 = 7.70771$, $m_2 = 0.53777$, $M_2 = 5.46221$, $m_3 = -14.15050$, $M_3 = -4.71071$, $m_4 = 12.91724$, $M_4 = 36.$, so $m_L = m_2$, $M_R = M_1$, $m = M_3$ and $M = m_4$ (rounded to *five decimal places). Also,*

$$\frac{1}{\alpha}\left(\Phi_1(A_1)+\Phi_2(A_2)\right)=\frac{1}{\beta}\left(\Phi_3(A_3)+\Phi_4(A_4)\right)=\begin{pmatrix}4 & 9/4\\ 9/4 & 3\end{pmatrix}$$

and

$$A_f \equiv \frac{1}{\alpha}\left(\Phi_1(A_1^4)+\Phi_2(A_2^4)\right)=\begin{pmatrix}989.00391 & 663.46875\\ 663.46875 & 526.12891\end{pmatrix}$$

$$C_f \equiv \Phi_1(A_1^4)+\Phi_2(A_2^4)+\Phi_3(A_3^4)+\Phi_4(A_4^4)=\begin{pmatrix}68093.14258 & 48477.98437\\ 48477.98437 & 51335.39258\end{pmatrix}$$

$$B_f \equiv \frac{1}{\beta}\left(\Phi_3(A_3^4)+\Phi_4(A_4^4)\right)=\begin{pmatrix}135197.28125 & 96292.5\\ 96292.5 & 102144.65625\end{pmatrix}$$

Then

$$A_f < C_f < B_f \tag{51}$$

holds (which is consistent with (31)*).*

We will choose three pairs of numbers $(\bar{m},\bar{M})$, $\bar{m}\in[-4.71071, 0.53777]$, $\bar{M}\in[7.70771, 12.91724]$ *as follows*

i) $\bar{m}=m_L=0.53777$, $\bar{M}=M_R=7.70771$, *then*

$$\widetilde{\Delta}_1=\beta\delta_f\widetilde{A}=0.5\cdot 2951.69249\cdot\begin{pmatrix}0.15678 & 0.09030\\ 0.09030 & 0.15943\end{pmatrix}=\begin{pmatrix}231.38908 & 133.26139\\ 133.26139 & 235.29515\end{pmatrix}$$

ii) $\bar{m}=m=-4.71071$, $\bar{M}=M=12.91724$, *then*

$$\widetilde{\Delta}_2=\beta\delta_f\widetilde{A}=0.5\cdot 27766.07963\cdot\begin{pmatrix}0.36022 & 0.03573\\ 0.03573 & 0.36155\end{pmatrix}=\begin{pmatrix}5000.89860 & 496.04498\\ 496.04498 & 5019.50711\end{pmatrix}$$

iii) $\bar{m}=-1$, $\bar{M}=10$, *then*

$$\widetilde{\Delta}_3=\beta\delta_f\widetilde{A}=0.5\cdot 9180.875\cdot\begin{pmatrix}0.28203 & 0.08975\\ 0.08975 & 0.27557\end{pmatrix}=\begin{pmatrix}1294.66 & 411.999\\ 411.999 & 1265.\end{pmatrix}$$

New, we obtain the following improvement of (51) *(see* (50)*)*

$$\textbf{\textit{i)}}\quad A_f < A_f+\widetilde{\Delta}_1=\begin{pmatrix}1220.39299 & 796.73014\\ 796.73014 & 761.42406\end{pmatrix}$$
$$< C_f < \begin{pmatrix}134965.89217 & 96159.23861\\ 96159.23861 & 101909.36110\end{pmatrix}=B_f-\widetilde{\Delta}_1<B_f$$

$$\textbf{\textit{ii)}}\quad A_f < A_f+\widetilde{\Delta}_2=\begin{pmatrix}5989.90251 & 1159.51373\\ 1159.51373 & 5545.63601\end{pmatrix}$$
$$< C_f < \begin{pmatrix}130196.38265 & 95796.45502\\ 95796.45502 & 97125.14914\end{pmatrix}=B_f-\widetilde{\Delta}_2<B_f$$

$$\textbf{\textit{iii)}}\quad A_f < A_f+\widetilde{\Delta}_3=\begin{pmatrix}2283.66362 & 1075.46746\\ 1075.46746 & 1791.12874\end{pmatrix}$$
$$< C_f < \begin{pmatrix}133902.62153 & 95880.50129\\ 95880.50129 & 100879.65641\end{pmatrix}=B_f-\widetilde{\Delta}_3<B_f$$

Using Theorem 21 we get the following result.

Corollary 23. *Let the assumptions of Theorem 21 hold. Then*

$$\frac{1}{\alpha}\sum_{i=1}^{n_1}\Phi_i(f(A_i)) \le \frac{1}{\alpha}\sum_{i=1}^{n_1}\Phi_i(f(A_i)) + \gamma_1\delta_f\widetilde{A} \le \frac{1}{\beta}\sum_{i=n_1+1}^{n}\Phi_i(f(A_i)) \tag{52}$$

and

$$\frac{1}{\alpha}\sum_{i=1}^{n_1}\Phi_i(f(A_i)) \le \frac{1}{\beta}\sum_{i=n_1+1}^{n}\Phi_i(f(A_i)) - \gamma_2\delta_f\widetilde{A} \le \frac{1}{\beta}\sum_{i=n_1+1}^{n}\Phi_i(f(A_i)) \tag{53}$$

holds for every γ_1, γ_2 in the close interval joining α and β, where δ_f and $\widetilde{A}$ are defined by (45).

Proof. Adding $\alpha\delta_f\widetilde{A}$ in (44) and noticing $\delta_f\widetilde{A} \ge 0$, we obtain

$$\frac{1}{\alpha}\sum_{i=1}^{n_1}\Phi_i(f(A_i)) \le \frac{1}{\alpha}\sum_{i=1}^{n_1}\Phi_i(f(A_i)) + \alpha\delta_f\widetilde{A} \le \frac{1}{\beta}\sum_{i=n_1+1}^{n}\Phi_i(f(A_i))$$

Taking into account the above inequality and the left hand side of (44) we obtain (52). Similarly, subtracting $\beta\delta_f\widetilde{A}$ in (44) we obtain (53). □

Remark 24. *We can obtain extensions of inequalities which are given in Remark 16 and 17. Also, we can obtain a special case of Theorem 21 with the convex combination of operators A_i putting $\Phi_i(B) = \alpha_i B$, for $i = 1, \dots, n$, similarly as in Corollary 19. Finally, applying this result, we can give another proof of Corollary 14. The interested reader can see the details in [30].*

Author details

Jadranka Mićić
Faculty of Mechanical Engineering and Naval Architecture, University of Zagreb, Ivana Lučića 5, 10000 Zagreb, Croatia

Josip Pečarić
Faculty of Textile Technology, University of Zagreb, Prilaz baruna Filipovića 30, 10000 Zagreb, Croatia

7. References

[1] Davis C (1957) A Schwarz inequality for convex operator functions. Proc. Amer. Math. Soc. 8: 42-44.
[2] Choi M.D (1974) A Schwarz inequality for positive linear maps on C*-algebras. Illinois J. Math. 18: 565-574.
[3] Hansen F, Pedersen G.K (1982) Jensen's inequality for operators and Löwner's theorem. Math. Ann. 258: 229-241.
[4] Mond B, Pečarić J (1995) On Jensen's inequality for operator convex functions. Houston J. Math. 21: 739-754.
[5] Hansen F, Pedersen G.K (2003) Jensen's operator inequality. Bull. London Math. Soc. 35: 553-564.

[6] Mond B, Pečarić J (1994) Converses of Jensen's inequality for several operators, Rev. Anal. Numér. Théor. Approx. 23: 179-183.

[7] Mond B, Pečarić J.E (1993) Converses of Jensen's inequality for linear maps of operators, An. Univ. Vest Timiş. Ser. Mat.-Inform. XXXI 2: 223-228.

[8] Furuta T (1998) Operator inequalities associated with Hölder-McCarthy and Kantorovich inequalities. J. Inequal. Appl. 2: 137-148.

[9] Mićić J, Seo Y, Takahasi S.E, Tominaga M (1999) Inequalities of Furuta and Mond-Pečarić. Math. Inequal. Appl. 2: 83-111.

[10] Mićić J, Pečarić J, Seo Y, Tominaga M (2000) Inequalities of positive linear maps on Hermitian matrices. Math. Inequal. Appl. 3: 559-591.

[11] Fujii M, Mićić Hot J, Pečarić J, Seo Y (2012) Recent Developments of Mond-Pečarić method in Operator Inequalities. Zagreb: Element. 320 p. In print.

[12] Fujii J.I (2011) An external version of the Jensen operator inequality. Sci. Math. Japon. Online e-2011: 59-62.

[13] Fujii J.I, Pečarić J, Seo Y (2012) The Jensen inequality in an external formula. J. Math. Inequal. Accepted for publication.

[14] Hansen F, Pečarić J, Perić I (2007) Jensen's operator inequality and it's converses. Math. Scand. 100: 61-73.

[15] Furuta T, Mićić Hot J, Pečarić J, Seo Y (2005) Mond-Pečarić Method in Operator Inequalities. Zagreb: Element. 262 p.

[16] Mićić J, Pečarić J, Seo Y (2010) Converses of Jensen's operator inequality. Oper. Matrices 4: 385-403.

[17] Mićić J, Pavić Z, Pečarić J (2011) Some better bounds in converses of the Jensen operator inequality. Oper. Matrices. In print.

[18] Mićić J, Pavić Z, Pečarić J (2011) Jensen's inequality for operators without operator convexity. Linear Algebra Appl. 434: 1228-1237.

[19] Mićić J, Pečarić J, Perić J (2012) Refined Jensen's operator inequality with condition on spectra. Oper. Matrices. Accepted for publication.

[20] Mitrinović D.S, Pečarić J.E, Fink A.M (1993) Classical and New Inequalities in Analysis. Dordrecht-Boston-London: Kluwer Acad. Publ. 740 p.

[21] Mićić J, Pavić Z, Pečarić J (2011) Extension of Jensen's operator inequality for operators without operator convexity. Abstr. Appl. Anal. 2011: 1-14.

[22] Abramovich S, Jameson G, Sinnamon G (2004) Refining Jensen's inequality, Bull. Math. Soc. Sci. Math. Roumanie (N.S.) 47: 3-14.

[23] Dragomir S.S (2010) A new refinement of Jensen's inequality in linear spaces with applications. Math. Comput. Modelling 52: 1497-1505.

[24] Khosravi M, Aujla J.S, Dragomir S.S, Moslehian M.S (2011) Refinements of Choi-Davis-Jensen's inequality. Bull. Math. Anal. Appl. 3: 127-133.

[25] Moslehian M.S (2009) Operator extensions of Hua's inequality. Linear Algebra Appl. 430: 1131-1139.

[26] Rooin J (2005) A refinement of Jensen's inequality, J. Ineq. Pure and Appl. Math. 6. 2. Art. 38: 4 p.

[27] Srivastava H.M, Xia Z.G, Zhang Z.H (2011) Some further refinements and extensions of the Hermite-Hadamard and Jensen inequalities in several variables. Math. Comput. Modelling 54: 2709-2717.

[28] Xiao Z.G, Srivastava H.M, Zhang Z.H (2010) Further refinements of the Jensen inequalities based upon samples with repetitions. Math. Comput. Modelling 51: 592-600.

[29] Wang L.C, Ma X.F, Liu L.H (2009) A note on some new refinements of Jensen's inequality for convex functions. J. Inequal. Pure Appl. Math. 10. 2. Art. 48: 6 p.

[30] Mićić J, Pečarić J, Perić J (2012) Extension of the refined Jensen's operator inequality with condition on spectra. Ann. Funct. Anal. 3: 67-85.

Chapter 10

A Linear System of Both Equations and Inequalities in Max-Algebra

Abdulhadi Aminu

Additional information is available at the end of the chapter

http://dx.doi.org/10.5772/48195

1. Introduction

The aim of this chapter is to present a system of linear equation and inequalities in max-algebra. Max-algebra is an analogue of linear algebra developed on the pair of operations $(\oplus, \otimes)$ extended to matrices and vectors, where $a \oplus b = max(a, b)$ and $a \otimes b = a + b$ for $a, b \in \mathbb{R}$. The system of equations $A \otimes x = c$ and inequalities $B \otimes x \leq d$ have each been studied in the literature. We will present necessary and sufficient conditions for the solvability of a system consisting of these two systems and also develop a polynomial algorithm for solving max-linear program whose constraints are max-linear equations and inequalities. Moreover, some solvability concepts of an inteval system of linear equations and inequalities will also be presented.

Max-algebraic linear systems were investigated in the first publications which deal with the introduction of algebraic structures called (max,+) algebras. Systems of equations with variables only on one side were considered in these publications [1, 2] and [3]. Other systems with a special structure were investigated in the context of solving eigenvalue problems in correspondence with algebraic structures or synchronisation of discrete event systems, see [4] and also [1] for additional information. Given a matrix A, a vector b of an appropriate size, using the notation $\oplus = \max$, $\otimes = \text{plus}$, the studied systems had one of the following forms: $A \otimes x = b$, $A \otimes x = x$ or $A \otimes x = x \oplus b$. An infinite dimensional generalisation can be found in [5].

In [1] Cuninghame-Green showed that the problem $A \otimes x = b$ can be solved using residuation [6]. That is the equality in $A \otimes x = b$ be relaxed so that the set of its sub-solutions is studied. It was shown that the greatest solution of $A \otimes x \leq b$ is given by $\bar{x}$ where

$$\bar{x}_j = \min_{i \in M}(b_i \otimes a_{ij}^{-1}) \text{ for all } j \in N$$

The equation $A \otimes x = b$ is also solved using the above result as follows: The equation $A \otimes x = b$ has solution if and only if $A \otimes \bar{x} = b$. Also, Gaubert [7] proposed a method for solving the one-sided system $x = A \otimes x \oplus b$ using rational calculus.

Zimmermann [3] developed a method for solving $A \otimes x = b$ by set covering and also presented an algorithm for solving max-linear programs with one sided constraints. This method is proved to has a computational complexity of $O(mn)$, where m, n are the number of rows and columns of input matrices respectively. Akian, Gaubert and Kolokoltsov [5] extended Zimmermann's solution method by set covering to the case of functional Galois connections.

Butkovic [8] developed a max-algebraic method for finding all solutions to a system of inequalities $x_i - x_j > b_{ij}, i, j = 1, ..., n$ using n generators. Using this method Butkovic [8] developed a pseudopolynomial algorithm which either finds a bounded mixed-integer solution, or decides that no such solution exists. Summary of these results can be found in [9] and [10]

Cechlárova and Diko [11] proposed a method for resolving infeasibility of the system $A \otimes x = b$. The techniques presented in this method are to modify the right-hand side as little as possible or to omit some equations. It was shown that the problem of finding the minimum number of those equations is NP-complete.

2. Max-algebra and some basic definitions

In this section we introduce max-algebra, give the essential definitions and show how the operations of max-algebra can be extended to matrices and vectors.

In max-algebra, we replace addition and multiplication, the binary operations in conventional linear algebra, by maximum and addition respectively. For any problem that involves adding numbers together and taking the maximum of numbers, it may be possible to describe it in max-algebra. A problem that is nonlinear when described in conventional terms may be converted to a max-algebraic problem that is linear with respect to $(\oplus, \otimes) = (\max, +)$.

Definition 1. The max-plus semiring $\overline{\mathbb{R}}$ is the set $\mathbb{R} \cup \{-\infty\}$, equipped with the addition $(a, b) \mapsto \max(a, b)$ and multiplication $(a, b) \mapsto a + b$ denoted by $\oplus$ and $\otimes$ respectively. That is $a \oplus b = \max(a, b)$ and $a \otimes b = a + b$. The identity element for the addition (or zero) is $-\infty$, and the identity element for the multiplication (or unit) is 0.

Definition 2. The min-plus semiring $\mathbb{R}_{\min}$ is the set $\mathbb{R} \cup \{+\infty\}$, equipped with the addition $(a, b) \mapsto \min(a, b)$ and multiplication $(a, b) \mapsto a + b$ denoted by $\oplus'$ and $\otimes'$ respectively. The zero is $+\infty$, and the unit is 0. The name tropical semiring is also used as a synonym of min-plus when the ground set is $\mathbb{N}$.

The completed max-plus semiring $\overline{\mathbb{R}}_{max}$ is the set $\mathbb{R} \cup \{\pm\infty\}$, equipped with the addition $(a, b) \mapsto \max(a, b)$ and multiplication $(a, b) \mapsto a + b$, with the convention that $-\infty + (+\infty) = +\infty + (-\infty) = -\infty$. The completed min-plus semiring $\overline{\mathbb{R}}_{\min}$ is defined in the dual way.

Proposition 1. The following properties hold for all $a, b, c \in \overline{\mathbb{R}}$:

$$a \oplus b = b \oplus a$$
$$a \otimes b = b \otimes a$$
$$a \oplus (b \oplus c) = (a \oplus b) \oplus c$$
$$a \otimes (b \otimes c) = (a \otimes b) \otimes c$$

$$a \otimes (b \oplus c) = a \otimes b \oplus a \otimes c$$

$$a \oplus (-\infty) = -\infty = (-\infty) \oplus a$$

$$a \otimes 0 = a = 0 \otimes a$$

$$a \otimes a^{-1} = 0, a, a^{-1} \in \mathbb{R}$$

Proof.
The statements follow from the definitions. □

Proposition 2. For all $a, b, c \in \overline{\mathbb{R}}$ the following properties hold:

$$a \leq b \Longrightarrow a \oplus c \leq b \oplus c$$

$$a \leq b \Longleftrightarrow a \otimes c \leq b \otimes c, c \in \mathbb{R}$$

$$a \leq b \Longleftrightarrow a \oplus b = b$$

$$a > b \Longleftrightarrow a \otimes c > b \otimes c, \ -\infty < c < +\infty$$

Proof. The statements follow from definitions. □

The pair of operations $(\oplus,\otimes)$ is extended to matrices and vectors as in the conventional linear algebra as follows: For $A = (a_{ij})$, $B = (b_{ij})$ of compatible sizes and $\alpha \in \mathbb{R}$ we have:

$$A \oplus B = (a_{ij} \oplus b_{ij})$$

$$A \otimes B = \left(\sum_k^{\oplus} a_{ik} \otimes b_{kj} \right)$$

$$\alpha \otimes A = (\alpha \otimes a_{ij})$$

Example 1.

$$\begin{pmatrix} 3 & 1 & 5 \\ 2 & 1 & 5 \end{pmatrix} \oplus \begin{pmatrix} -1 & 0 & 2 \\ 6 & -5 & 4 \end{pmatrix} = \begin{pmatrix} 3 & 1 & 5 \\ 6 & 1 & 5 \end{pmatrix}$$

Example 2.

$$\begin{pmatrix} -4 & 1 & -5 \\ 3 & 0 & 8 \end{pmatrix} \otimes \begin{pmatrix} -1 & 2 \\ 1 & 7 \\ 3 & 1 \end{pmatrix}$$
$$= \begin{pmatrix} (-4+(-1)) \oplus (1+1) \oplus (-5+3) & (-4+2) \oplus (1+7) \oplus (-5+1) \\ (3+(-1)) \oplus (0+1) \oplus (8+3) & (3+2) \oplus (0+7) \oplus (8+1) \end{pmatrix} = \begin{pmatrix} 2 & 8 \\ 11 & 9 \end{pmatrix}$$

Example 3.

$$10 \otimes \begin{pmatrix} 7 & -3 & 2 \\ 6 & 1 & 0 \end{pmatrix} = \begin{pmatrix} 17 & 7 & 12 \\ 16 & 11 & 10 \end{pmatrix}$$

Proposition 3.
For $A, B, C \in \overline{\mathbb{R}}^{m\times n}$ of compatible sizes, the following properties hold:

$$\begin{aligned} A \oplus B &= B \oplus A \\ A \oplus (B \oplus C) &= (A \oplus B) \oplus C \\ A \otimes (B \otimes C) &= (A \otimes B) \otimes C \\ A \otimes (B \oplus C) &= A \otimes B \oplus A \otimes C \\ (A \oplus B) \otimes C &= A \otimes C \oplus B \otimes C \end{aligned}$$

Proof.
The statements follow from the definitions. □

Proposition 4.
The following hold for A, B, C, a, b, c, x, y of compatible sizes and $\alpha, \beta \in \mathbb{R}$:

$$\begin{aligned} A \otimes (\alpha \otimes B) &= \alpha \otimes (A \otimes B) \\ \alpha \otimes (A \oplus B) &= \alpha \otimes A \oplus \alpha \otimes B \\ (\alpha \oplus \beta) \otimes A &= \alpha \otimes A \oplus \beta \otimes B \\ x^T \otimes \alpha \otimes y &= \alpha \otimes x^T \otimes y \\ a \leq b &\Longrightarrow c^T \otimes a \leq c^T \otimes b \\ A \leq B &\Longrightarrow A \oplus C \leq B \oplus C \\ A \leq B &\Longrightarrow A \otimes C \leq B \otimes C \\ A \leq B &\Longleftrightarrow A \oplus B = B \end{aligned}$$

Proof. The statements follow from the definition of the pair of operations $(\oplus, \otimes)$. □

Definition 3. Given real numbers $a, b, c, \ldots$, a max-algebraic *diagonal matrix* is defined as:

$$\mathrm{diag}(a, b, c, \ldots) = \begin{pmatrix} a & & & & \\ & b & & -\infty & \\ & & c & & \\ & -\infty & & \ddots & \\ & & & & \ddots \end{pmatrix}$$

Given a vector $d = (d_1, d_2, \ldots, d_n)$, the *diagonal of the vector* d is denoted as $\mathrm{diag}(d) = \mathrm{diag}(d_1, d_2, \ldots, d_n)$.

Definition 4. Max-algebraic *identity matrix* is a diagonal matrix with all diagonal entries zero. We denote by I an identity matrix. Therefore, *identity matrix* $I = \mathrm{diag}(0, 0, 0, \ldots)$.

It is obvious that $A \otimes I = I \otimes A$ for any matrices A and I of compatible sizes.

Definition 5. Any matrix that can be obtained from the identity matrix, I, by permuting its rows and or columns is called a *permutation matrix*. A matrix arising as a product of a diagonal matrix and a permutation matrix is called a *generalised permutation matrix* [12].

Definition 6. A matrix $A \in \overline{\mathbb{R}}^{n \times n}$ is *invertible* if there exists a matrix $B \in \overline{\mathbb{R}}^{n \times n}$, such that $A \otimes B = B \otimes A = I$. The matrix B is unique and will be called the *inverse* of A. We will henceforth denote B by A^{-1}.

It has been shown in [1] that a matrix is *invertible* if and only if it is a generalised permutation matrix. If $x = (x_1, \ldots, x_n)$ we will denote $x^{-1} = (x_1^{-1}, \ldots, x_n^{-1})$, that is $x^{-1} = -x$, in conventional notation.

Example 4.
Consider the following matrices

$$A = \begin{pmatrix} -\infty & -\infty & 3 \\ 5 & -\infty & -\infty \\ -\infty & 8 & -\infty \end{pmatrix} \text{ and } B = \begin{pmatrix} -\infty & -5 & -\infty \\ -\infty & -\infty & -8 \\ -3 & -\infty & -\infty \end{pmatrix}$$

The matrix B is an inverse of A because,

$$A \otimes B = \begin{pmatrix} -\infty & -\infty & 3 \\ 5 & -\infty & -\infty \\ -\infty & 8 & -\infty \end{pmatrix} \otimes \begin{pmatrix} -\infty & -5 & -\infty \\ -\infty & -\infty & -8 \\ 3 & -\infty & -\infty \end{pmatrix} = \begin{pmatrix} 0 & -\infty & -\infty \\ -\infty & 0 & -\infty \\ -\infty & -\infty & 0 \end{pmatrix}$$

Given a matrix $A = (a_{ij}) \in \overline{\mathbb{R}}$, the *transpose* of A will be denoted by A^T, that is $A^T = (a_{ji})$. Structures of discrete-event dynamic systems may be represented by square matrices A over the semiring:

$$\overline{\mathbb{R}} = (\{-\infty\} \cup \mathbb{R}, \oplus, \otimes) = (\{-\infty\} \cup \mathbb{R}, max, +)$$

The system $\mathfrak{R}$ is embeddable in the self-dual system:

$$\overline{\overline{\mathbb{R}}} = (\{-\infty\} \cup \mathbb{R}\{+\infty\}, \oplus, \otimes, \oplus', \otimes') = (\{-\infty\} \cup \mathbb{R}\{+\infty\}, max, +, min, +)$$

Basic algebraic properties for $\oplus'$ and $\otimes'$ are similar to those of $\oplus$ and $\otimes$ described earlier. These are obtained by swapping $\leq$ and $\geq$. Extension of the pair $(\oplus', \otimes')$ to matrices and vectors is as follows:
Given A, B of compatible sizes and $\alpha \in \mathbb{R}$, we define the following:

$$A \oplus' B = (a_{ij} \oplus' b_{ij})$$

$$A \otimes' B = \left(\sum_k^{\oplus'} a_{ik} \otimes' b_{kj} \right) = \min_k (a_{ik} + b_{kj})$$

$$\alpha \otimes' A = (\alpha \otimes' a_{ij})$$

Also, properties of matrices for the pair $(\oplus', \otimes')$ are similar to those of $(\oplus, \otimes)$, just swap $\leq$ and $\geq$. For any matrix $A = [a_{ij}]$ over $\overline{\overline{\mathbb{R}}}$, the *conjugate* matrix is $A^* = [-a_{ji}]$ obtained by negation and transposition, that is $A = -A^T$.

Proposition 5. The following relations hold for any matrices U, V, W over $\overline{\overline{\mathbb{R}}}$.

$$(U \otimes' V) \otimes W \leq U \otimes' (V \otimes W) \tag{1}$$

$$U \otimes (U^* \otimes' W) \leq W \tag{2}$$

$$U \otimes (U^* \otimes' (U \otimes W)) = U \otimes W \tag{3}$$

Proof. Follows from the definitions. □

3. The Multiprocessor Interactive System (MPIS): A practical application

Linear equations and inequalities in max-algebra have a considerable number of applications, the model we present here is called the *multiprocessor interactive system (MPIS)* which is formulated as follows:

Products $P_1, \ldots, P_m$ are prepared using n processors, every processor contributing to the completion of each product by producing a partial product. It is assumed that every processor can work on all products simultaneously and that all these actions on a processor start as soon as the processor is ready to work. Let a_{ij} be the duration of the work of the jth processor needed to complete the partial product for P_i $(i = 1, \ldots, m; j = 1, \ldots, n)$. Let us denote by x_j the starting time of the jth processor $(j = 1, \ldots, n)$. Then, all partial products for P_i $(i = 1, \ldots, m; j = 1, \ldots, n)$ will be ready at time $\max(a_{i1} + x_1, \ldots, a_{in} + x_n)$. If the completion times $b_1, \ldots, b_m$ are given for each product then the starting times have to satisfy the following system of equations:

$$\max(a_{i1} + x_1, \ldots, a_{in} + x_n) = b_i \text{ for all } i \in M$$

Using the notation $a \oplus b = max(a, b)$ and $a \otimes b = a + b$ for $a, b \in \mathbb{R}$ extended to matrices and vectors in the same way as in linear algebra, then this system can be written as

$$A \otimes x = b \tag{4}$$

Any system of the form (4) is called 'one-sided max-linear system'. Also, if the requirement is that each product is to be produced on or before the completion times $b_1, \ldots, b_m$, then the starting times have to satisfy

$$\max(a_{i1} + x_1, \ldots, a_{in} + x_n) \leq b_i \text{ for all } i \in M$$

which can also be written as

$$A \otimes x \leq b \tag{5}$$

The system of inequalities (5) is called 'one-sided max-linear system of inequalities'.

4. Linear equations and inequalities in max-algebra

In this section we will present a system of linear equation and inequalities in max-algebra. Solvability conditions for linear system and inequalities will each be presented. A system consisting of max-linear equations and inequalities will also be discussed and necessary and sufficient conditions for the solvability of this system will be presented.

4.1. System of equations

In this section we present a solution method for the system $A \otimes x = b$ as given in [1, 3, 13] and also in the monograph [10]. Results concerning the existence and uniqueness of solution to the system will also be presented.

Given $A = (a_{ij}) \in \overline{\mathbb{R}}^{m \times n}$ and $b = (b_1, \ldots, b_m)^T \in \overline{\mathbb{R}}^m$, a system of the form

$$A \otimes x = b \tag{6}$$

is called a *one-sided max-linear system*, some times we may omit 'max-linear' and say one-sided system. This system can be written using the conventional notation as follows

$$\max_{j=1,\ldots,n} (a_{ij} + x_j) = b_i,\ i \in M \tag{7}$$

The system in (7) can be written after subtracting the right-hand sides constants as

$$\max_{j=1,\ldots,n} (a_{ij} \otimes b_i^{-1} + x_j) = 0,\ i \in M$$

A one-sided max-linear system whose all right hand side constants are zero is called *normalised max-linear system* or just *normalised* and the process of subtracting the right-hand side constants is called *normalisation*. Equivalently, *normalisation* is the process of multiplying the system (6) by the matrix B' from the left. That is

$$B' \otimes A \otimes x = B' \otimes b = 0$$

where,

$$B' = diag(b_1^{-1}, b_2^{-1}, \ldots, b_m^{-1}) = \mathrm{diag}(b^{-1})$$

For instance, consider the following one-sided system:

$$\begin{pmatrix} -2 & 1 & 3 \\ 3 & 0 & 2 \\ 1 & 2 & 1 \end{pmatrix} \otimes \begin{pmatrix} x_1 \\ x_2 \\ x_3 \end{pmatrix} = \begin{pmatrix} 5 \\ 6 \\ 3 \end{pmatrix} \tag{8}$$

After normalisation, this system is equivalent to

$$\begin{pmatrix} -7 & -4 & -2 \\ -3 & -6 & -4 \\ -2 & -1 & -2 \end{pmatrix} \otimes \begin{pmatrix} x_1 \\ x_2 \\ x_3 \end{pmatrix} = \begin{pmatrix} 0 \\ 0 \\ 0 \end{pmatrix}$$

That is after multiplying the system (8) by

$$\begin{pmatrix} -5 & -\infty & -\infty \\ -\infty & -6 & -\infty \\ -\infty & -\infty & -3 \end{pmatrix}$$

Consider the first equation of the normalised system above, that is $max(x_1 - 7, x_2 - 4, x_3 - 2) = 0$. This means that if $(x_1, x_2, x_3)^T$ is a solution to this system then $x_1 \leq 7$,$x_2 \leq 4$, $x_3 \leq 2$ and at least one of these inequalities will be satisfied with equality. From the other equations of the system, we have for $x_1 \leq 3$, $x_1 \leq 2$, hence we have $x_1 \leq min(7,3,2) = -max(-7,-3,-2) = -\bar{x}_1$ where $-\bar{x}_1$ is the column 1 maximum. It is clear that for all j then $x_j \leq \bar{x}_j$, where $-\bar{x}_j$ is the column j maximum. At the same time equality must be attained in some of these inequalities so that in every row there is at least one column maximum which is attained by x_j. This observation was made in [3].

Definition 7. A matrix A is called *doubly* $\mathbb{R}$*-astic* [14, 15], if it has at least one finite element on each row and on each column.

We introduce the following notations

$$S(A,b) = \{x \in \overline{\mathbb{R}}^n; A \otimes x = b\}$$
$$M_j = \{k \in M; b_k \otimes a_{kj}^{-1} = \max_i (b_i \otimes a_{ij}^{-1})\} \text{ for all } j \in N$$
$$\bar{x}(A,b)_j = \min_{i \in M} (b_i \otimes a_{ij}^{-1}) \text{ for all } j \in N$$

We now consider the cases when $A = -\infty$ and/or $b = -\infty$. Suppose that $b = -\infty$. Then $S(A,b)$ can simply be written as

$$S(A,b) = \{x \in \overline{\mathbb{R}}^n; x_j = -\infty, \text{ if } A_j \neq -\infty, j \in N\}$$

Therefore if $A = -\infty$ we have $S(A,b) = \overline{\mathbb{R}}^n$. Now, if $A = -\infty$ and $b \neq -\infty$ then $S(A,b) = \emptyset$. Thus, we may assume in this section that $A = -\infty$ and $b \neq -\infty$. If $b_k = -\infty$ for some $k \in M$ then for any $x \in S(A,b)$ we have $x_j = -\infty$ if $a_{kj} \neq -\infty$, $j \in N$, as a result the kth equation could be removed from the system together with every column j in the matrix A where $a_{kj} \neq -\infty$ (if any), and set the corresponding $x_j = -\infty$. Consequently, we may assume without loss of generality that $b \in \mathbb{R}^m$.

Moreover, if $b \in \mathbb{R}^m$ and A has an $-\infty$ row then $S(A,b) = \emptyset$. If there is an $-\infty$ column j in A then x_j may take on any value in a solution x. Thus, in what follows we assume without loss of generality that A is doubly $\mathbb{R} - astic$ and $b \in \mathbb{R}^m$.

Theorem 1. Let $A = (a_{ij}) \in \overline{\mathbb{R}}^{m \times n}$ be doubly $\mathbb{R} - astic$ and $b \in \mathbb{R}^m$. Then $x \in S(A,b)$ if and only if

i) $x \leq \bar{x}(A,b)$ and

ii) $\bigcup_{j \in N_x} M_j = M$ where $N_x = \{j \in N; x_j = \bar{x}(A,b)_j\}$

Proof. Suppose $x \in S(A,b)$. Thus we have,

$$\begin{aligned} & A \otimes x = b \\ \Longleftrightarrow\ & \max_j (a_{ij} + x_j) = b_i \text{ for all } i \in M \\ \Longleftrightarrow\ & a_{ij} + x_j = b_i \text{ for some } j \in N \\ \Longleftrightarrow\ & x_j \leq b_i \otimes a_{ij}^{-1} \text{ for all } i \in M \\ \Longleftrightarrow\ & x_j \leq \min_{i \in M} (b_i \otimes a_{ij}^{-1}) \text{ for all } j \in N \end{aligned}$$

Hence, $x \leq \bar{x}$.

Now that $x \in S(A,b)$. Since $M_j \subseteq M$ we only need to show that $M \subseteq \bigcup_{j \in N_x} M_j$. Let $k \in M$. Since $b_k = a_{kj} \otimes x_j > -\infty$ for some $j \in N$ and $x_j^{-1} \geq \bar{x}_j^{-1} \geq a_{ij} \otimes b_i^{-1}$ for every $i \in M$ we have $x_j^{-1} = a_{kj} \otimes b_k^{-1} = \max_{i \in M} a_{ij} \otimes b_i^{-1}$. Hence $k \in M_j$ and $x_j = \bar{x}_j$.

Suppose that $x \leq \bar{x}$ and $\bigcup_{j \in N_x} M_j = M$. Let $k \in M$, $j \in N$. Then $a_{kj} \otimes x_j \leq b_k$ if $a_{kj} = -\infty$. If $a_{kj} \neq -\infty$ then

$$a_{kj} \otimes x_j \leq a_{kj} \otimes \bar{x}_j \leq a_{kj} \otimes b_k \otimes a_{kj}^{-1} = b_k \tag{9}$$

Therefore $A \otimes x \leq b$. At the same time $k \in M_j$ for some $j \in N$ satisfying $x_j = \bar{x}_j$. For this j both inequalities in (9) are equalities and thus $A \otimes x = b$. □

The following is a summary of prerequisites proved in [1] and [12]:

Theorem 2. Let $A = (a_{ij}) \in \overline{\mathbb{R}}^{m\times n}$ be doubly $\mathbb{R}-astic$ and $b \in \mathbb{R}^m$. The system $A \otimes x = b$ has a solution if and only if $\bar{x}(A,b)$ is a solution.

Proof.
Follows from Theorem 1. □

Since $\bar{x}(A,b)$ has played an important role in the solution of $A \otimes x = b$. This vector $\bar{x}$ is called the *principal solution* to $A \otimes x = b$ [1], and we will call it likewise. The principal solution will also be used when studying the systems $A \otimes x \leq b$ and also when solving the one-sided system containing both equations and inequalities. The one-sided systems containing both equations and inequalities have been studied in [16] and the result will be presented later in this chapter.

Note that the principal solution may not be a solution to the system $A \otimes x = b$. More precisely, the following are observed in [12]:

Corollary 1. Let $A = (a_{ij}) \in \overline{\mathbb{R}}^{m\times n}$ be doubly $\mathbb{R}-astic$ and $b \in \mathbb{R}^m$. Then the following three statements are equivalent:

i) $S(A,b) \neq \emptyset$

ii) $\bar{x}(A,b) \in S(A,b)$

iii) $\bigcup_{j\in N} M_j = M$

Proof.
The statements follow from Theorems 1 and 2. □

For the existence of a unique solution to the max-linear system $A \otimes x = b$ we have the following corollary:

Corollary 2. Let $A = (a_{ij}) \in \overline{\mathbb{R}}^{m\times n}$ be doubly $\mathbb{R}-astic$ and $b \in \mathbb{R}^m$. Then $S(A,b) = \{\bar{x}(A,b)\}$ if and only if

i) $\bigcup_{j\in N} M_j = M$ and

ii) $\bigcup_{j\in N} M_j \neq M$ for any $N' \subseteq N, N' \neq N$

Proof.
Follows from Theorem 1. □

The question of solvability and unique solvability of the system $A \otimes x = b$ was linked to the set covering and minimal set covering problem of combinatorics in [12].

4.2. System of inequalities

In this section we show how a solution to the one-sided system of inequalities can be obtained. Let $A = (a_{ij}) \in \mathbb{R}^{m\times n}$ and $b = (b_1, \ldots, b_m)^T \in \mathbb{R}$. A system of the form:

$$A \otimes x \leq b \tag{10}$$

is called *one-sided max-linear system of inequalities* or just *one-sided system of inequalities*. The one-sided system of inequalities has received some attention in the past, see [1, 3] and [17] for more information. Here, we will only present a result which shows that the principal solution, $\bar{x}(A,b)$ is the greatest solution to (10). That is if (10) has a solution then $\bar{x}(A,b)$ is the greatest of all the solutions. We denote the solution set of (10) by $S(A,b,\leq)$. That is

$$S(A,b,\leq) = \{x \in \mathbb{R}^n; A \otimes x \leq b\}$$

Theorem 3. $x \in S(A,b,\leq)$ if and only if $x \leq \bar{x}(A,b)$.

Proof. Suppose $x \in S(A,b,\leq)$. Then we have

$$\begin{aligned}
& A \otimes x \leq b \\
\Longleftrightarrow\ & \max_j(a_{ij} + x_j) \leq b_i \text{ for all } i \\
\Longleftrightarrow\ & a_{ij} + x_j \leq b_i \text{ for all } i,j \\
\Longleftrightarrow\ & x_j \leq b_i \otimes a_{ij}^{-1} \text{ for all } i,j \\
\Longleftrightarrow\ & x_j \leq \min_i(b_i \otimes a_{ij}^{-1}) \text{ for all } j \\
\Longleftrightarrow\ & x \leq \bar{x}(A,b)
\end{aligned}$$

and the proof is now complete. □

The system of inequalities

$$\begin{aligned} A \otimes x &\leq b \\ C \otimes x &\geq d \end{aligned} \tag{11}$$

was discussed in [18] where the following result was presented.

Lemma 1. A system of inequalities (11) has a solution if and only if $C \otimes \bar{x}(A,b) \geq d$

4.3. A system containing of both equations and inequalities

In this section a system containing both equations and inequalities will be presented, the results were taken from [16]. Let $A = (a_{ij}) \in \mathbb{R}^{k\times n}$, $C = (c_{ij}) \in \mathbb{R}^{r\times n}$, $b = (b_1, \ldots, b_k)^T \in \mathbb{R}^k$ and $d = (d_1, \ldots, d_r)^T \in \mathbb{R}^r$. A *one-sided max-linear system with both equations and inequalities* is of the form:

$$\begin{aligned} A \otimes x &= b \\ C \otimes x &< d \end{aligned} \qquad (12)$$

We shall use the following notation throughout this paper

$$R = \{1, 2, ..., r\}$$

$$S(A, C, b, d) = \{x \in \mathbb{R}^n; A \otimes x = b \text{ and } C \otimes x \leq d\}$$
$$S(C, d, \leq) = \{x \in \mathbb{R}^n; C \otimes x \leq d\}$$

$$\bar{x}_j(C, d) = \min_{i \in R}(d_i \otimes c_{ij}^{-1}) \text{ for all } j \in N$$

$$K = \{1, \ldots, k\}$$

$$K_j = \left\{k \in K; b_k \otimes a_{kj}^{-1} = \min_{i \in K}\left(b_i \otimes a_{ij}^{-1}\right)\right\} \text{ for all } j \in N$$

$$\bar{x}_j(A, b) = \min_{i \in K}(b_i \otimes a_{ij}^{-1}) \text{ for all } j \in N$$

$$\bar{x} = (\bar{x}_1, ..., \bar{x}_n)^T$$

$$J = \{j \in N; \bar{x}_j(C, d) \geq \bar{x}_j(A, b)\} \text{ and}$$

$$L = N \setminus J$$

We also define the vector $\hat{x} = (\hat{x}_1, \hat{x}_2, ..., \hat{x}_n)^T$, where

$$\hat{x}_j(A, C, b, d) \equiv \begin{cases} \bar{x}_j(A, b) & \text{if } j \in J \\ \bar{x}_j(C, d) & \text{if } j \in L \end{cases} \qquad (13)$$

and $N_{\hat{x}} = \{j \in N; \hat{x}_j = \bar{x}_j\}$.

Theorem 4. Let $A = (a_{ij}) \in \mathbb{R}^{k \times n}$, $C = (c_{ij}) \in \mathbb{R}^{r \times n}$, $b = (b_1, \ldots, b_k)^T \in \mathbb{R}^k$ and $d = (d_1, \ldots, d_r)^T \in \mathbb{R}^r$. Then the following three statements are equivalent:

(i) $S(A, C, b, d) \neq \emptyset$
(ii) $\hat{x}(A, C, b, d) \in S(A, C, b, d)$
(iii) $\bigcup_{j \in J} K_j = K$

Proof. *(i)* $\Longrightarrow$ *(ii)*. Let $x \in S(A, C, b, d)$, therefore $x \in S(A, b)$ and $x \in S(C, d, \leq)$. Since $x \in S(C, d, \leq)$, it follows from Theorem 3 that $x \leq \bar{x}(C, d)$. Now that $x \in S(A, b)$ and also $x \in S(C, d, \leq)$, we need to show that $\bar{x}_j(C, d) \geq \bar{x}_j(A, b)$ for all $j \in N_x$ (that is $N_x \subseteq J$). Let $j \in N_x$ then $x_j = \bar{x}_j(A, b)$. Since $x \in S(C, d, \leq)$ we have $x \leq \bar{x}(C, d)$ and therefore $\bar{x}_j(A, b) \leq \bar{x}_j(C, d)$ thus $j \in J$. Hence, $N_x \subseteq J$ and by Theorem 1 $\bigcup_{j \in J} K_j = K$. This also proves

$(i) \Longrightarrow (iii)$
$(iii) \Longrightarrow (i)$. Suppose $\bigcup_{j\in J} K_j = K$. Since $\hat{x}(A,C,b,d) \le \bar{x}(C,d)$ we have $\hat{x}(A,C,b,d) \in S(C,d,\le)$. Also $\hat{x}(A,C,b,d) \le \bar{x}(A,b)$ and $N_{\hat{x}} \supseteq J$ gives $\bigcup_{j\in N_{\hat{x}(A,C,b,d)}} K_j \supseteq \bigcup_{j\in J} K_j = K$. Hence $\bigcup_{j\in N_{\hat{x}(A,C,b,d)}} K_j = K$, therefore $\hat{x}(A,C,b,d) \in S(A,b)$ and $\hat{x}(A,C,b,d) \in S(C,d,\le)$. Hence $\hat{x}(A,C,b,d) \in S(A,C,b,d)$ (that is $S(A,C,b,d) \neq \emptyset$) and this also proves $(iii) \Longrightarrow (ii)$. □

Theorem 5. Let $A = (a_{ij}) \in \mathbb{R}^{k\times n}$, $C = (c_{ij}) \in \mathbb{R}^{r\times n}$, $b = (b_1,\ldots,b_k)^T \in \mathbb{R}^k$ and $d = (d_1,\ldots,d_r)^T \in \mathbb{R}^r$. Then $x \in S(A,C,b,d)$ if and only if

(i) $x \le \hat{x}(A,C,b,d)$ and

(ii) $\bigcup_{j\in N_x} K_j = K$ where $N_x = \{j \in N \; ; \; x_j = \bar{x}_j(A,b)\}$

Proof. ($\Longrightarrow$) Let $x \in S(A,C,b,d)$, then $x \le \bar{x}(A,b)$ and $x \le \bar{x}(C,d)$. Since $\hat{x}(A,C,b,d) = \bar{x}(A,b) \oplus' \bar{x}(C,d)$ we have $x \le \hat{x}(A,C,b,d)$. Also, $x \in S(A,C,b,d)$ implies that $x \in S(C,d,\le)$. It follows from Theorem 1 that $\bigcup_{j\in N_x} K_j = K$.
($\Longleftarrow$) Suppose that $x \le \hat{x}(A,C,b,d) = \bar{x}(A,b) \oplus' \bar{x}(C,d)$ and $\bigcup_{j\in N_x} K_j = K$. It follows from Theorem 1 that $x \in S(A,b)$, also by Theorem 3 $x \in S(C,d,\le)$. Thus $x \in S(A,b) \cap S(C,d,\le) = S(A,C,b,d)$. □

We introduce the symbol $|X|$ which stands for the number of elements of the set X.

Lemma 2. Let $A = (a_{ij}) \in \mathbb{R}^{k\times n}$, $C = (c_{ij}) \in \mathbb{R}^{r\times n}$, $b = (b_1,\ldots,b_k)^T \in \mathbb{R}^k$ and $d = (d_1,\ldots,d_r)^T \in \mathbb{R}^r$. If $|S(A,C,b,d)| = 1$ then $|S(A,b)| = 1$.

Proof. Suppose $|S(A,C,b,d)| = 1$, that is $S(A,C,b,d) = \{x\}$ for an $x \in \mathbb{R}^n$. Since $S(A,C,b,d) = \{x\}$ we have $x \in S(A,b)$ and thus $S(A,b) \neq \emptyset$. For contradiction, suppose $|S(A,b)| > 1$. We need to check the following two cases: (i) $L \neq \emptyset$ and (ii) $L = \emptyset$ where $L = N \setminus J$, and show in each case that $|S(A,C,b,d)| > 1$.

Proof of Case (i), that is $L \neq \emptyset$: Suppose that L contains only one element say $n \in N$ i.e $L = \{n\}$. Since $x \in S(A,C,b,d)$ it follows from Theorem 4that $\hat{x}(A,C,b,d) \in S(A,C,b,d)$. That is $x = \hat{x}(A,C,b,d) =$
$(\bar{x}_1(A,b), \bar{x}_2(A,b), \ldots, \bar{x}_{n-1}(A,b), \bar{x}_n(C,d)) \in S(A,C,b,d)$. It can also be seen that, $\bar{x}(C,d)_n < \bar{x}_n(A,b)$ and any vector of the form $z =$
$(\bar{x}_1(A,b), \bar{x}_2(A,b), \ldots, \bar{x}_{n-1}(A,b), \alpha) \in S(A,C,b,d)$, where $\alpha \le \bar{x}_n(C,d)$. Hence $|S(A,C,b,d)| > 1$. If L contains more than one element, then the proof is done in a similar way.

Proof of Case (ii), that is $L = \emptyset$ ($J = N$): Suppose that $J = N$. Then we have $\hat{x}(A,C,b,d) = \bar{x}(A,b) \le \bar{x}(C,d)$. Suppose without loss of generality that $x, x' \in S(A,b)$ such that $x \neq x'$. Then $x \le \bar{x}(A,b) \le \bar{x}(C,d)$ and also $x' \le \bar{x}(A,b) \le \bar{x}(C,d)$. Thus, $x, x' \in S(C,d,\le)$. Consequently, $x, x' \in S(A,C,b,d)$ and $x \neq x'$. Hence $|S(A,C,b,d)| > 1$. □

Theorem 6. Let $A = (a_{ij}) \in \mathbb{R}^{k\times n}$, $C = (c_{ij}) \in \mathbb{R}^{r\times n}$, $b = (b_1,\ldots,b_k)^T \in \mathbb{R}^k$ and $d = (d_1,\ldots,d_r)^T \in \mathbb{R}^r$. If $|S(A,C,b,d)| = 1$ then $J = N$.

Proof. Suppose $|S(A,C,b,d)| = 1$. It follows from Theorem 4 that $\bigcup_{j\in J} K_j = K$. Also, $|S(A,C,b,d)| = 1$ implies that $|S(A,b)| = 1$ (Lemma 2). Moreover, $|S(A,b)| = 1$ implies that $\bigcup_{j\in N} K_j = K$ and $\bigcup_{j\in N'} K_j \neq K, N' \subseteq N, N' \neq N$ (Theorem 2). Since $J \subseteq N$ and $\bigcup_{j\in J} K_j = K$, we have $J = N$. □

Corollary 3. Let $A = (a_{ij}) \in \mathbb{R}^{k\times n}$, $C = (c_{ij}) \in \mathbb{R}^{r\times n}$, $b = (b_1,\ldots,b_k)^T \in \mathbb{R}^k$ and $d = (d_1,\ldots,d_r)^T \in \mathbb{R}^r$. If $|S(A,C,b,d)| = 1$ then $S(A,C,b,d) = \{\bar{x}(A,b)\}$.

Proof. The statement follows from Theorem 6 and Lemma 2. □

Corollary 4. Let $A = (a_{ij}) \in \mathbb{R}^{k\times n}$, $C = (c_{ij}) \in \mathbb{R}^{r\times n}$, $b = (b_1,\ldots,b_k)^T \in \mathbb{R}^k$ and $d = (d_1,\ldots,d_r)^T \in \mathbb{R}^k$. Then, the following three statements are equivalent:

(i) $|S(A,C,b,d)| = 1$

(ii) $|S(A,b)| = 1$ and $J = N$

(iii) $\bigcup\limits_{j\in J} K_j = K$ and $\bigcup\limits_{j\in J'} K_j \neq K$, for every $J' \subseteq J, J' \neq J$, and $J = N$

Proof. $(i) \Longrightarrow (ii)$ Follows from Lemma 2 and Theorem 6.
$(ii) \Longrightarrow (i)$ Let $J = N$, therefore $\bar{x} \leq \bar{x}(C,d)$ and thus $S(A,b) \subseteq S(C,d,\leq)$. Therefore we have $S(A,C,b,d) = S(A,b) \cap S(C,d,\leq) = S(A,b)$. Hence $|S(A,C,b,d)| = 1$.
$(ii) \Longrightarrow (iii)$ Suppose that $S(A,b) = \{x\}$ and $J = N$. It follows from Theorem 2 that $\bigcup_{j\in N} K_j = K$ and $\bigcup_{j\in N'} K_j \neq K, N' \subseteq N, N' \neq N$. Since $J = N$ the statement now follows from Theorem 2.
$(iii) \Longrightarrow (ii)$ It is immediate that $J = N$ and the statement now follows from Theorem 2. □

Theorem 7. Let $A = (a_{ij}) \in \mathbb{R}^{k\times n}$, $C = (c_{ij}) \in \mathbb{R}^{r\times n}$, $b = (b_1,\ldots,b_k)^T \in \mathbb{R}^k$ and $d = (d_1,\ldots,d_r)^T \in \mathbb{R}^k$. If $|S(A,C,b,d)| > 1$ then $|S(A,C,b,d)|$ is infinite .

Proof. Suppose $|S(A,C,b,d)| > 1$. By Corollary 4 we have $\bigcup_{j\in J} K_j = K$, for some $J \subseteq N$, $J \neq N$(that is $\exists j \in N$ such that $\bar{x}_j(A,b) > \bar{x}_j(C,d)$). Now $J \subseteq N$ and $\bigcup_{j\in J} K_j = K$, Theorem 5 implies that any vector $x = (x_1, x_2, ..., x_n)^T$ of the form

$$x_j \equiv \begin{cases} \bar{x}_j(A,b) & \text{if } j \in J \\ y \leq \bar{x}_j(C,d) & \text{if } j \in L \end{cases}$$

is in $S(A,C,b,d)$, and the statement follows. □

Remark 1. From Theorem 7 we can say that the number of solutions to the one-sided system containing both equations and inequalities can only be $0, 1$, or ∞.

The vector $\hat{x}(A,C,b,d)$ plays an important role in the solution of the one-sided system containing both equations and inequalities. This role is the same as that of the principal solution $\bar{x}(A,b)$ to the one-sided max-linear system $A \otimes x = b$, see [19] for more details.

5. Max-linear program with equation and inequality constraints

Suppose that the vector $f = (f_1, f_2, ..., f_n)^T \in \mathbb{R}^n$ is given. The task of minimizing [maximizing]the function $f(x) = f^T \otimes x = \max(f_1 + x_1, f_1 + x_2 ..., f_n + x_n)$ subject to (12) is called max-linear program with one-sided equations and inequalities and will be denoted by $\text{MLP}_{\leq}^{\min}$ and $[\text{MLP}_{\leq}^{\max}]$. We denote the sets of optimal solutions by $S^{\min}(A, C, b, d)$ and $S^{\max}(A, \bar{C}, b, d)$, respectively.

Lemma 3. Suppose $f \in \mathbb{R}^n$ and let $f(x) = f^T \otimes x$ be defined on $\mathbb{R}^n$. Then,
(i) $f(x)$ is max-linear, i.e. $f(\lambda \otimes x \oplus \mu \otimes y) = \lambda \otimes f(x) \oplus \mu \otimes f(y)$ for every $x, y \in \mathbb{R}^n$.
(ii) $f(x)$ is isotone, i.e. $f(x) \leq f(y)$ for every $x, y \in \mathbb{R}^n$, $x \leq y$.

Proof.
(i) Let $\alpha \in \mathbb{R}$. Then we have

$$\begin{aligned} f(\lambda \otimes x \oplus \mu \otimes y) &= f^T \otimes \lambda \otimes x \oplus f^T \otimes \mu \otimes y \\ &= \lambda \otimes f^T \otimes x \oplus \mu \otimes f^T \otimes y \\ &= \lambda \otimes f(x) \oplus \mu \otimes f(y) \end{aligned}$$

and the statement now follows.

(ii) Let $x, y \in \mathbb{R}^n$ such that $x \leq y$. Since $x \leq y$, we have

$$\begin{aligned} & \max(x) \leq \max(y) \\ \Longleftrightarrow \; & f^T \otimes x \leq f^T \otimes y, \text{ for any, } f \in \mathbb{R}^n \\ \Longleftrightarrow \; & f(x) \leq f(y). \end{aligned}$$

□

Note that it would be possible to convert equations to inequalities and conversely but this would result in an increase of the number of constraints or variables and thus increasing the computational complexity. The method we present here does not require any new constraint or variable.

We denote by

$$(A \otimes x)_i = \max_{j \in N}(a_{ij} + x_j)$$

A variable x_j will be called *active* if $x_j = f(x)$, for some $j \in N$. Also, a variable will be called *active* on the constraint equation if the value $(A \otimes x)_i$ is attained at the term x_j respectively. It follows from Theorem 5 and Lemma 3 that $\hat{x}(A, C, b, d) \in S^{\max}(A, C, b, d)$. We now present a polynomial algorithm which finds $x \in S^{\min}(A, C, b, d)$ or recognizes that $S^{\min}(A, B, c, d) = \emptyset$. Due to Theorem 4 either $\hat{x}(A, C, b, d) \in S(A, C, b, d)$ or $S(A, C, b, d) = \emptyset$. Therefore, we assume in the following algorithm that $S(A, C, b, d) \neq \emptyset$ and also $S^{\min}(A, C, b, d) \neq \emptyset$.

Theorem 8. The algorithm ONEMLP-EI is correct and its computational complexity is $O((k + r)n^2)$.

Algorithm 1 ONEMLP-EI(Max-linear program with one-sided equations and inequalities)

Input: $f = (f_1, f_2, ..., f_n)^T \in \mathbb{R}^n$, $b = (b_1, b_2, ...b_k)^T \in \mathbb{R}^k$, $d = (d_1, d_2, ...d_r)^T \in \mathbb{R}^r$, $A = (a_{ij}) \in R^{k \times n}$ and $C = (c_{ij}) \in R^{r \times n}$.

Output: $x \in S^{\min}(A, C, b, d)$.

1. Find $\bar{x}(A,b)$, $\bar{x}(C,d)$, $\hat{x}(A,C,b,d)$ and $K_j, j \in J; J = \{j \in N; \bar{x}_j(C,d) \geq \bar{x}_j(A,b)\}$
2. $x := \hat{x}(A,C,b,d)$
3. $H(x) := \{j \in N; f_j + x_j = f(x)\}$
4. $J := J \setminus H(x)$
5. If
$$\bigcup_{j \in J} K_j \neq K$$
then stop ($x \in S^{\min}(A,C,b,d)$)
6. Set x_j small enough (so that it is not active on any equation or inequality) for every $j \in H(x)$
7. Go to 3

Proof. The correctness follows from Theorem 5 and the computational complexity is computed as follows. In Step 1 $\bar{x}(A,b)$ is $O(kn)$, while $\bar{x}(C,d)$, $\hat{x}(A,C,b,d)$ and K_j can be determined in $O(rn)$, $O(k+r)n$ and $O(kn)$ respectively. The loop 3-7 can be repeated at most $n-1$ times, since the number of elements in J is at most n and in Step 4 at least one element will be removed at a time. Step 3 is $O(n)$, Step 6 is $O(kn)$ and Step 7 is $O(n)$. Hence loop 3-7 is $O(kn^2)$. □

5.1. An example

Consider the following system max-linear program in which $f = (5, 6, 1, 4, -1)^T$,

$$A = \begin{pmatrix} 3 & 8 & 4 & 0 & 1 \\ 0 & 6 & 2 & 2 & 1 \\ 0 & 1 & -2 & 4 & 8 \end{pmatrix}, b = \begin{pmatrix} 7 \\ 5 \\ 7 \end{pmatrix},$$

$$C = \begin{pmatrix} -1 & 2 & -3 & 0 & 6 \\ 3 & 4 & -2 & 2 & 1 \\ 1 & 3 & -2 & 3 & 4 \end{pmatrix} \text{ and } d = \begin{pmatrix} 5 \\ 5 \\ 6 \end{pmatrix}.$$

We now make a record run of Algorithm ONEMLP-EI. $\bar{x}(A,b) = (5, -1, 3, 3, -1)^T$, $\bar{x}(C,d) = (2, 1, 7, 3, -1)^T$, $\hat{x}(A,C,b,d) = (2, -1, 3, 3, -1)^T$, $J = \{2, 3, 4, 5\}$ and $K_2 = \{1, 2\}$, $K_3 = \{1, 2\}$, $K_4 = \{2, 3\}$ and $K_5 = \{3\}$. $x := \hat{x}(A,C,b,d) = (2, -1, 3, 3, -1)^T$ and $H(x) = \{1, 4\}$ and $J \not\subseteq H(x)$. We also have $J := J \setminus H(x) = \{2, 3, 5\}$ and $K_2 \cup K_3 \cup K_5 = K$. Then set $x_1 = x_4 = 10^{-4}$ (say) and $x = (10^{-4}, -1, 3, 10^{-4}, -1)^T$. Now $H(x) = \{2\}$ and $J := J \setminus H(x) = \{3, 5\}$. Since $K_3 \cup K_5 = K$ set $x_2 = 10^{-4}$(say) and we have $x = (10^{-4}, 10^{-4}, 3, 10^{-4}, -1)^T$. Now $H(x) = \{3\}$ and $J := J \setminus H(x) = \{5\}$. Since $K_5 \neq K$ then we stop and an optimal solution is $x = (10^{-4}, 10^{-4}, 3, 10^{-4}, -1)^T$ and $f^{\min} = 4$.

6. A special case of max-linear program with two-sided constraints

Suppose $c = (c_1, c_2, ..., c_m)^T, d = (d_1, d_2, ..., d_m)^T \in \mathbb{R}^m, A = (a_{ij})$ and $B = (b_{ij}) \in \mathbb{R}^{m\times n}$ are given matrices and vectors. The system

$$A \otimes x \oplus c = B \otimes x \oplus d \tag{14}$$

is called non-homogeneous two-sided max-linear system and the set of solutions of this system will be denoted by S. Two-sided max-linear systems have been studied in [20], [21], [22] and [23].

Optimization problems whose objective function is max-linear and constraint (14) are called max-linear programs (MLP). Max-linear programs are studied in [24] and solution methods for both minimization and maximization problems were developed. The methods are proved to be pseudopolynomial if all entries are integer. Also non-linear programs with max-linear constraints were dealt with in [25], where heuristic methods were develeoped and tested for a number of instances.

Consider max-linear programs with two-sided constraints (minimization), $\text{MLP}^{\min}$

$$\begin{aligned} &f(x) = f^T \otimes x \longrightarrow \min \\ &\text{subject to} \\ &A \otimes x \oplus c = B \otimes x \oplus d \end{aligned} \tag{15}$$

where $f = (f_1, \ldots, f_n)^T \in \mathbb{R}^n$, $c = (c_1, \ldots, c_m)^T$, $d = (d_1, \ldots, d_m)^T \in \mathbb{R}^m$, $A = (a_{ij})$ and $B = (b_{ij}) \in \mathbb{R}^{m\times n}$ are given matrices and vectors. We introduce the following:

$$\begin{aligned} y &= (f_1 \otimes x_1, f_2 \otimes x_2, \ldots, f_n \otimes x_n) \\ &= \text{diag}(f) \otimes x \end{aligned} \tag{16}$$

$\text{diag}(f)$ means a diagonal matrix whose diagonal elements are $f_1, f_2, ..., f_n$ and off diagonal elements are $-\infty$. It therefore follows from (16) that

$$\begin{aligned} &f^T \otimes x = 0^T \otimes y \\ \Longleftrightarrow\ &x = (f_1^{-1} \otimes y_1, f_2^{-1} \otimes y_2, \ldots, f_n^{-1} \otimes y_n) \\ &\quad = (\text{diag}(f))^{-1} \otimes y \end{aligned} \tag{17}$$

Hence, by substituting (16) and (17) into (15) we have

$$\begin{aligned} &0^T \otimes y \longrightarrow \min \\ &\text{subject to} \\ &A' \otimes y \oplus c = B' \otimes y \oplus d, \end{aligned} \tag{18}$$

where 0^T is transpose of the zero vector, $A' = A \otimes (\text{diag}(f))^{-1}$ and $B' = B \otimes (\text{diag}(f))^{-1}$

Therefore we assume without loss of generality that $f = 0$ and hence (15) is equivalent to

$$\begin{aligned} &f(x) = \sum_{j=1,\ldots,n}^{\oplus} x_j \longrightarrow \min \\ &\text{subject to} \\ &A \otimes x \oplus c = B \otimes x \oplus d \end{aligned} \tag{19}$$

The set of feasible solutions for (19) will be denoted by S and the set of optimal solutions by $S^{\min}$. A vector is called *constant* if all its components are equal. That is a vector $x \in \mathbb{R}^n$ is constant if $x_1 = x_2 = \cdots = x_n$. For any $x \in S$ we define the set $Q(x) = \{i \in M; (A \otimes x)_i > c_i\}$. We introduce the following notation of matrices. Let $A = (a_{ij}) \in R^{m\times n}$, $1 \leq i_1 < i_2 < \cdots < i_q \leq m$ and $1 \leq j_1 < j_2 < \cdots < j_r \leq n$. Then,

$$A\begin{pmatrix} i_1, i_2, \dots, i_q \\ j_1, j_2, \dots, j_r \end{pmatrix} = \begin{pmatrix} a_{i_1j_1} a_{i_1j_2} \dots a_{i_1j_r} \\ a_{i_2j_1} a_{i_2j_2} \dots a_{i_2j_r} \\ \dots \\ a_{i_qj_1} a_{i_qj_2} \dots a_{i_qj_r} \end{pmatrix} = A(Q,R)$$

where, $Q = \{i_1, \dots, i_q\}$, $R = \{j_1, \dots, j_r\}$. Similar notation is used for vectors $c(i_1, \dots, i_r) = (c_{i_1} \dots c_{i_r})^T = c(R)$. Given MLP$^{\min}$ with $c \geq d$, we define the following sets

$$M^{=} = \{i \in M; c_i = d_i\} \text{ and}$$
$$M^{>} = \{i \in M; c_i > d_i\}$$

We also define the following matrices:

$$\begin{aligned} A_{=} &= A(M^{=}, N),\ A_{>} = A(M^{>}, N) \\ B_{=} &= B(M^{=}, N),\ B_{>} = B(M^{>}, N) \\ c_{=} &= c(M^{=}),\ c_{>} = c(M^{>}) \end{aligned} \tag{20}$$

An easily solvable case arises when there is a constant vector $x \in S$ such that the set $Q(x) = \emptyset$. This constant vector x satisfies the following equations and inequalities

$$\begin{aligned} A_{=} \otimes x &\leq c_{=} \\ A_{>} \otimes x &\leq c_{>} \\ B_{=} \otimes x &\leq c_{=} \\ B_{>} \otimes x &= c_{>} \end{aligned} \tag{21}$$

where $A_{=}, A_{>}, B_{=}, B_{>}$, $c_{=}$ and $c_{>}$ are defined in (20). The one-sided system of equation and inequalities (21) can be written as

$$\begin{aligned} G \otimes x &= p \\ H \otimes x &\leq q \end{aligned} \tag{22}$$

where,

$$G = (B_{>}),\ H = \begin{pmatrix} A_{=} \\ A_{>} \\ B_{=} \end{pmatrix}$$
$$p = c_{>} \text{ and } q = \begin{pmatrix} c_{=} \\ c_{>} \\ c_{=} \end{pmatrix} \tag{23}$$

Recall that $S(G, H, p, q)$ is the set of solutions for (22).

Theorem 9. Let $Q(x) = \emptyset$ for some constant vector $x = (\alpha, \ldots, \alpha)^T \in S$. If $z \in S^{\min}$ then $z \in S(G, H, p, q)$.

Proof. Let $x = (\alpha, \ldots, \alpha)^T \in S$. Suppose $Q(z) = \emptyset$ and $z \in S^{\min}$. This implies that $f(z) \leq f(x) = \alpha$. Therefore we have, $\forall j \in N$, $z \leq \alpha$. Consequently, $z \leq x$ and $(A \otimes z)_i \leq (A \otimes x)_i$ for all $i \in M$. Since, $Q(z) = \emptyset$ and $z \in S(G, H, p, q)$. □

Corollary 5. If $Q(x) = \emptyset$ for some constant vector $x \in S$ then $S^{\min} \subseteq S^{\min}(G, H, p, q)$.

Proof. The statement follows from Theorem 9. □

7. Some solvability concepts of a linear system containing of both equations and inequalities

System of max-separable linear equations and inequalities arise frequently in several branches of Applied Mathematics: for instance in the description of discrete-event dynamic system [1, 4] and machine scheduling [10]. However, choosing unsuitable values for the matrix entries and right-handside vectors may lead to unsolvable systems. Therefore, methods for restoring solvability suggested in the literature could be employed. These methods include modifying the input data [11, 26] or dropping some equations [11]. Another possibility is to replace each entry by an interval of possible values. In doing so, our question will be shifted to asking about weak solvability, strong solvability and control solvability.

Interval mathematics was championed by Moore [27] as a tool for bounding errors in computer programs. The area has now been developed in to a general methodology for investigating numerical uncertainty in several problems. System of interval equations and inequalities in max-algebra have each been studied in the literature. In [26] weak and strong solvability of interval equations were discussed, control sovability, weak control solvability and universal solvability have been dealt with in [28]. In [29] a system of linear inequality with interval coefficients was discussed. In this section we consider a system consisting of interval linear equations and inequalities and present solvability concepts for such system.

An algebraic structure $(B, \oplus, \otimes)$ with two binary operations $\oplus$ and $\otimes$ is called max-plus algebra if

$$B = \mathbb{R} \cup \{-\infty\},\ a \oplus b = \max\{a, b\},\ a \otimes b = a + b$$

for any $a, b \in \mathbb{R}$.

Let m, n, r be given positive integers and $a \in \mathbb{R}$, we use throughout the paper the notation $M = \{1, 2, ..., m\}$, $N = \{1, 2, ..., n\}, R = \{1, 2, ..., r\}$ and $a^{-1} = -a$. The set of all $m \times n$, $r \times n$ matrices over B is denoted by $B(m, n)$ and $B(r, n)$ respectively. The set of all n-dimensional vectors is denoted by $B(n)$. Then for each matrix $A \in B(n, m)$ and vector $x \in B(n)$ the product $A \otimes x$ is define as

$$(A \otimes x) = \max_{j \in N} \left(a_{ij} + x_j \right)$$

For a given matrix interval $\boldsymbol{A} = [\underline{A}, \overline{A}]$ with $\underline{A}, \overline{A} \in B(k, n), \underline{A} \leq \overline{A}$ and given vector interval $\boldsymbol{b} = [\underline{b}, \overline{b}]$ with $\underline{b}, \overline{b} \in B(n), \underline{b} \leq \overline{b}$ the notation

$$\boldsymbol{A} \otimes x = \boldsymbol{b} \tag{24}$$

represents an interval system of linear max-separable equations of the form

$$A \otimes x = b \tag{25}$$

Similarly, for a given matrix interval $\boldsymbol{C} = [\underline{A}, \overline{A}]$ with $\underline{C}, \overline{C} \in B(r,n), \underline{C} \leq \overline{C}$ and given vector interval $\boldsymbol{d} = [\underline{d}, \overline{d}]$ with $\underline{d}, \overline{d} \in B(n), \underline{b} \leq \overline{b}$ the notation

$$\boldsymbol{C} \otimes x \leq \boldsymbol{d} \tag{26}$$

represents an interval system of linear max-separable inequalities of the form

$$C \otimes x \leq d \tag{27}$$

Interval system of linear max-separable equations and inequalities have each been studied in the literature, for more information the reader is reffered to . The following notation

$$\begin{aligned} \boldsymbol{A} \otimes x &= \boldsymbol{b} \\ \boldsymbol{C} \otimes x &\leq \boldsymbol{d} \end{aligned} \tag{28}$$

represents an interval system of linear max-separable equations and inequalities of the form

$$\begin{aligned} A \otimes x &= b \\ C \otimes x &\leq d \end{aligned} \tag{29}$$

where $A \in \boldsymbol{A}, C \in \boldsymbol{C}, b \in \boldsymbol{b}$ and $d \in \boldsymbol{d}$.

The aim of this section is to consider a system consisting of max-separable linear equations and inequalities and presents some solvability conditions of such system. Note that it is possible to convert equations to inequalities and conversely, but this would result in an increase in the number of equations and inequalities or an increase in the number of unknowns thus increasing the computational complexity when testing the solvability conditions. Each system of the form (29) is said to be a subsystem of (28). An interval system (29) has constant matrices if $\underline{A} = \overline{A}$ and $\underline{C} = \overline{C}$. Similarly, an interval system has constant right hand side if $\underline{b} = \overline{b}$ and $\underline{d} = \overline{d}$. In what follows we will consider $A \in \mathbb{R}(k,n)$ and $C \in \mathbb{R}(r,n)$.

7.1. Weak solvability

Definition 8. A vector y is a weak solution to an interval system (29) if there exists $A \in \boldsymbol{A}, C \in \boldsymbol{C}, b \in \boldsymbol{b}$ and $d \in \boldsymbol{d}$ such that

$$\begin{aligned} A \otimes y &= b \\ C \otimes y &\leq d \end{aligned} \tag{30}$$

Theorem 10. A vector $x \in \mathbb{R}^n$ is a weak solution of (29) if and only if

$$x = \bar{x}\begin{pmatrix} \underline{A} & \overline{b} \\ \underline{C} & \overline{d} \end{pmatrix}$$

and

$$\overline{A} \otimes \bar{x}\begin{pmatrix} \underline{A} & \overline{b} \\ \underline{C} & \overline{d} \end{pmatrix} \geq \underline{b}$$

Proof. Let $i = \{1, ..., m\}$ be an arbitrary chosen index and $x = (x_1, x_2, ..., x_n)^T \in \mathbb{R}^n$ fixed. If $A \in \boldsymbol{A}$ then $(A \otimes x)_i$ is isotone and we have

$$(A \otimes x)_i \in [(\underline{A} \otimes x)_i, (\overline{A} \otimes x)_i] \subseteq \mathbb{R}$$

Hence, x is a weak solution if and only if

$$[(\underline{A} \otimes x)_i, (\overline{A} \otimes x)_i] \cap [\underline{b}_i, \overline{b}_i] \tag{31}$$

Similarly, if $\underline{C} \otimes x \leq \overline{d}$ then x is obviously a weak solution to

$$\begin{aligned} \underline{A} \otimes x \leq \overline{b} \\ \underline{C} \otimes x \leq \overline{d} \end{aligned} \tag{32}$$

That is

$$x = \bar{x}\begin{pmatrix} \underline{A} & \overline{b} \\ \underline{C} & \overline{d} \end{pmatrix}$$

Also from (31) x is a weak solution if and only if

$$[(\underline{A} \otimes x)_i, (\overline{A} \otimes x)_i] \cap [\underline{b}_i, \overline{b}_i] \neq \emptyset, \forall i = 1, 2, ..., m$$

That is

$$\overline{A} \otimes \bar{x}\begin{pmatrix} \underline{A} & \overline{b} \\ \underline{C} & \overline{d} \end{pmatrix} \geq \underline{b}$$

$\square$

Definition 9. An interval system (29) is weakly solvable if there exists $A \in \boldsymbol{A}, C \in \boldsymbol{C}, b \in \boldsymbol{b}$ and $d \in \boldsymbol{d}$ such that (29) is solvable.

Theorem 11. An interval system (29) with constant matrix $A = \underline{A} = \overline{A}, C = \underline{C} = \overline{C}$ is weakly solvable if and only if

$$A \otimes \bar{x}\begin{pmatrix} A & \overline{b} \\ C & \overline{d} \end{pmatrix} \geq \underline{b}$$

Proof. The (if) part follows from the definition. Conversely, Let

$$A \otimes \bar{x}\begin{pmatrix} A & b \\ C & d \end{pmatrix} = b$$

be solvable subsystem for $b \in [\underline{b}_i, \overline{b}_i]$. Then we have

$$A \otimes \bar{x}\begin{pmatrix} A & \overline{b} \\ C & \overline{d} \end{pmatrix} \geq A \otimes \bar{x}\begin{pmatrix} A & b \\ C & d \end{pmatrix} = b \geq \underline{b}$$

$\square$

7.2. Strong solvability

Definition 10. A vector x is a strong solution to an interval system (29) if for each $A \in \boldsymbol{A}, C \in \boldsymbol{C}$ and each $b \in \boldsymbol{b}, d \in \boldsymbol{d}$ there is an $x \in \mathbb{R}$ such that (29) holds.

Theorem 12. a vector x is a strong solution to (29) if and only if it is a solution to

$$E \otimes x = f$$
$$\overline{C} \otimes x \leq \underline{d}$$

where

$$E = \begin{pmatrix} \overline{A} \\ \underline{A} \end{pmatrix}, f = \begin{pmatrix} \underline{b} \\ \overline{b} \end{pmatrix} \tag{33}$$

Proof. If x is a strong solution of (29), it obviously satisfies (33). Conversely, suppose x satisfies (33) and let $\tilde{A} \in \mathbf{A}, \tilde{C} \in \mathbf{C}, \tilde{b} \in \mathbf{b}, \tilde{d} \in \mathbf{d}$ such that $\tilde{A} \otimes x \neq \tilde{b}$ and $\tilde{C} \otimes x > \tilde{d}$. Then $\exists i \in (1, 2, ..., m)$ such that either $(\tilde{A} \otimes x)_i < \tilde{b}_i$ or $(\tilde{A} \otimes x)_i > \underline{\tilde{b}_i}$ and $(\tilde{C} \otimes x)_i > \tilde{d}_i$. Therefore, $(\underline{A} \otimes x)_i < (\tilde{A} \otimes x)_i < b_i$, $(\overline{A} \otimes x)_i \geq (\tilde{A} \otimes x)_i > b_i$ and $(\overline{C} \otimes x)_i > (\tilde{C} \otimes x)_i > d_i$ and the theorem statement follows. □

Acknowledgement

The author is grateful to the Kano University of Science and Technology, Wudil for paying the publication fee.

Author details

Abdulhadi Aminu
Department of Mathematics, Kano University of Science and Technology, Wudil, P.M.B 3244, Kano, Nigeria

8. References

[1] R. A. Cuninghame-Green, Minimax Algebra, *Lecture Notes in Economics and Mathematical Systems*, vol.166, Springer, Berlin (1979).
[2] N. N. Vorobyov, Extremal algebra of positive matrices, *Elektronische Datenverarbeitung und Kybernetik* 3 (1967) 39-71 (in Russian).
[3] K. Zimmermann, Extremální algebra, *Výzkumná publikace Ekonomicko-matematické laboratoře při Ekonomickém ústave ČSAV*, 46, Praha (1976) (in Czech).
[4] F. L. Bacelli, G. Cohen, G. J. Olsder, J. P. Quadrat, *Synchronization and Linearity, An Algebra for Discrete Event Systems*, Wiley, Chichester (1992).
[5] M. Akian, S. Gaubert, V. Kolokoltsov, Set covering and invertibility of functional Galois connections, in: *G. L. Litvinov, V. P. Maslov (Ed.), Idempotent Mathematics and Mathematical Physics*, American Mathematical Society (2005) 19-51.
[6] T. S. Blyth, M. F. Janowitz, *Residuation Theory*, Pergamon press (1972).
[7] S. Gaubert, Methods and Application of (max,+) Linear Algebra, *INRIA* (1997).
[8] P. Butkovič, Finding a bounded mixed-integer solution to a system of dual inequalities, *Operations Research Letters*. 36 (2008) 623-627.

[9] M. Akian, R. Bapat, S. Gaubert, Max-plus algebra, in: *L. Hogben (Ed.), Handbook of Linear algebra: Discrete Mathematics and its Application*, Chapman & Hall/CRC, Baton Rouge, L.A (2007).

[10] P. Butkovič, *Max-linear Systems: Theory and Algorithms.*, Springer Monographs in Mathematics, Springer-Verlag (2010).

[11] K. Cechlárová, P. Diko, Resolving infeasibility in extremal algebras, *Linear Algebra & Appl.* 290 (1999) 267-273.

[12] P. Butkovič, Max-algebra: the linear algebra of combinatorics?, *Linear Algebra & Appl.* 367 (2003) 313-335.

[13] R. A. Cuninghame-Green, Minimax Algebra and applications, in: *Advances in Imaging and Electron Physics*, vol. 90, Academic Press, New York (1995) 1-121.

[14] R. A. Cuninghame-Green, K. Zimmermann, Equation with residual functions, *Comment. Math. Uni. Carolina* 42(2001) 729-740.

[15] P. D. Moral, G. Salut, Random particle methods in (max,+) optimisation problems in: *Gunawardena (Ed.), Idempotency*, Cambridge (1988) 383-391.

[16] A. Aminu, Simultaneous solution of linear equations and inequalities in max-algebra *Kybernetika*, 47, 2, (2011),241-250.

[17] P. Butkovič, Necessary solvability conditions of linear extremal equations, *Discrete Applied Mathematics* 10 (1985) 19-26, North-Holland.

[18] K. Cechlárová, Solution of interval systems in max-algebra, in: *V. Rupnik, L. Zadnik-stirn, S. Drobne(Eds.)*,Proc. SOR (2001) Preddvor, Slonvenia, 321-326.

[19] A. Aminu, Max-algebraic linear systems and programs, *PhD Thesis, University of Birmingham, UK* , (2009).

[20] P. Butkovič, G. Hegedüs, An elemination method for finding all solutions of the system of linear equations over an extremal algebra, *Ekonom. mat. Obzor* 20 (1984) 2003-215.

[21] R. A. Cuninghame-Green, P. Butkovič, The equation $A \otimes x = B \otimes y$ over (max,+), *Theoret. Comput. Sci.* 293 (1991) 3-12.

[22] B. Heidergott, G. J. Olsder, J. van der Woude, *Max-plus at work, Modelling and Analysis of Synchronized Systems: A course on Max-Plus Algebra and Its Applications*, Princeton University Press, New Jersey (2006).

[23] E. A. Walkup, G. Boriello, A general linear max-plus solution technique, in: *Gunawardena(Ed.), Idempotency*, Cambridge (1988) 406-415.

[24] P. Butkovič, A. Aminu, Introduction to Max-linear programming, *IMA Journal of Management Mathematics* (2008), 20, 3 (2009) 233-249.

[25] A. Aminu, P. Butkovič, Non-linear programs with max-linear constraints: a heuristic approach, *IMA Journal of Management Mathematics* 23, 1, (2012), 41-66.

[26] R.A Cuninghame-Green, K. Cechlárová *Residuationin fuzzy algebra and some appplications*, Fuzzy Sets and Systems 71 (1995) 227-239.

[27] R.E Moore *Methods and application of interval analysis*, SIAM, (1979).

[28] H. Myšová, Control solvability of interval systems in max-separable linear equations, Linear algebra and its Application, (2006) 416, 215-223.

[29] M. Fielder, J. Nedoma, J. Ramik, J.Rohn, K. Zimmermann, *Linear optimization problems with inexact data*, Springer, Berlin, (2006).

Chapter 11

Efficient Model Transition in Adaptive Multi-Resolution Modeling of Biopolymers

Mohammad Poursina, Imad M. Khan and Kurt S. Anderson

Additional information is available at the end of the chapter

http://dx.doi.org/10.5772/48196

1. Introduction

Multibody dynamics methods are being used extensively to model biomolecular systems to study important physical phenomena occurring at different spatial and temporal scales [10, 13]. These systems may contain thousands or even millions of degrees of freedom, whereas, the size of the time step involved during the simulation is on the order of femto seconds. Examples of such problems may include proteins, DNAs, and RNAs. These highly complex physical systems are often studied at resolutions ranging from a fully atomistic model to coarse-grained molecules, up to a continuum level system [4, 19, 20]. In studying these problems, it is often desirable to change the system definition during the course of the simulation in order to achieve an optimal combination of accuracy and speed. For example, in order to study the overall conformational motion of a bimolecular system, a model based on super-atoms (beads) [18, 22] or articulated multi-rigid and/or flexible body [21, 23] can be used. Whereas, localized behavior has to be studied using fully atomistic models. In such cases, the need for the transition from a fine-scale to a coarse model and vice versa arises. Illustrations of a fully atomistic model of a molecule, and its coarse-grained model are shown in Fig. (1-a) and Fig. (1-b).

Given the complexity and nonlinearity of challenging bimolecular systems, it is expected that different physical parameters such as dynamic boundary conditions and applied forces will have a significant affect on the behavior of the system. It is shown in [16] that time-invariant coarse models may provide inadequate or poor results and as such, an adaptive framework to model these systems should be considered [14]. Transitions between different system models can be achieved by intelligently removing or adding certain degrees of freedom (dof). This change occurs instantaneously and as such, may be viewed as model changes as a result of impulsively applied constraints. For multi-rigid and flexible body systems, the transition from a higher fidelity (fine-scale) model to a lower fidelity model (coarse-scale) using divide-and-conquer algorithm (DCA) has been studied previously in [8, 12]. DCA efficiently provides the unique states of the system after this transition. In this chapter, we focus on the transition from a coarse model to a fine-scale model. When the system is modeled in an articulated multi-flexible-body framework, such transitions may be achieved by two

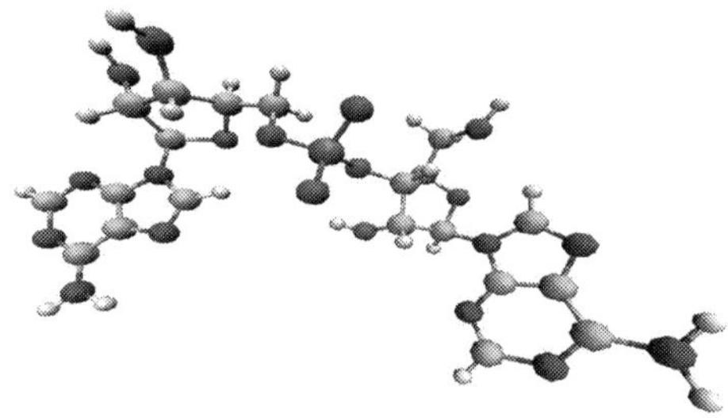

(a) Fully atomistic model

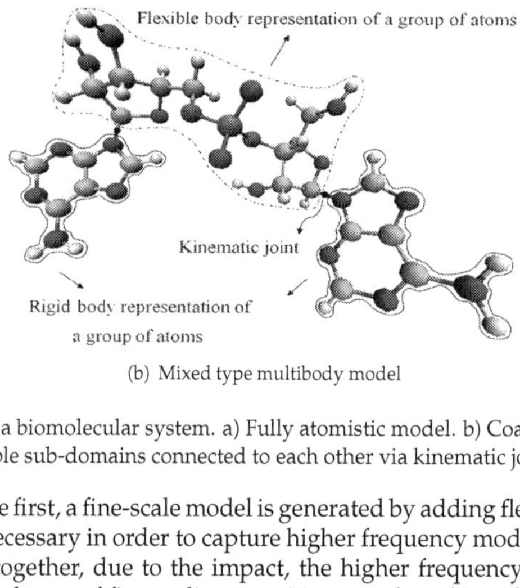

(b) Mixed type multibody model

Figure 1. Illustration of a biomolecular system. a) Fully atomistic model. b) Coarse grain model with different rigid and flexible sub-domains connected to each other via kinematic joints.

different means. In the first, a fine-scale model is generated by adding flexible dof. This type of fine scaling may be necessary in order to capture higher frequency modes. For instance, when two molecules bind together, due to the impact, the higher frequency modes of the system are excited. The second type of fine scaling transition may be achieved through releasing the connecting joints in the multi-flexible-body system. In other words, certain constraints on joints are removed to introduce new dof in the model.

In contrast to those types of dynamic systems in which the model definition is persistent, and the total energy of the system is conserved, the class of problems discussed here experiences discontinuous changes in the model definition and hence, the energy of the system must also change (nominally increase) in a discontinuous fashion. During the coarse graining process, based on a predefined metric, one may conclude that naturally existing higher modes are less relevant and can be ignored. As such, the kinetic energy associated with those modes must be estimated and properly accounted for, when transitioning back to the fine-scale model. Moreover, any change in the system model definition is assumed to occur as a result of impulsively applied constraints, without the influence of external loads. As such, the generalized momentum of the system must also be conserved [6]. In other words, the momentum of each differential element projected onto the space of admissible motions permitted by the more restrictive model (whether pre- or post-transition) when integrated over the entire system must be conserved across the model transition. If the generalized momentum is not conserved during the transition, the results are non-physical, and the new initial conditions for the rest of the simulation of the system are invalid.

In the next section, a brief overview of the DCA and analytical preliminaries necessary to the algorithm development are presented. The optimization problem associated with the coarse to fine scale transitioning is discussed next. Then the impulse-momentum formulation for transitioning from coarse models to fine-scale models in the articulated flexible-body scheme is presented. Finally conclusions are made.

2. Theoretical background

In this section, a brief introduction of the basic divide-and-conquer algorithm is presented. The DCA scheme has been developed for the simulation of general multi-rigid and multi-flexible-body systems [5, 8, 9], and systems with discontinuous changes [11, 12]. The basic algorithm described here is independent of the type of problem and is presented so that the chapter might be more self contained. In other words, it can be used to study the behavior of any rigid- and flexible-body system, even if the system undergoes a discontinuous change. Some mathematical preliminaries are also presented in this section which are important to the development of the algorithm.

2.1. Basic divide-and-conquer algorithm

The basic DCA scheme presented in this chapter works in a similar manner described in detail in [5, 9]. Consider two representative flexible bodies k and $k+1$ connected to each other by a joint J^k as shown in Fig. (2-a). The two points of interest, H_1^k and H_2^k, on body k are termed *handles*. A handle is any selected point through which a body interacts with the environment. In this chapter, we will limit our attention to each body having two handles, and each handle coincides with the joint location on the body, i.e. joint locations J^{k-1} and J^k in case of body k. Similarly, for body $k+1$, the points H_1^{k+1} and H_2^{k+1} are located at the joint locations J^k and J^{k+1}, respectively. Furthermore, large rotations and translations in the flexible bodies are modeled as rigid body dof. Elastic deformations in the flexible bodies are modeled through the use of modal coordinates and admissible shape functions.

DCA is implemented using two main processes, hierarchic assembly and disassembly. The goal of the assembly process is to find the equations describing the dynamics of each body in the hierarchy at its two handles. This process begins at the level of individual bodies and adjacent bodies are assembled in a binary tree configuration. Using recursive formulations, this process couples the two-handle equations of successive bodies to find the two-handle equations of the resulting assembly. For example, body k and body $k+1$ are coupled together to form the assembly shown in Fig. (2-b). At the end of the assembly process, the two-handle equations of the entire system are obtained.

The hierarchic disassembly process begins with the solution of the two-handle equations associated with the primary node of the binary tree. The process works from this node to the individual sub-domain nodes of the binary tree to solve for the two-handle equations of the constituent subassemblies. This process is repeated until all unknowns (e.g., spatial constraint forces, spatial constraint impulses, spatial accelerations, jumps in the spatial velocities) of the bodies at the individual sub-domain level of the binary tree are known. The assembly and disassembly processes are illustrated in Fig. (3).

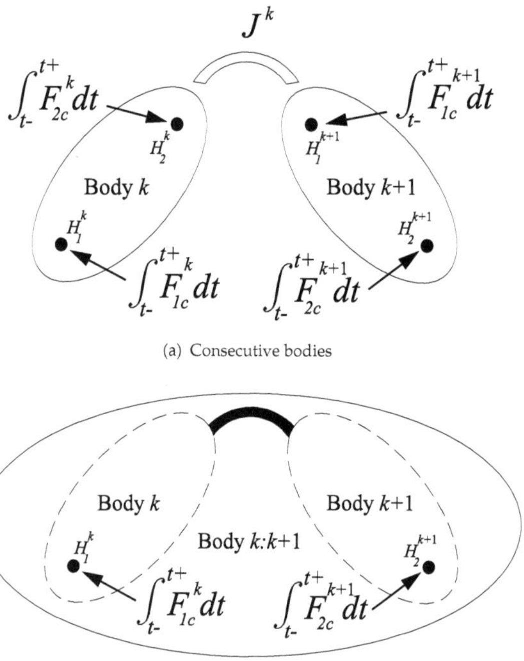

(a) Consecutive bodies

(b) Assembly $k : k+1$

Figure 2. Assembling of the two bodies to form a subassembly. a) Consecutive bodies k and $k+1$. b) A fictitious subassembly formed by coupling bodies k and $k+1$.

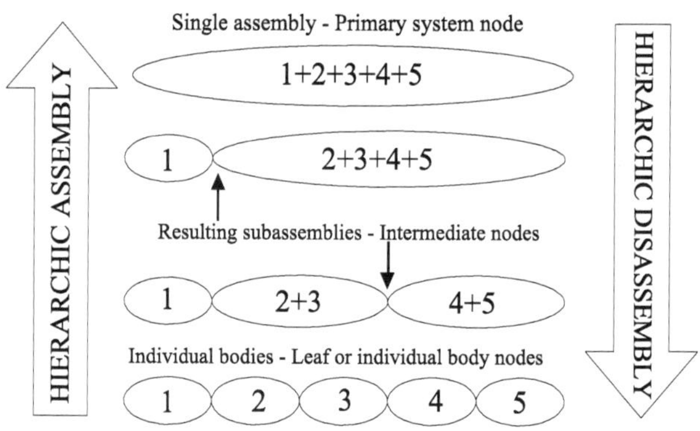

Figure 3. The hierarchic assembly-disassembly process in DCA.

2.2. Analytical preliminaries

For convenience, the superscript c shows that a quantity of interest is associated with the coarse model, while f denotes that it is associated with the fine model. For example, the column matrix $\begin{bmatrix} v_1 \\ \dot{q}_1 \end{bmatrix}^c$ represents the velocity of handle-1 in the coarse model, and $\begin{bmatrix} v_1 \\ \dot{q}_1 \end{bmatrix}^f$ represents the velocity of the same handle in the fine-scale model. In these matrices, v_1 and $\dot{q}_1$ are the spatial velocity vector of handle-1 and the associated generalized modal speeds, respectively.

As discussed previously, the change in the system model definition may occur by changing the number of flexible modes used to describe the behavior of flexible bodies, and/or the number of dof of the connecting joints. To implement these changes in the system model mathematically, the joint free-motion map is defined as follows.

The joint free-motion map $P_R^{J^k}$ can be interpreted as the $6 \times \nu^k$ matrix of the free-modes of motion permitted by the ν^k degree-of-freedom joint, J^k. In other words, $P_R^{J^k}$ maps $\nu^k \times 1$ generalized speeds associated with relative free motion permitted by the joint into a 6×1 spatial relative velocity vector which may occur across the joint, J^k [5]. For instance, consider a transition in which a spherical joint in the system is altered, where only one dof is locked about the first axis. The joint free-motion maps of the fine and coarse models in this case are shown in the following:

$$P_R^{J^k f} = \begin{bmatrix} 1\,0\,0 \\ 0\,1\,0 \\ 0\,0\,1 \\ 0\,0\,0 \\ 0\,0\,0 \\ 0\,0\,0 \end{bmatrix}, P_R^{J^k c} = \begin{bmatrix} 0\,0 \\ 1\,0 \\ 0\,1 \\ 0\,0 \\ 0\,0 \\ 0\,0 \end{bmatrix} \tag{1}$$

We define the orthogonal complement of the joint free-motion map, D_R^k. As such, by definition one arises at the following

$$(D_R^k)^T P_R^{J^k} = 0 \tag{2}$$

3. Optimization problem

Any violation in the conservation of the generalized momentum of the system in the transition between different models leads to non-physical results since the instantaneous switch in the system model definition is incurred without the influence of any external load. In other words, the momentum of each differential element projected onto the space of the admissible motions permitted by the more restrictive model (whether pre- or post-transition) when integrated over the entire system must be conserved across the model transition [6]. Jumps in the system partial velocities due to the sudden change in the model resolution result in the jumps in the generalized speeds corresponding to the new set of degrees of freedom. Since the model is instantaneously swapped, the position of the system does not change. Hence, the position dependent forces acting on the system do not change, and do not affect the generalized speeds. Any change in the applied loads (e.g., damping terms) which might occur due to the change in the model definition and the associated velocity jumps do not contribute to the impulse-momentum equations which describe the model transition. This is because these changes in the applied loads are bounded, and integrated over the infinitesimally short transition time.

Consider a fine-scale model with n dof. Let the dof of the model reduce to $n - m$ after the imposition of certain instantaneous constraints. In this case, the conservation of the generalized momentum of the system is expressed as

$$L^{c/c} = L^{f/c} \tag{3}$$

In the above equation, $L^{c/c}$ and $L^{f/c}$ represent the momenta of the coarse and fine models, respectively, projected on to the space of partial velocity vectors of the coarse model. Equation (3) provides a set of $n-m$ equations which are linear in the generalized speeds of the coarse model and solvable for the unique and physically meaningful states of the system after the transition to the coarser model.

Now consider the case in which, the coarse model is transitioned back to the fine-scale model. Equation (3) is still valid, and provides $n-m$ equations with n unknown generalized speeds of the finer model. Furthermore, during the coarsening process, the level of the kinetic energy also drops because we chose to ignore certain modes of the system. However, in actual biomolecular systems such a decrease in energy does not happen. Consequently, it is important to realize the proper kinetic energy when transitioning back to the fine-scale model. Therefore, the following equation must be satisfied

$$KE^f = \frac{1}{2}(u^f)^T \mathcal{M} u^f \tag{4}$$

In the above equation u^f is the $n \times 1$ column matrix containing the generalized speeds of the fine model, and $\mathcal{M}$ represents the generalized inertia matrix of the fine model. It is clear that Eqs. (3) and (4) provide $n-m+1$ equations with n unknowns. This indicates that the problem is under-determined when multiple dof of the system are released. We may arrive at a unique or finite number of solutions, solving the following optimization problem

$$\text{Optimize} \quad J(u^f, t) \tag{5}$$

$$\text{Subjected to} \quad \Theta_i(u^f, t) = 0,\ i = 1, \cdots, k \tag{6}$$

In the above equation, J is the physics- or knowledge- or mathematics-based objective function to be optimized (nominally minimized) subjected to the constraint equations Θ_i. In [1, 15], different objective functions are proposed for coarse to fine-scale transition problems. For instance, in order to prevent the generalized speeds of the new fine-scale model from deviating greatly from those of the coarse scale model, we may minimize the L^2 norm of the difference between the generalized speeds of the coarse and fine scale models as follows

$$J = (u^f - u^c)^T (u^f - u^c) \tag{7}$$

As indicated previously, $(n-m)$ constraint equations governing the optimization problem are obtained from the conservation of the generalized momentum of the system within the transition. The rest of the constraint equations are obtained from other information about the system, such as the specific value of kinetic energy or the temperature of the system.

The generalized momenta balance equations from Eq. (3) are expressed as

$$Au^f = b \tag{8}$$

where A and b are $(n-m) \times n$ and n known matrices, respectively, and u^f is an $n \times 1$ column matrix of the generalized speeds of the fine-scale system model. As a part of the optimization problem, one must solve this linear system for $n - m$ dependent generalized speeds in terms of m independent generalized speeds. Therefore, the optimization is performed on a much smaller number (m) variables, with a cost of $O(m^3)$. For a complex molecular system, n could be very large, and $n >> m$, hence a significant reduction is achieved in the overall cost of optimization as compared to other traditional techniques, such as Lagrange multipliers [3]. However, the computations required to find the relations between dependent and independent generalized speeds can impose a significant burden on these simulations. It is shown in [2] that if traditional methods such as Gaussian elimination or LU factorization are used to find these relations, this cost tends to be $O(n(n-m)^2)$. The DCA scheme provided here finds these relations at the end of the hierarchic disassembly process with computational complexity of almost $O(n)$ in serial implementation. In other words, in this strategy, DCA formulates the impulse-momentum equations of the system which is followed by providing the relations between dependent and independent generalized speeds of the system in a timely manner. As such, this significantly reduces the costs associated with forming and solving the optimization problem in the transitions to the finer models.

4. DCA-based momenta balance for multi-flexible bodies

In this section, two-handle impulse-momentum equations of flexible bodies are derived. Mathematical modeling of the transition from a coarse model to a fine-scale model is discussed. For the fine-scale to coarse-scale model transition in multi-flexible-body system the reader is referred to [7, 17]. We will now derive the two-handle impulse-momentum equations when flexible degrees of freedom of a flexible body or the joints in the system are released. Then, the assembly of two consecutive bodies for which the connecting joint is unlocked is discussed. Finally, the hierarchic assembly-disassembly process for the multi-flexible-body system is presented.

4.1. Two-handle impulse-momentum equations in coarse to fine transitions

Now, we develop the two-handle impulse-momentum equations for consecutive flexible bodies in the transition from a coarse model to a fine-scale model. It is desired to develop the handle equations which express the spatial velocity vectors of the handles after the transition to the finer model as explicit functions of only newly introduced modal generalized speeds of the fine model. For this purpose, we start from the impulse-momentum equation of the flexible body as

$$\Gamma^f \mathbf{v}_1^f - \Gamma^c \mathbf{v}_1^c = \begin{bmatrix} \gamma_R \\ \gamma_F \end{bmatrix}_1^c \int_{t_c}^{t_f} F_{1c} dt + \begin{bmatrix} \gamma_R \\ \gamma_F \end{bmatrix}_2^c \int_{t_c}^{t_f} F_{2c} dt \tag{9}$$

where Γ^f and Γ^c are the inertia matrices associated with the fine-scale and coarse models, respectively. Also, t_c and t_f represent the time right before and right after the transition. The quantities $\int_{t_c}^{t_f} F_{1c} dt$ and $\int_{t_c}^{t_f} F_{2c} dt$ are the spatial impulsive constraint forces on handle-1 and handle-2 of the flexible body. The matrices $\begin{bmatrix} \gamma_R \\ \gamma_F \end{bmatrix}_1^c$ and $\begin{bmatrix} \gamma_R \\ \gamma_F \end{bmatrix}_2^c$ are the coefficients resulting from the generalized constraint force contribution at handle-1 and handle-2, respectively. Moreover, in Eq. (9), the impulses due to the applied loads are not considered since they represent a bounded loads integrated over an infinitesimal time interval. For detailed derivation of these quantities the reader is referred to [8]. It is desired to develop the handle equations which provide the spatial velocity vectors of the handles right after the transition to the fine-scale model in terms of newly added modal generalized speeds. Therefore, in Eq. (9), the inertia matrix of the flexible body is represented by its components corresponding to rigid and flexible modes, as well as the coupling terms

$$\begin{bmatrix} \Gamma_{RR} & \Gamma_{RF} \\ \Gamma_{FR} & \Gamma_{FF} \end{bmatrix}^f \begin{bmatrix} v_1 \\ \dot{q} \end{bmatrix}^f = \begin{bmatrix} \gamma_R \\ \gamma_F \end{bmatrix}_1^c \int_{t_c}^{t_f} F_{1c} dt + \begin{bmatrix} \gamma_R \\ \gamma_F \end{bmatrix}_2^c \int_{t_c}^{t_f} F_{2c} dt + \begin{bmatrix} \Gamma_{RR} & \Gamma_{RF} \\ \Gamma_{FR} & \Gamma_{FF} \end{bmatrix}^c \begin{bmatrix} v_1 \\ \dot{q} \end{bmatrix}^c \tag{10}$$

which is decomposed to the following relations

$$\Gamma^f_{FF}\dot{q}^f = \gamma^c_{F1} \int_{t_c}^{t_f} F_{1c} dt + \gamma^c_{F2} \int_{t_c}^{t_f} F_{2c} dt + \Gamma^c_{FR} v^c_1 + \Gamma^c_{FF}\dot{q}^c - \Gamma^f_{FR} v^f_1 \tag{11}$$

$$\Gamma^f_{RR} v^f_1 = \gamma^c_{R1} \int_{t_c}^{t_f} F_{1c} dt + \gamma^c_{R2} \int_{t_c}^{t_f} F_{2c} dt + \Gamma^c_{RR} v^c_1 + \Gamma^c_{RF}\dot{q}^c - \Gamma^f_{RF}\dot{q}^f \tag{12}$$

Since the generalized momentum equations are calculated based on the projection onto the space of the coarser model, the matrix Γ^f is not a square matrix and thus Γ^f_{FF} is not invertible. However, we can partition Eq. (11) in terms of dependent (those associated with the coarser model) and independent (newly introduced) generalized speeds as

$$\begin{bmatrix} \Gamma^{f_d}_{FF} \vdots \Gamma^{f_i}_{FF} \end{bmatrix} \begin{bmatrix} \dot{q}^{f_d} \\ \cdots \\ \dot{q}^{f_i} \end{bmatrix} = \gamma^c_{F1} \int_{t_c}^{t_f} F_{1c} dt + \gamma^c_{F2} \int_{t_c}^{t_f} F_{2c} dt + \Gamma^c_{FR} v^c_1 + \Gamma^c_{FF}\dot{q}^c - \Gamma^f_{FR} v^f_1 \tag{13}$$

Using the above relation, the expression for the dependent generalized modal speeds is written as

$$\dot{q}^{f_d} = (\Gamma^{f_d}_{FF})^{-1} [\gamma^c_{F1} \int_{t_c}^{t_f} F_{1c} dt + \gamma^c_{F2} \int_{t_c}^{t_f} F_{2c} dt + \Gamma^c_{FR} v^c_1 + \Gamma^c_{FF}\dot{q}^c - \Gamma^f_{FR} v^f_1 - \Gamma^{f_i}_{FF}\dot{q}^{f_i}] \tag{14}$$

Defining

$$\Gamma^f_{RF} = \begin{bmatrix} \Gamma^{f_d}_{RF} \vdots \Gamma^{f_i}_{RF} \end{bmatrix} \tag{15}$$

$$\Lambda = [\Gamma^f_{RR} - \Gamma^{f_d}_{RF} (\Gamma^{f_d}_{FF})^{-1} \Gamma^f_{FR}]^{-1} \tag{16}$$

$$\zeta_1 = [\gamma^c_{R1} - \Gamma^{f_d}_{RF} (\Gamma^{f_d}_{FF})^{-1} \gamma^c_{F1}] \tag{17}$$

$$\zeta_2 = [\gamma^c_{R2} - \Gamma^{f_d}_{RF} (\Gamma^{f_d}_{FF})^{-1} \gamma^c_{F2}] \tag{18}$$

$$\zeta_3 = [\Gamma^c_{RR} - \Gamma^{f_d}_{RF} (\Gamma^{f_d}_{FF})^{-1} \Gamma^c_{FR}] v^c_1 + [\Gamma^c_{RF} - \Gamma^{f_d}_{RF} (\Gamma^{f_d}_{FF})^{-1} \Gamma^c_{FF}] \dot{q}^c \tag{19}$$

$$\zeta_4 = [\Gamma^{f_d}_{RF} (\Gamma^{f_d}_{FF})^{-1} \Gamma^{f_i}_{FF} - \Gamma^{f_i}_{RF}] \tag{20}$$

$$\lambda_{1i} = \Lambda \zeta_i, \quad (i = 1, 2, 3, 4) \tag{21}$$

and using Eqs. (12) and (14), the spatial velocity vector of handle-1 can be written in terms of the independent modal speeds

$$v_1^f = \lambda_{11} \int_{t_c}^{t_f} F_{1c} dt + \lambda_{12} \int_{t_c}^{t_f} F_{2c} dt + \lambda_{13} + \lambda_{14} \dot{q}^{f_i} \tag{22}$$

As such, the spatial velocity vector of handle-2 becomes

$$v_2^f = (S^{k1k2})^T v_1^f + \phi_2^f \dot{q}^f \tag{23}$$

Employing the same partitioning technique, Eqs. (23) can be written as

$$v_2^f = (S^{k1k2})^T v_1^f + \begin{bmatrix} \phi_2^{f_d} \vdots \phi_2^{f_i} \end{bmatrix} \begin{bmatrix} \dot{q}^{f_d} \\ \cdots \\ \dot{q}^{f_i} \end{bmatrix} \tag{24}$$

$$\Rightarrow v_2^f = (S^{k1k2})^T v_1^f + \phi_2^{f_d} \dot{q}^{f_d} + \phi_2^{f_i} \dot{q}^{f_i} \tag{25}$$

Using

$$\lambda_{21} = [(S^{k1k2})^T \lambda_{11} + \phi_2^{f_d} (\Gamma_{FF}^{f_d})^{-1} \gamma_{F1}^c - \phi_2^{f_d} (\Gamma_{FF}^{f_d})^{-1} \Gamma_{FR}^f \lambda_{11}] \tag{26}$$

$$\lambda_{22} = [(S^{k1k2})^T \lambda_{12} + \phi_2^{f_d} (\Gamma_{FF}^{f_d})^{-1} \gamma_{F2}^c - \phi_2^{f_d} (\Gamma_{FF}^{f_d})^{-1} \Gamma_{FR}^f \lambda_{12}] \tag{27}$$

$$\begin{aligned} \lambda_{23} = [S^{k1k2})^T \lambda_{13} + \phi_2^{f_d} (\Gamma_{FF}^{f_d})^{-1} \Gamma_{FR}^c v_1^c + \phi_2^{f_d} (\Gamma_{FF}^{f_d})^{-1} \Gamma_{FF}^c \dot{q}^c \\ - \phi_2^{f_d} (\Gamma_{FF}^{f_d})^{-1} \Gamma_{FR}^f \lambda_{13}] \end{aligned} \tag{28}$$

$$\lambda_{24} = [S^{k1k2})^T \lambda_{14} + \phi_2^{f_i} - \phi_2^{f_d} (\Gamma_{FF}^{f_d})^{-1} \Gamma_{FF}^{f_i} - \phi_2^{f_d} (\Gamma_{FF}^{f_d})^{-1} \Gamma_{FR}^f \lambda_{14}] \tag{29}$$

and Eq. (25), the spatial velocity vector of handle-2 can be written as

$$v_2^f = \lambda_{21} \int_{t_c}^{t_f} F_{1c} dt + \lambda_{22} \int_{t_c}^{t_f} F_{2c} dt + \lambda_{23} + \lambda_{24} \dot{q}^{f_i} \tag{30}$$

Equations (22) and (30) are now in two-handle impulse-momentum form and along with Eq. (14), give the new velocities associated with each handle after the transition. These equations express the spatial velocity vectors of the handles of the body as well as the modal generalized speeds which have not changed within the transition in terms of the newly added modal generalized speeds. This important property will be used in the optimization problem to provide the states of the system after the transition to the finer models.

As such, the two-handle equations describing the impulse-momentum of two consecutive bodies, body k and body $k+1$ are expressed as

$$v_1^{(k)f} = \lambda_{11}^{(k)} \int_{t_c}^{t_f} F_{1c}^{(k)} dt + \lambda_{12}^{(k)} \int_{t_c}^{t_f} F_{2c}^{(k)} dt + \lambda_{13}^{(k)} + \lambda_{14}^{(k)} \dot{q}^{(k)f_i} \tag{31}$$

$$v_2^{(k)f} = \lambda_{21}^{(k)} \int_{t_c}^{t_f} F_{1c}^{(k)} dt + \lambda_{22}^{(k)} \int_{t_c}^{t_f} F_{2c}^{(k)} dt + \lambda_{23}^{(k)} + \lambda_{24}^{(k)} \dot{q}^{(k)f_i} \tag{32}$$

$$v_1^{(k+1)f} = \lambda_{11}^{(k+1)} \int_{t_c}^{t_f} F_{1c}^{(k+1)} dt + \lambda_{12}^{(k+1)} \int_{t_c}^{t_f} F_{2c}^{(k+1)} dt + \lambda_{13}^{(k+1)} + \lambda_{14}^{(k+1)} \dot{q}^{(k+1)f_i} \tag{33}$$

$$v_2^{(k+1)f} = \lambda_{21}^{(k+1)} \int_{t_c}^{t_f} F_{1c}^{(k+1)} dt + \lambda_{22}^{(k+1)} \int_{t_c}^{t_f} F_{2c}^{(k+1)} dt + \lambda_{23}^{(k+1)} + \lambda_{24}^{(k+1)} \dot{q}^{(k+1)f_i} \tag{34}$$

4.2. Assembly process and releasing the joint between two consecutive bodies

In this section, a method to combine the two-handle equations of individual flexible bodies to form the equations of the resulting assembly is presented. Herein, the assembly process of the consecutive bodies is discussed only within the transition from a coarse model to a finer model. This transition is achieved by releasing the joint between two consecutive bodies. Clearly, this would mean a change in the joint free-motion map $P_R^{J^k}$ and its orthogonal complement $D_R^{J^k}$. It will become evident that the assembly process of the consecutive bodies for the fine to coarse transition is similar, and the associated equations can be easily derived by following the given procedure.

From the definition of joint free-motion map, the relative spatial velocity vector at the joint between two consecutive bodies is expressed by the following kinematic constraint

$$v_1^{(k+1)f} - v_2^{(k)f} = P_R^{J^k f} u^{(k/k+1)f} \tag{35}$$

In the above equation, $u^{(k/k+1)f}$ is the relative generalized speed defined at the joint of the fine model. From Newton's third law of motion, the impulses at the intermediate joint are related by

$$\int_{t_c}^{t_f} F_{2c}^{(k)} dt = -\int_{t_c}^{t_f} F_{1c}^{(k+1)} dt \tag{36}$$

Substituting Eqs. (32), (33) and (36) into Eq. (35) results in

$$\begin{aligned}(\lambda_{11}^{(k+1)} + \lambda_{22}^{(k)}) \int_{t_c}^{t_f} F_{1c}^{(k+1)} dt &= \lambda_{21}^{(k)} \int_{t_c}^{t_f} F_{1c}^{(k)} dt - \lambda_{12}^{(k+1)} \int_{t_c}^{t_f} F_{2c}^{(k+1)} dt \\ &+ \lambda_{23}^{(k)} - \lambda_{13}^{(k+1)} + \lambda_{24}^{(k)} \dot{q}^{(k)f_i} - \lambda_{14}^{(k+1)} \dot{q}^{(k+1)f_i} + P_R^{J^k f} u^{(k/k+1)f}\end{aligned} \tag{37}$$

Using the definition of the joint free-motion map, the spatial constraint impulses lie exactly in the space spanned by the orthogonal complement of joint free-motion map of the *coarser* model. These constraint impulses can be expressed as

$$\int_{t_c}^{t_f} F_{1c}^{(k+1)} dt = D_R^{J^k c} \int_{t_c}^{t_f} \mathbf{F}_{1c}^{(k+1)} dt \tag{38}$$

In the above equation, $\int_{t_c}^{t_f} \mathbf{F}_{1c}^{(k+1)} dt$ is an ordered measure number of the impulsive constraint torques and forces. Pre-multiplying Eq. (37) by $(D_R^{J^k c})^T$, one arrives at the expression for $\int_{t_c}^{t_f} F_{1c}^{(k+1)} dt$ as

$$\begin{aligned}\int_{t_c}^{t_f} F_{1c}^{(k+1)} dt &= X\lambda_{21}^{(k)} \int_{t_c}^{t_f} F_{1c}^{(k)} dt - X\lambda_{12}^{(k+1)} \int_{t_c}^{t_f} F_{2c}^{(k+1)} dt \\ &+ XY + X\lambda_{24}^{(k)} \dot{q}^{(k)f_i} - X\lambda_{14}^{(k+1)} \dot{q}^{(k+1)f_i} + XP_R^{J^k f} u^{(k/k+1)f}\end{aligned} \tag{39}$$

where

$$X = D_R^{J^k c}[(D_R^{J^k c})^T(\lambda_{11}^{(k+1)} + \lambda_{22}^{(k)})D_R^{J^k c}]^{-1}(D_R^{J^k c})^T \tag{40}$$

$$Y = \lambda_{23}^{(k)} - \lambda_{13}^{(k+1)} \tag{41}$$

Using Eqs. (31), (34), and (39), we write the two-handle equations for the assembly $k : k+1$

$$
\begin{aligned}
v_1^{(k:k+1)f} &= \Psi_{11}^{(k:k+1)} \int_{t_c}^{t_f} F_{1c}^{(k)} dt + \Psi_{12}^{(k:k+1)} \int_{t_c}^{t_f} F_{2c}^{(k+1)} dt \\
&+ \Psi_{13}^{(k:k+1)} + \Psi_{14}^{(k:k+1)} \dot{q}^{(k)f_i} + \Psi_{15}^{(k:k+1)} \dot{q}^{(k+1)f_i} + \Psi_{16}^{(k:k+1)} u^{(k/k+1)f} \qquad (42)
\end{aligned}
$$

$$
\begin{aligned}
v_2^{(k:k+1)f} &= \Psi_{21}^{(k:k+1)} \int_{t_c}^{t_f} F_{1c}^{(k)} dt + \Psi_{22}^{(k:k+1)} \int_{t_c}^{t_f} F_{2c}^{(k+1)} dt \\
&+ \Psi_{23}^{(k:k+1)} + \Psi_{24}^{(k:k+1)} \dot{q}^{(k)f_i} + \Psi_{25}^{(k:k+1)} \dot{q}^{(k+1)f_i} + \Psi_{26}^{(k:k+1)} u^{(k/k+1)f} \qquad (43)
\end{aligned}
$$

where:

$$\Psi_{11}^{(k:k+1)} = \lambda_{11}^{(k)} - \lambda_{12}^{(k)} X \lambda_{21}^{(k)} \qquad (44)$$

$$\Psi_{12}^{(k:k+1)} = \lambda_{12}^{(k)} X \lambda_{12}^{(k+1)} \qquad (45)$$

$$\Psi_{13}^{(k:k+1)} = \lambda_{13}^{(k)} - \lambda_{12}^{(k)} XY \qquad (46)$$

$$\Psi_{14}^{(k:k+1)} = \lambda_{14}^{(k)} - \lambda_{12}^{(k)} X \lambda_{24}^{(k)} \qquad (47)$$

$$\Psi_{15}^{(k:k+1)} = \lambda_{12}^{(k)} X \lambda_{14}^{(k+1)} \qquad (48)$$

$$\Psi_{16}^{(k:k+1)} = -\lambda_{12}^{(k)} X P_R^{J^k f} \qquad (49)$$

$$\Psi_{21}^{(k:k+1)} = \lambda_{21}^{(k+1)} X \lambda_{21}^{(k)} \qquad (50)$$

$$\Psi_{22}^{(k:k+1)} = \lambda_{22}^{(k+1)} - \lambda_{21}^{(k+1)} X \lambda_{12}^{(k+1)} \qquad (51)$$

$$\Psi_{23}^{(k:k+1)} = \lambda_{21}^{(k+1)} XY + \lambda_{23}^{(k+1)} \qquad (52)$$

$$\Psi_{24}^{(k:k+1)} = \lambda_{21}^{(k+1)} X \lambda_{24}^{(k)} \qquad (53)$$

$$\Psi_{25}^{(k:k+1)} = \lambda_{24}^{(k+1)} - \lambda_{21}^{(k+1)} X \lambda_{14}^{(k+1)} \qquad (54)$$

$$\Psi_{26}^{(k:k+1)} = \lambda_{21}^{(k+1)} X P_R^{J^k f} \qquad (55)$$

The two-handle equations of the resultant assembly express the spatial velocity vectors of the terminal handles of the assembly in terms of the spatial constraint impulses on the same handles, as well as the newly added modal generalized speeds of each constituent flexible body, and the newly introduced dof at the connecting joint. These are the equations which address the dynamics of the assembly when both types of transitions occur simultaneously. In other words, they are applicable when new flexible modes are added to the flexible constituent subassemblies and new degrees of freedom are released at the connecting joint. If there is no change in the joint free-motion map, the spatial partial velocity vector associated with $u^{(k/k+1)f}$ does not appear in the handle equations of the resulting assembly.

5. Hierarchic assembly-disassembly

The DCA is implemented in two main passes: assembly and disassembly [8, 9]. As mentioned previously, two consecutive bodies can be combined together to recursively form the handle equations of the resulting assembly. As such, the assembly process starts at the individual sub-domain level of the binary tree to combine the adjacent bodies and form the equations of motion of the resulting assembly. This process is recursively implemented as that of the binary tree to find the impulse-momentum equations of the new assemblies. In this process, the spatial velocity vector (after transition) and impulsive load of the handles at the common joint of the consecutive bodies are eliminated. The handle equations of the resulting assembly are expressed in terms of the constraint impulses and spatial velocities of the terminal handles, as well as the newly introduce modal generalized speeds and generalized speeds associated with the newly added degrees of freedom at the connecting joints. This process stops at the top level of the binary tree in which the impulse-momentum equations of the entire system are expressed by the following two-handle equations

$$\begin{aligned} v_1^{1f} &= \Psi_{11}^{(1:n)} \int_{t_c}^{t_f} F_{1c}^1 dt + \Psi_{12}^{(1:n)} \int_{t_c}^{t_f} F_{2c}^n dt + \Psi_{13}^{(1:n)} \\ &+ \Psi_{14}^{(1:n)} \dot{q}^{(1:n)f_i} + \Psi_{15}^{(1:n)} u^{(1:n)f} \end{aligned} \tag{56}$$

$$\begin{aligned} v_2^{nf} &= \Psi_{21}^{(1:n)} \int_{t_c}^{t_f} F_{1c}^1 dt + \Psi_{22}^{(1:n)} \int_{t_c}^{t_f} F_{2c}^n dt + \Psi_{23}^{(1:n)} \\ &+ \Psi_{24}^{(1:n)} \dot{q}^{(1:n)f_i} + \Psi_{25}^{(1:n)} u^{(1:n)f} \end{aligned} \tag{57}$$

Note that through the partial velocity vectors $\Psi_{ij}^{(1:n)}, (i = 1,2 \text{ and } j = 4,5)$, these equations are linear in terms of the newly added generalized modal speeds as well as the generalized speeds associated with the released dof at the joints of the system.

The two-handle equations for the assembly at the primary node is solvable by imposing appropriate boundary conditions. Solving for the unknowns of the terminal handles initiates the disassembly process [1, 11]. In this process, the known quantities of the terminal handles of each assembly are used to solve for the spatial velocities and the impulsive loads at the common joint of the constituent subassemblies using the handle equations of each individual subassembly. This process is repeated in a hierarchic disassembly of the binary tree where the known boundary conditions are used to solve the impulse-momentum equations of the subassemblies, until the spatial velocities of the fine model and impulses on all bodies in the system are determined as a known linear function of the newly introduced generalized speeds of the fine model.

6. Conclusion

The method presented in this chapter is able to efficiently simulate discontinuous changes in the model definitions for articulated multi-flexible-body systems. The impulse-momentum equations govern the dynamics of the transitions when the number of deformable modes changes and the joints in the system are locked or released. The method is implemented in a divide-and-conquer scheme which provides linear and logarithmic complexity when implemented in serial and parallel, respectively. Moreover, the transition from a coarse-scale to a fine-scale model is treated as an optimization problem to arrive at a finite number of solutions or even a unique one. The divide-and-conquer algorithm is able to efficiently produce equations to express the generalized speeds of the system after the transition to the finer models in terms of the newly added generalized speeds. This allows the reduction in computational expenses associated with forming and solving the optimization problem.

Acknowledgment

Support for this work received under National Science Foundation through award No. 0757936 is gratefully acknowledged.

Author details

Mohammad Poursina, Imad M. Khan and Kurt S. Anderson
Department of Mechanical, Aeronautics, and Nuclear Engineering, Rensselaer Polytechnic Institute

7. References

[1] Anderson, K. S. & Poursina, M. [2009a]. Energy concern in biomolecular simulations with transition from a coarse to a fine model, *Proceedings of the Seventh International Conference on Multibody Systems, Nonlinear Dynamics and Control, ASME Design Engineering Technical Conference*, number IDETC2009 in *MSND-87297*, San Diego, CA.

[2] Anderson, K. S. & Poursina, M. [2009b]. Optimization problem in biomolecular simulations with DCA-based modeling of transition from a coarse to a fine fidelity, *Proceedings of the Seventh International Conference on Multibody Systems, Nonlinear Dynamics and Control, ASME Design Engineering Technical Conference 2009, (IDETC09)*, number IDETC2009/MSND-87319, San Diego, CA.

[3] Demel, J. W. [1997]. *Applied Numerical Linear Algebra*, SIAM, Philadelphia, PA.

[4] Dill, K. A., Ozkan, S. B., Shell, M. S. & Weikl, T. R. [2008]. The protein folding problem, *Annual Review of Biophysics* 37(1): 289–316.

[5] Featherstone, R. [1999]. A divide-and-conquer articulated body algorithm for parallel $O(\log(n))$ calculation of rigid body dynamics. Part 1: Basic algorithm, *International Journal of Robotics Research* 18(9): 867–875.

[6] Kane, T. R. & Levinson, D. A. [1985]. *Dynamics: Theory and Application*, Mcgraw-Hill, NY.

[7] Khan, I., Poursina, M. & Anderson, K. S. [2011]. DCA-based optimization in transitioning to finer models in articulated multi-flexible-body modeling of biopolymers, *Proceedings of the ECCOMAS Thematic Conference - Multibody Systems Dynamics*, Brussels, Belgium.

[8] Mukherjee, R. & Anderson, K. S. [2007a]. A logarithmic complexity divide-and-conquer algorithm for multi-flexible articulated body systems, *Computational and Nonlinear Dynamics* 2(1): 10–21.

[9] Mukherjee, R. & Anderson, K. S. [2007b]. An orthogonal complement based divide-and-conquer algorithm for constrained multibody systems, *Nonlinear Dynamics* 48(1-2): 199–215.

[10] Mukherjee, R., Crozierb, P. S., Plimptonb, S. J. & Anderson, K. S. [2008]. Substructured molecular dynamics using multibody dynamics algorithms, *International Journal of Non-Linear Mechanics* 43: 1040–1055.

[11] Mukherjee, R. M. & Anderson, K. S. [2007c]. Efficient methodology for multibody simulations with discontinuous changes in system definition, *Multibody System Dynamics* 18: 145–168.

[12] Mukherjee, R. M. & Anderson, K. S. [2008]. A generalized momentum method for multi-flexible body systems for model resolution change, *Proceedings of the 12th conference on nonlinear vibrations, dynamics, and multibody systems*, Blacksburg, VA.

[13] Norberg, J. & Nilsson, L. [2003]. Advances in biomolecular simulations: methodology and recent applications, *Quarterly Reviews of Biophysics* 36(3): 257–306.

[14] Poursina, M. [2011]. *Robust Framework for the Adaptive Multiscale Modeling of Biopolymers*, PhD thesis, Rensselaer Polytechnic Institute, Troy.

[15] Poursina, M., Bhalerao, K. D. & Anderson, K. S. [2009]. Energy concern in biomolecular simulations with discontinuous changes in system definition, *Proceedings of the ECCOMAS Thematic Conference - Multibody Systems Dynamics*, Warsaw, Poland.

[16] Poursina, M., Bhalerao, K. D., Flores, S., Anderson, K. S. & Laederach, A. [2011]. Strategies for articulated multibody-based adaptive coarse grain simulation of RNA, *Methods in Enzymology* 487: 73–98.

[17] Poursina, M., Khan, I. & Anderson, K. S. [2011]. Model transitions and optimization problem in multi-flexible-body modeling of biopolymers, *Proceedings of the Eighths International Conference on Multibody Systems, Nonlinear Dynam. and Control, ASME Design Engineering Technical Conference 2011, (IDETC11)*, number DETC2011-48383, Washington, DC.

[18] Praprotnik, M., Site, L. & Kremer, K. [2005]. Adaptive resolution molecular-dynamics simulation: Changing the degrees of freedom on the fly, *J. Chem. Phys.* 123(22): 224106–224114.

[19] Scheraga, H. A., Khalili, M. & Liwo, A. [2007]. Protein-folding dynamics: Overview of molecular simulation techniques, *Annu. Rev. Phys. Chem.* 58(1): 57–83.

[20] Shahbazi, Z., Ilies, H. & Kazerounian, K. [2010]. Hydrogen bonds and kinematic mobility of protein molecules, *Journal of Mechanisms and Robotics* 2(2): 021009–9.

[21] Turner, J. D., Weiner, P., Robson, B., Venugopal, R., III, H. S. & Singh, R. [1995]. Reduced variable molecular dynamics, *Journal of Computational chemistry* 16: 1271–1290.

[22] Voltz, K., Trylska, J., Tozzini, V., Kurkal-Siebert, V., Langowski, J. & Smith, J. [2008]. Coarse-grained force field for the nucleosome from self-consistent multiscaling, *Journal of Computational chemistry* 29(9): 1429–1439.

[23] Wu, X. W. & Sung, S. S. [1998]. Constraint dynamics algorithm for simulation of semiflexible macromolecules, *Journal of Computational chemistry* 19(14): 1555–1566.